# Archaeological Landscape and Settlement II

# Archaeological Landscape and Settlement II

Guest Editors

**Paolo Biagi**
**Elisabetta Starnini**

MDPI

Basel • Beijing • Wuhan • Barcelona • Belgrade • Novi Sad • Cluj • Manchester

*Guest Editors*

Paolo Biagi
Department of Asian and
North African Studies
Ca' Foscari University of Venice
Venezia
Italy

Elisabetta Starnini
Department of Civilizations
and Forms of Knowledge
University of Pisa
Pisa
Italy

*Editorial Office*

MDPI AG
Grosspeteranlage 5
4052 Basel, Switzerland

This is a reprint of the Special Issue, published open access by the journal *Land* (ISSN 2073-445X), freely accessible at: https://www.mdpi.com/journal/land/special_issues/X0P9F0B844.

For citation purposes, cite each article independently as indicated on the article page online and as indicated below:

Lastname, A.A.; Lastname, B.B. Article Title. *Journal Name* **Year**, *Volume Number*, Page Range.

**ISBN 978-3-7258-7184-1 (Hbk)**
**ISBN 978-3-7258-7185-8 (PDF)**
**https://doi.org/10.3390/books978-3-7258-7185-8**

Cover image courtesy of Paolo Biagi
A historical corral photographed from a Bronze Age kurgan on Mt. Paravani in the Georgian Caucasus

# Contents

# About the Editors

**Paolo Biagi**

Paolo Biagi was born in Brescia, Italy, on January 30th, 1948. He is a Senior Researcher at the Ca' Foscari University of Venice, Italy. He was formerly a Full Professor of Prehistory and Protohistory at the same University. His interests include prehistoric archaeology, the Mesolithic and Neolithic eras, Highland Zone archaeology in Europe, shell middens and coastal archaeology in South Asia, chert and obsidian mining and procurement, lithic technology, and radiocarbon dating. He has carried out research and excavations in Oman, Pakistan, Kuwait, Italy, Sardinia, Romania, Greece, and Georgia. He is the co-director of the Italian Archaeological Expedition in the Caucasus of Georgia (with V. Licheli) and Director of the Italian Archaeological Expedition in Lower Sindh and Las Bela (Pakistan).

**Elisabetta Starnini**

Elisabetta Starnini is an Associate Professor of Early and Late Prehistory at the University of Pisa, Italy. Her research interests are quite broad. Her research deals with the Neolithization of the Balkan Peninsula and Northern Italy, raw material sourcing, procurement, and use in prehistoric times, the archaeometry of ceramics and stone artefacts, and the Prehistory and Bronze Age of the Indus Valley. She is also working on the Paleolithic and Mesolithic eras in Greece and Northern Italy. She has directed a national research project on obsidian sourcing in the central Mediterranean as a PI. She is the author/co-author of more than 300 publications, including monographs, peer-reviewed articles in international journals, and book chapters.

# Preface

This Reprint of the second volume of the *Land* Special Issue "Archaeological Landscape and Settlement" consists of eleven research papers providing landscape archaeology insights independently of the geomorphological and altitude locations of a wide variety of sites, their characteristics, functions, the construction methods of their structures, and their chronology in a global perspective.

The Reprint consists of multidisciplinary papers that present and discuss many different topics related to prehistoric and historical archaeology in countries that range from Mediterranean Europe to Africa, Central and Eastern Asia and South America, from the Middle Palaeolithic to Medieval and subrecent times, and from seaside landscapes to the highland zones of Mongolian Central Asia and the Andean chain, which were exploited for economic subsistence and defensive purposes. Considering such a wide range of topics, the Reprint provides cases that show how landscape archaeological methods have been employed by different authors to interpret different events that took place throughout the world in a variety of highly different territories.

**Paolo Biagi and Elisabetta Starnini**
*Guest Editors*

land

Article

# Beyond Colonial Boundaries: Reimagining the Rozvi through Landscapes, Identities and Indigenous Epistemologies

**Lesley Hatipone Machiridza [1,2] and Russell Kapumha [3,*]**

[1] Alexander Von Humboldt Post-Doctoral Fellow, Institute of African Studies and Egyptology, University of Cologne Albertus-Magnus-Platz, D-50923 Cologne, Germany; lesley.machiridza@gmail or lmachiri@uni-koeln.de
[2] Department of Development Studies, History and Archaeology, Simon Muzenda School of Arts, Culture and Heritage Great Zimbabwe University, Masvingo P.O. Box 1235, Zimbabwe
[3] School of Archaeology, University of Oxford, Oxford OX1 3AZ, UK
* Correspondence: russell.kapumha@arch.ox.ac.uk

**Abstract:** The land, 'things'/objects, and memory in the form of narratives and metaphors are intricately bound together. They all constitute the iconography of a shared set of ideas, beliefs, feelings, values, practices, and performances that objectify collective identities. Respectively, these complex entangled tangible and spiritual/invisible indices of identities situated in places deserve special archaeological devotion. However, since African archaeology and history remains trapped in Eurocentric colonial metanarratives, indigenous epistemologies and ontologies have somehow remained on the margins of knowledge production processes. This deliberate erasure and silencing continues to impede archaeology's capacity to explore hidden meanings and values that people imbue to places and landscapes through time. Owing to this setback, multiple precolonial group identities in parts of Zimbabwe, South Africa, Botswana, and Mozambique such as Torwa, Twamamba, Rozvi, Singo, and Venda, among others, remain vague and subjectively tied to the archaeology of Butua/Torwa (AD 1400–1644) and Rozvi (AD 1685–1830) state systems. The failure to read the landscape as both a repository of memory and an agent for collective identities continues to compound our archaeological challenges. Against this background, Rozvi oral narratives and the Insiza cluster Khami-phase sites in southwestern Zimbabwe are subjected to renewed scrutiny. Following a critical review of colonial archives and Rozvi traditions, it turned out that instead of contradicting 'science', oral traditions actually amplify our reading of the archaeological record, only if handled properly.

**Keywords:** land; landscapes; identities; metaphors; Rozvi; Khami phase

## 1. Introduction

The Rozvi, a dynamic historical political formation, first rose to prominence in northeastern Zimbabwe around the 1670s before migrating southwestwards to further consolidate their power under a major pre-colonial state system (1685–1830), variably called Rozvi, Vurozvi, Ukalanga, Gore, or Goremukuru [1–3]. This development, coupled with other incessant Rozvi dispersals stirred by internal succession disputes, further entangled their past with a very wide landscape dotted by numerous Khami-phase sites. Through time, extensive Rozvi movements, contact, and integration with other sociopolitical groupings, as well as their adoption and use of several Khami-phase sites, has triggered immense scholarly debate [1,2,4–14]. Their extensive spatial distribution through time has created a complex cultural web that continues to cause archaeological interpretive challenges. At one point, a prominent historian, ref. [1] was prompted to say:

> "...the Rozvi even more than the rest of the Shona have acquired a historiography without a history..."

Perhaps, the greatest challenge bedeviling the Rozvi past stems from the gross mistakes that were perpetuated by antiquarians during the colonial era. These were recklessly tolerated and reproduced in corpus literature until the Rozvi past was completely submerged in multiple layers of myths [1]. These antiquarian misrepresentations, coupled with the negative influence of Western discourse about Africans, typical of the era of colonialism and imperialism, inflicted long-lasting damage on our ability to understand and interpret this intriguing past. Amidst these colonial ideologies, Rozvi origins were externalized, their cultural ingenuity and complexity denied, and their oral histories dismissed and ignored as myth while their heritage and land were systematically seized and renamed.

Of course, it would be misleading to sorely blame antiquarians and colonial settlers for Rozvi disconnections from their core state, land, and territories. Rather, owing to incessant succession disputes that ravaged their system of governance, numerous Rozvi groups had already started fragmenting from the core state as early as the 1690s [9,11,12]. In addition, the Nguni-inspired incursions in southern Africa of the 1820s to 1860s, known as the *mfecane,* further spurred more Rozvi dispersals before the colonial land seizures finally forced the last remnants from core ancestral lands. Generally, Rozvi territories overlap with numerous Khami-phase sites (see Figures 1–4 below), which are also politically connected with the Torwa/Butua state (1400–1644). As such, the actual role of Rozvi dynasties in the rise and spread of these settlements remains contested. Although there has been significant progress in historicizing the Rozvi, accounting for their archeological footprints within this massive landscape has remained quite problematic [8,10,13–15]. Rozvi identities have largely remained somewhat rhetorical, well encapsulated by Rozvi and Shona descendants, despite their evidently being subjects of scientific enquiry.

To this end, Shona oral traditions are quite emphatic and explicit that all *madzimbahwe* (dry stone-walled monuments) of the Zimbabwe Culture were built and occupied by the Rozvi. In fact, while the Rozvi have always argued that these sites and associated landscapes directly belong to their former ruling ancestors, historians and archaeologists have remained very skeptical of such claims. The archaeological record too, has also been quite stubborn by either revealing very little or actually contradicting the Rozvi oral traditions in certain ways [2]. Perhaps it is high time that we took heed of the advice never to use archaeology as a counter-narrative to indigenous stories [16–21]. Therefore, this article engages a constructive interaction process that is quite sensitive to indigenous metaphors and historical identities. It does so by particularly drawing from Tilley's [21] observation that the invisible aspects of a thing are as essential to its meaning and significance as those that are visible. Thus, Rozvi historical connections to Khami-phase sites and associated landscapes are subjected to renewed scrutiny and interpretation under these premises.

**Figure 1.** Locational map of Zimbabwe and associated Khami-phase sites.

## 2. The Colonial Archive: Metanarratives and the Construction of Dominant Memory

The world over, colonialism and imperialism ushered in a broad range of oppressive scientific research practices that denigrated, marginalized, tamed, and silenced Africans and the recognition of their indigenous narratives [22,23]. As a result, literature that was generated back then is awash with layered ideologies, biases, and stereotypes about Africans [24,25]. Additionally, it is imperative to note that most of these colonial distortions were particularly aimed at dispossessing and claiming indigenous lands through misrepresentation, denying and rejecting their cultural ingenuity and complexity. By projecting everything indigenous as backward, inferior, and uncivilized, colonial agents sought to justify imperialism and pave the way for it. As such, the 'colonial library' crafted very subjective portrayals of indigenous histories through the eyes of the West, which is problematic. For instance, Smith [23] argues that, from the perspective of the once-colonized peoples across the world, the word 'research' remains one of the 'dirtiest' terms in their vocabulary because it was inextricably bound with European imperialism and colonialism. In this respect, most colonial archives should be treated with great caution. Although they are products of their time and still form a large corpus of the historical literature today, they must be subjected to critical review in order to avoid recycling inherent biases. Such an approach effectively arms us to minimize those veiled distortions that were deliberately perpetuated by antiquarians and their like-minded colleagues as we strive to objectively reconstruct past realities. To this effect, research became a site of struggle between the interests and ways of knowing of the West and the interests and ways of knowing of the 'Other' [23].

The colonial administration and its associates are also guilty of elevating their own worldview by treating indigenous ways of knowing as unscientific, naïve, contradictory, illogical, and invalid [22,23,26]. Respectively, it is not surprising that when European explorers toured indigenous lands across the world, they perceived them as undocumented, untamed, available, and devoid of indigenous people's memories and histories [23,27]. This was a deliberate way of denying the physical existence and validity of indigenous people's claims to their ancestral lands. As Dei [28] observes, indigenous or 'heritage' knowledge has always been characterized by the realities of the cosmos and the nexus of nature, society, and culture. Thus, indigenous ways of knowing are holistic and expressed through stories and practices tied to the land. Instead of investing time and effort to fully understand these indigenous knowledge systems (IKS), most colonialists chose to ignore and dismiss these as senseless 'nativist' discourse' [23]. At least, we now know that, as part of attempts to remove and dispossess indigenous people of their land and heritage, colonialists resorted to denying all forms of local memory. Some natives were even labeled as recent immigrants with tenuous claims to their land. According to Russell [27], this 'lack of landscape connection', or rather the lack of memory, or the disruption or deliberate absenting of it was used to distance locals from their heritage. This way, they effectively dispossessed them of their land and legitimized all forms of European land invasions. Similar 'dirty tactics' were progressively used to externalize Rozvi origins and distance their descendants from ancestral lands and heritage sites [2].

Closely aligned to these colonial prejudices was the promotion of binary oppositions, dualisms, and hierarchical orderings of the world. More precisely, back then, Western thought emphasized distinctions between the primitive and civilized, prehistoric and historic, and literate and non-literate societies [24,29]. Schmidt and Pikirayi [30] further note that numerous exclusionary and unnecessary arbitrary boundaries underwrite and amplify colonial experiences. More bluntly, the collective memory of imperialism was premised on the random collection, documentation, classification, and (mis)representation of indigenous people to the West as a bunch of ignorant, lazy, hopeless, and culturally frozen people. Oftentimes, antiquarians and their colonial counterparts expressed personal views, omissions, and commissions as official knowledge, yet these all strived to dehumanize indigenous people through discourses of race and gender [23]. For decades, indigenous people and their histories were under siege, and most of these colonial distortions went unchecked and continue to be accepted as gospel by a wide range of unsuspecting readers/audiences. Once again, we note with deep regret that indigenous histories still continue to be sidelined and ignored. Colonial historians, for example, alienated and denigrated indigenous histories through Eurocentric historiographies, which encapsulated problematic concepts such as prehistory. Today, prehistory continues to be uncritically used, despite the implications for African history [30–32]. This way, African history has extensively suffered because it was unjustly relegated to the margins of knowledge as primitive knowledge that naturally ceased to exist at the onset of native people's contact with European civilization and modernity [23].

There is scholarly evidence that speaks to the view that indigenous communities never lost their history and heritage; rather, they were simply projected in colonial archives as people without a history. Even in historical archaeology, indigenous narratives were never given as much value as documentary evidence [24]. It therefore goes without saying that the bulk of research conducted under the banner of colonial ideology was prejudicial to indigenous people and their cultures and dehumanized them [27,33]. In like manner, Rozvi oral narratives were often ignored and dismissed in historical and archaeological writings as mere myths. Despite numerous claims made by local people living near Khami-phase sites that these settlements were indeed remnants of a glorious Rozvi past, such claims only received piecemeal scholarly attention (Machiridza 2005; [13,34]). Unfortunately, these kinds of publications now constitute the bulk of the academic literature and have kept indigenous views on the Rozvi in perpetual obscurity. This challenge is certainly not unique to the Rozvi past but affects several other indigenous histories in Zimbabwe and further beyond. Thus, Smith [23] laments this 'policing' research practice in the following words:

> "Authorities and outside experts are often called to verify, comment upon, and give judgement about the validity of indigenous claims to cultural beliefs, values, ways of knowing and historical accounts...".

Apart from disrespecting and denying indigenous people's claims to ingenuity and complexity, white settlers also sought to portray indigenous people as very recent immigrants. This way associated ancient indigenous cultural heritage was purported to belong to another ancient racial group that was distinctly different from the locals but racially closer to the colonizers. This brazen lie was widely and extensively used to legitimize all forms of colonial ideologies or efforts to repossess their so-called 'lost ancestral inheritance' [16,30,35–37]. Naturally, archaeology and other related scientific disciplines of the Western academy got entangled with these colonial paradigms and epistemologies, which still continue to constrain our full appreciation of the dynamics of space, place, time, being, and materiality in the African context [17,18]. In attempts to systematically counter these hostile colonial practices, the succeeding section delves deep into contested conceptions of the land, landscapes, placemaking, memory, metaphors, and identities.

Any attempt to challenge and eliminate grand colonial ideologies and epistemologies culminates in a number of exciting theoretical and methodological insights often referred to as postcolonial theory, indigenous revisionist models, anti-colonialist practice, decolonization, deconstructionist epistemologies, Afrocentricity, and post-modernist frameworks [17,22]. These theoretical orientations have also inspired pluralistic approaches to envisage the past using alternative historical sources, silenced voices, and the co-creation of knowledge about space and time [18,19]. Closely tied to this are discourses on decolonization that continue to trigger immense scholarly interest and debate. In brief, decolonization is the process of unmasking all forms of colonial exploitation of the 'Other' while systematically re-centering or empowering the once marginalized worldviews, including local ontologies, epistemologies, and axiologies [25,28,38]. For example, and as part of this trend and paradigm shift, historical archaeology, and many other historical sub-disciplines, have rapidly blossomed and adopted an amplified sense of history that emphasizes critical reflection upon processes of continuity and change, history, ideology, naming, dwelling, storytelling and identity construction, negotiation, and maintenance. All this is aimed at fully understanding and interpreting how people in the past and present experience landscapes.

### 3. A Landscape Approach

It is imperative, at this juncture, to examine the basic constituents of a landscape approach, which includes notions of the land, space, place, memory, identity, and metaphors. Although several analyses of land, space, place, and time exist, there remains relative confusion on their specific meanings and differences. Some scholars even go so far as using these terms interchangeably. This is problematic. To start with, land is the physical backdrop of all human action, it is the common base upon which people inscribe meanings so that, in a way, it is the equivalent of space. According to Henry [39], space is generally conceived of as the continuous extension in length, breadth, and width that can be mapped by the three-dimensional coordinate system of rational geometry. While relevant and acceptable, the challenge with this definition is that it overlaps with our understanding of place and sites. Smith [23] robustly argues that space is often viewed as empty, static, or divorced from time; it is something that can be well-defined, fixed, and devoid of politics. A key takeaway from this definition is the idea of space as land devoid of politics. Anschuetz et al. [40] also argues that space only gains importance through attachment of economic values and its strategic use by actors. This profoundly implies that when land is devoid of people and related sociocultural meanings, it remains as space. Thus, space only exists as an empty expanse that transforms into something else when given meaning by human beings. Again, this explains why imperialists and colonialists were bent on describing indigenous or native land as untamed and undocumented space, meaning it was readily available for their possession [23,27]. In this way, colonial agents notoriously justified colonialism, invasion, and control of native lands by claiming that these were once open spaces devoid of indigenous people, memory, and history, which of course was not true. Once in charge, they began to imbue their own meanings into these so-called spaces, thereby transforming them into colonial places, which takes us to 'place', yet another complex term worth unpacking.

Unlike space, place, which in essence is a 'space of meeting' is always in a state of becoming, it is never bound, fixed, grounded, and stable [21]. This implies that place is more abstract than space. It is best conceptualized as a spatial site, position, or social unit that is either consciously or politically produced by people. In layman's terms, when land is politically charged or made to belong to a particular group of people, it becomes place. For instance, Relph [41] proffers an interesting conceptualization of place in the following manner:

> "Each place is a territory of significance, distinguished from larger or smaller areas by its name, by its particular environmental qualities, by the stories and shared memories connected to it, and by the intensity of meanings people give to or derive from it...".

It therefore follows that when land is imbued with a diverse range of meanings by a particular group of people through various sociocultural processes in time, it transforms into place. Once imbued with meanings, space becomes an active and powerful agent of enforcing collective identities as a place. When this happens, the place itself may not even change physically; rather, what alters are the attached social meanings, imaginations, and invisible and spiritual attachments. In view of this, it is worth reiterating that at times material forms may only offer us a brief snapshot into past realities; otherwise, much deeper meanings always extend much further into the invisible realms/dimensions [21,42]. Sinamai [18] further cements this reasoning by arguing that archaeologists should now wake up from their prolonged slumber by turning away from their sole obsession with materiality/things and beginning to also pay much closer attention to attached intangible meanings. This clarion call brings us to yet another interesting and complex concept that is often confused with place, which is the landscape.

Just like place, landscape is a culturally fashioned creation that is imbued with diverse meanings, hotly contested, and variably characterized as 'text, artefact, and language' itself [21,43]. In attempts to clarify the conceptual difference between place and landscape, Tilley [21] avers that people in places, moving within and between these places, constitute landscapes. He further adds that landscape is not defined by spaces but by places: it sits within places, implying a network of places constitute landscape(s). On a related note, Anschuetz et al. [40] adds that a combination of the physical environment, the character of the people at a particular time, and associated meanings yields landscapes. This definition is broad enough to draw our attention to the tangible/scientific dimensions as well as the social/spiritual dimensions of landscapes. Owing to colonial ideologies and their projected monopoly over science and the benefits it brought forth by ignoring and distancing the 'Other' from their science and lands, the historical approach to landscapes suffered great neglect, particularly within Africa. In this regard, the politics of colonial research silenced indigenous communities by invalidating their spiritual ways of interacting with landscape. It is therefore crucial to note that through a diverse range of cultural processes, landscapes became extensions of group identities. As different agents participated in place-making through traditional practices and performances, naming processes, storytelling, and dwelling, they activated landscapes as metaphors of cultural meaning [18,21,23,27,39].

Tilley [21] stresses that places and landscapes anchor memories because we do not remember in a disembodied and placeless manner. Memory is hereby conceptualized as a personal activity that is subject to biases, quirks, and rhythms of the individual mind; it is a temporal matter of the past but its spatialization transforms it into something more stable and durable [21,39,42]. When this happens, especially among indigenous communities, memory becomes part of their history, but this is often misunderstood. Henry [39] further explains the distinction between memory and history by stipulating that history is conceptualized as a factual record of a given past, an objectively recorded account of past realities;

hence, it is often given authority over memory. Vansina [44] further informs us that remembering differs from memory in the sense that remembering is action and its mechanisms of cueing and scanning are meant to recall specific memories. This implies that among indigenous communities that still rely on memory as part of recalling history, remembering is often enhanced by attaching a cue to every item memorialized. This way, music, objects, and landscapes then function as cues, labels, or mnemotechnic devises that help recall specific memories (Vansina [44]). It is also interesting to note that what Jan Vansina calls cues are in fact metaphors that manifest in multiple forms culturally. In this case, metaphors are highly complex forms of communication that are used by people to achieve meaning through association, analogy, and allegory (Sinamai [18]). Principally, metaphors usually feature as part of language, naming patterns, stories, narratives, folklores, places, and nature, just as long as they express some form of cultural communication.

On the whole, places and landscapes are pivotal in the creation and consolidation of memories that foster identity sentiments. In fact, as people store their memories in places, those places also become part of people's memory [39]. Thus, as part of identity politics, the land was systematically imbibed with a network of memories as an active sociopolitical archive. Through repeated works termed placemaking, the social construction of space or the development of sense of place, the land became an active extension of people's identities. Hence, there cannot be identity without the land because the land and identities are inextricably bound [18,21,27,39]. Tilley [21] further argues that ideas and feelings about identities are located in the specificities of places and landscapes. However, most of such 'specificities' are metaphoric and invisible to archaeologists who largely rely on what they can actually see, touch, measure, and verify. For this reason, tracing identities and agency through the archaeological record alone has remained quite difficult and problematic [2,6,8,18,42]. Consequently, it is high time that we adopted a landscape approach that harmonizes both the spiritual and scientific dimensions of past realities. Since the Rozvi were a product of complex historical processes, their identities were self-proclaimed, imagined, worked/re-worked through time and subjectively linked to a diverse range of ancestral and historical landscape symbols. All ideas, feelings, and values that shaped and still shape Rozvi identities were situationally objectified through strategic material references to the past. Thus, all fluid and spiritual indices of Rozvi identities were systematically objectified through the landscape, including selected Khami-phase sites, which are briefly and closely explored below.

## 4. Contextualizing the Archaeology and Metanarratives: The Khami Phase

Historically, the Khami phase encompasses the Torwa/Butua and Rozvi state systems as well as several other chiefdoms that occupied the northern fringes of South Africa [2,45,46]. The archaeological term Khami-phase dating (1400–1830) derives from the main site of Khami, a precolonial Torwa/Butua state capital situated in southwestern Zimbabwe. Thus, all related sites with terraced monumental architecture, some of which accommodated massive elite *dhaka* huts on top, decorated dry stone walls with checker, herringbone, dentelle, cord, and chevron patterns, and ceramic wares with polychrome and graphite burnishing designs typically belong to this phase. Unlike their Zimbabwe tradition counterparts, elites at Khami-phase sites publicly displayed their wealth by exposing their massive huts at the apex of massive artificial platforms. Some scholars deliberately overlooked these cultural differences in order to advance the failed argument that it was dynastic migrations from either Great Zimbabwe or the Mutapa state that directly led to Khami-phase site origins [2,8]. Recent empirical data is increasingly pointing towards the independent development of the Khami phase [2,47]. Despite these research advances, we however remain uncertain about the actual relationship between the Khami-phase sites and earlier Leopard's Kopje/Woolandale tradition sites [2]. Nonetheless, without digressing much from the core research area, it suffices to briefly examine the broader context of cultural complexity processes in southern Africa.

Accordingly, the Khami phase is best understood and appreciated as a mere extension of long-established processes of complexity development that were happening across southern Africa [8,45,48]. In fact, the development of complexity in southern Africa goes as far back as AD 800 and presents an unbroken cultural sequence right into the historical period, generally accepted as the last 500 years. Although the broader southwestern regions are hot, dry, and barren, certain localities within offered immense opportunities for cultural progression. Typical examples include the Shashe-Limpopo confluence, the northeastern Tati area of Botswana and the central watershed regions of Zimbabwe, territories coinciding with the Butua/Guruuswa regions [2,14,48]. At certain times, associated strategic mineral resources, fertile soils, and floral and faunal species in these localities immensely attracted diverse communities. Spasms of favorable climatic conditions also brought forth immense economic and sociocultural benefits. It was through such sporadic and spasmodic processes that the Shashe-Limpopo basin eventually emerged as the first epicenter of precolonial power in southern Africa. The earliest Zimbabwe Culture state systems of Mapela and Mapungubwe evolved therein. In particular, these polities directly evolved from numerous competing Leopard's Kopje chiefdoms that systematically capitalized on the ideology of sacred leadership and exploited long-distance trade networks and associated local economic opportunities [47,48].

By taking full advantage of the ideology of sacred leadership, tribute and taxation, hunting, local industries and crafts manufacture, iron, copper, and gold mining, crop and livestock (particularly cattle) production, local and long-distance trade networks, and numerous chiefdoms and states gradually evolved through time. The term Zimbabwe Culture was thus adopted to broadly refer to numerous precolonial Shona states (900–1900) that exhibit evidence of class distinction through prestigious dry stone walls, massive *dhaka* huts, craft specialization, external trade, intensive cattle herding, and nucleation of activities around elite centers [2,5,14,46,49]. Against this background, the first phase of the Zimbabwe Culture comprises the competing polities of Mapela (1055–1400) and Mapungubwe (1220–2290). This was followed by the Zimbabwe phase constituting Great Zimbabwe (1300–1550) and the Mutapa state (1450–1900). Typically, Zimbabwe tradition sites are characterized by both plain and decorated free-standing dry stone-walled enclosures that secluded elites inside. These sites are widely spread across the Zimbabwean plateau and adjacent regions where they significantly overlap with the Khami-phase sites (see Figure 2 below). The extensive distribution of the Zimbabwe tradition sites and their overlap with both the first and third phases of the Zimbabwe Culture has constantly attracted immense scholarly attention. However, divergent theoretical orientations and emerging field data is increasingly pointing towards a multilinear development trajectory.

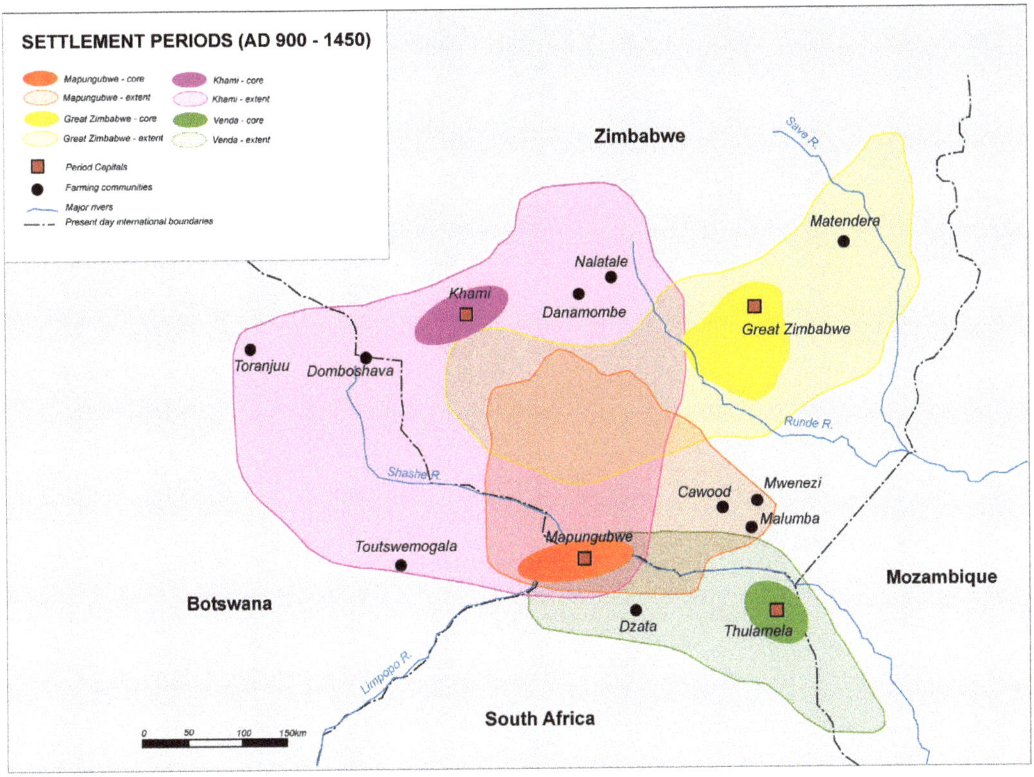


**Figure 2.** Geographical distribution of the Mapungubwe, Great Zimbabwe, and Khami phases [49].

Earlier on, the evolutionary model was widely popularized, thereby giving the impression that there was a neat linear succession of state systems (Main and Huffman [48]). Implicitly, when Mapungubwe was the center of power, all contemporary Zimbabwe Culture settlements of varying sizes and distance from this center were perceived as mere supporting administrative units [5,48,49]. This model further argues that the same structure was systematically repeated when Great Zimbabwe, Torwa/Butua, and Mutapa states later rose to prominence. However, in recent years this model has come under fire, particularly owing to the lack of fit with new or emerging field evidence [2,14,47,49]. In line with this, van Waarden [14] has also argued that before the fall of Mapungubwe, a Gumanye chiefdom was already on the rise at Great Zimbabwe. Further afield, around the Tati cluster of Botswana, numerous Zimbabwe tradition sites dating between the 10th and 11th centuries were also thriving independently as polities [14]. This therefore implies that individual agency was a key factor in the spontaneous development of multiple contemporaneous and overlapping complex societies that extensively scarred landscapes coinciding with the broader Guruuswa/Butua (see Figure 3) regions [2,49]. As already indicated above, the Khami phase was already developing by the turn of the 15th century, but who exactly was historically responsible for this extensive and complex cultural development?

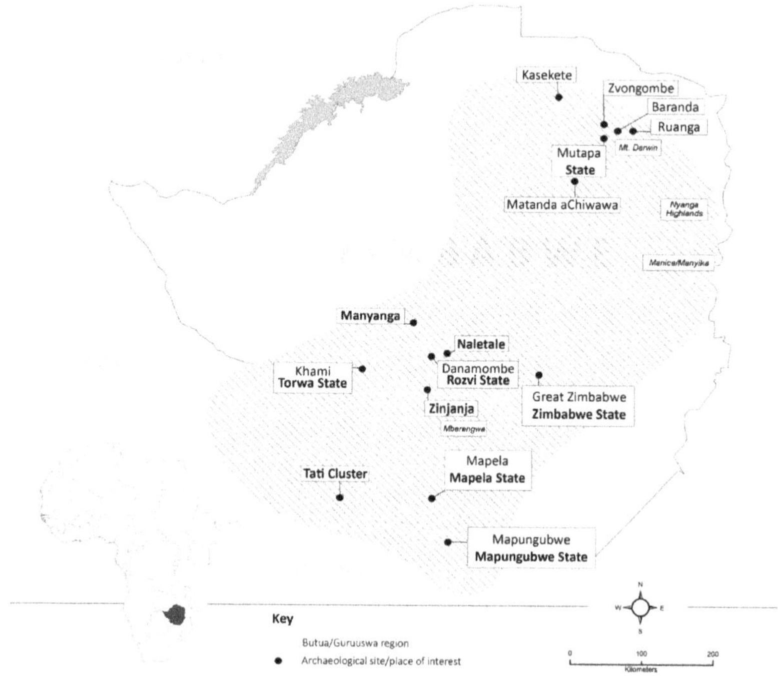


**Figure 3.** The highveld Butua/Guruuswa regions and associated Zimbabwe Culture states.

## 5. Early Dynastic History Mysteries' and the Khami Phase

Although there is a shared belief that the Torwa were the first dynasties to emerge and govern from Khami-phase sites, it remains quite difficult to tell whether they were also the sole founders of these settlements. Until now, the Torwa origin and historical identity remains a scholarly mystery [2,8]. More so, attributing all Khami-phase sites in southern Africa to Torwa dynasties would be farfetched, especially considering the extensive distribution of these sites. Respectively, the argument once advanced to challenge the predominance of Mapungubwe during the 13th century also comes to the fore here [49]. It would have been very difficult if not impossible for a few ruling elites based at the Khami capital to control such vast territories. Moreover, most of these Khami-phase sites are yet to be fully investigated, so claims that rulers from the Torwa state capital presided over the rest of the region are problematic. Be that as it may, one thing remains certain: before the Rozvi migrated into southwestern Zimbabwe, the Torwa were the ones in charge. As Randles [3] puts it, in 1512 Gaspar Veloso wrote a letter to the King of Portugal stating the following:

> "... between the land of the Monomotapa and Sofala all kings bore allegiance to the Monomotapa, but further inland (still on the plateau, of course) there was another king who had risen in revolt and with whom he was at war: the king of Butua. The latter was as powerful as the Monomotapa and his country held great reserves of gold".

In a different but related Portuguese account written by João de Barros in 1552, the Torwa regions were further described as follows:

> "They have other mines in a region named Toroa, also called the Kingdom of Butua, ruled by a prince [chief] named Burrom, a vassal of Benomotapa" [3].

These accounts constitute the earliest documentary references to southwestern Zimbabwe but they somehow mixed up the facts. Firstly, when the Portuguese arrived in southern Africa, they were not really interested in local historical details but economic trading opportunities. Hence, their references to local historical affairs should be considered as merely coincidental. Secondly, upon their arrival, they largely settled along the Indian Ocean coast, using Swahili traders as middlemen before venturing closer to their sources of ivory and gold around 1512 [1,3,50]. Therefore, their tendency to link everything internal with the Mutapa state should be appreciated in the context of their vested economic interests there. While it is true that the Butua state was a rival of the Mutapa state, it is not necessarily correct that it directly evolved from that state. In fact, von Sicard (1948:13) buttresses this point as follows:

> They must have been connected in one way or another not only with the first known Shangamire of the 15th century, but also with the south-western Butwa Kingdom under Boroma, heard of by Fernandes in 1513 and mentioned again by de Barros, in 1552. They were firmly established in Butwa Torwa long before Mutapa conquered the Dema or Limi tribes in the north-east... When at the beginning of the 18th century the great Shangamire conquered Butwa Torwa, most of the Twamamba still moved further south, and about half a century later they were more disintegrated by the advancing "Venda".

Still describing the same regions and people, he further comments:

> "There remains a last consideration with regards to the stone structures in the Twamamba country, including Great Zimbabwe. It must be remembered that Butwa Torwa is older than Monomotapa Kingdom in the north. Consequently, it existed before 1460... There is no reason why we should combine the Mutapa overthrow with the settlement of the Twamamba and kindred tribes in the southern parts of Butwa Torwa. On the other side there are indications that the Twamamba had occupied at least parts of it a long time before Mutapa established his rule north of the Zambezi. It seems, therefore, advisable to re-examine the whole question of the origin of our Rhodesian ruins... The material at my disposal is not sufficient to arrive at any definite conclusions as to the Twamamba building activities. But I should like to point out one special feature at several ruins, where it is most probable that the Twamamba have been at work ... probably a great number of our ruins are of Twamamba origin, since they are not built by the 18th century Rozwi..." [51].

These extracts clearly show the current confusion surrounding the historical identity of the Khami phase, especially when it comes to Torwa, Twamamba, Singo, and Rozvi identities. While others credit the Torwa as the first builders, von Sicard [51] gives credit to the Twamamba and disregards the Rozvi completely. However, Sicard [51] does not clearly qualify the relationship between the Torwa, Twamamba, and Singo dynasties. This prompts us to ask, were these the same people and, if not, how exactly were they different? Perhaps we may never fully get to the bottom of all this, however, we take solace in the fact that the Torwa/Butua state was quite peaceful right until AD 1644 when a civil war broke out following a succession dispute between two brothers aspiring for the same throne. Although this war did not mark the final end of the Torwa/Butua state, it severely weakened it right until AD 1685 when the Rozvi eventually usurped power. Portuguese meddling in the internal affairs of precolonial African states also continued right until the Rozvi finally put an end to all this between AD 1684 and 1693 [2]. Therefore, the next section shifts attention towards Rozvi origins and how they systematically naturalized their association with Khami-phase sites in southwestern Zimbabwe and broader landscapes.

## 6. The Power of Landscape Grammar: Rozvi Origins and Ethnicity

In terms of Rozvi origins, Mudenge [10] makes the following interesting comments:

> "In short there is no evidence from Portuguese sources that the Rozvi as a nation bearing the name 'Rozvi' existed before the advent of Dombo. Previous writers have at times unfortunately used eighteenth-century Portuguese documents clearly referring to Dombo (1684–95) as if they refer to the Changamire of c.1480–c.1492. It is partly this error which has given the impression that there are Portuguese documents which show the existence of the designation Rozvi prior to Dombo's advent. At present, however, we simply do not know of the existence of such sources".

Respectively, prior to Portuguese arrivals and the political turmoil that later ensued within the Mutapa state from the 1650s to 1684, the name Rozvi and the people bearing that name were simply non-existent. The earliest known and majority highveld plateau inhabitants were simply called Karanga and Kalanga speakers, while other diverse minority ethnic groupings were confined to the lowveld margins of the Zimbabwean plateau [1,2,12]. Thus, it was only after a series of historical events, particularly the Portuguese *prazo* system (farmland grabbing in the Mutapa state), *kuruva* tax collection (meant to force Africans into Portuguese farms and gold mines), and the numerous Portuguese political interferences in African states that the Rozvi identity eventually emerged [2,8,52]. However, these Portuguese interferences severely weakened the Mutapa state and inspired the rise of numerous land barons that aggressively exploited the traditional *Nyai* clientele system in order to build power [9]. As early as the 1670s, a wealthy provincial Mutapa cattle baron called Changamire Dombolakonachimwango or Dombo for short was on the rise. By the 1680s, he was well advanced, so he first rebelled against the Mutapa state in 1684 before launching other aggressive military campaigns between 1684 and 1693 [2,52]. His impressive record at the battlefront eventually earned him and his immediate Karanga followers the nickname *Varozvi,* which literary translates to 'the destroyers' [7].

Through space and time, the nickname *Varozvi* was systematically negotiated and transformed from its military connotations into a ruling elite and fluid 'imagined community' simply identified as 'Rozvi' [2,7,33]. In this respect, naming was strategically weaponized to produce and promote an ethnic community that resonated well with both past and emerging networks of memory. Therefore, Changamire Dombo successfully maximized on naming patterns to create a strong sense of belonging and cultural continuity, which challenged and changed old lines of memory and identity. Once the name Rozvi was fully established, it extensively fluctuated in meaning through space and time [2,7,8]. For some, Rozvi was just a name for Changamire Dombo and his immediate military followers, while for others it meant anyone historically, politically, and territorially connected to this Changamire and his state. Thus, Dionizio de Mello e Castro and Antonio Pinto de Miranda had this to say:

> While in charge of King Mucombue's herds, Changamire was instructed to drive them into the land of Orobze [Mello e Castro; Orobu in Pinto de Miranda]; there he rose in rebellion against the king having gathered together a great army; in consequence he is today the most powerful, the most dreaded and the most respected of all the chiefs who refuse to obey the emperors of Monomotapa, whom he has on several occasions defeated and routed in pitched battle. He is the absolute ruler over Abutua... [3].

Randles [3] further cites another Portuguese source that describes this Changamire rebel and his associated lands as follows:

> "... Urobze lies a long way from Manica, it would take a month to reach it, it is said to be nearer the Cape of Correntes. The Africans call it Goromucuro; it lies to the west of Manica. The Kingdom abounds in the rolling veld plains [campinas vistozas planices]. There are huge herds... There are many birds called 'emas'

[ostriches]. There are also very few trees, if at all, and firewood is replaced by cow-dung dried in the sun...".

Despite the variable spelling errors of Rozvi as Orobze, Orobu and Urobze, the Portuguese still reasonably managed to trace Rozvi origins, territories and shifts in name meanings. Most importantly, through the cultural practice of naming, Rozvi elites gradually imbued their constructed or invented identities on all former Torwa territories until they completely submerged them.

Therefore, Changamire Dombo and his immediate followers should be credited for their wisdom to exploit ordinary names as metaphors for their new identity and to claim space. Soon after their northeastern military campaigns around the Mutapa state, the Rozvi quickly retreated southwestwards into Torwa territories, also called Abutua, Goromucuro/Guruuswa, or Butua. The actual dates for their arrival and political takeover are contested but they generally fall between AD 1685 and 1696 [1,2,52]. At least these dates provide us rough estimates of the time period that Khami-phase sites became entangled with the Rozvi. As shown in the Portuguese extracts highlighted above, once the Rozvi overpowered the Torwa dynasties, everything associated with the former gradually became Rozvi. In fact, whenever dominant groups choose to use their own toponyms as part of new land claims, the histories of associated subordinate groups ultimately suffer. This is because toponyms are notorious for melding history with geography and conflating place with dominant group identity [53,54]. Thus, names are not just passive 'artefacts'; rather, they are active forces that shape processes of claiming and constructing landscape around specific ideological versions of the past [53]. Yi [55] further elaborates that naming entails taking possession, in fact, it is a cultural method of remapping indigenous geography. Therefore, just by renaming themselves and associated places as 'Rozvi', their identities simultaneously became part of the landscape. A detailed appreciation of this critical political strategy enables us to meaningfully explain how Khami-phase sites suddenly became Rozvi symbols even though the sites clearly predate and partly overlap with this historical group's political formation.

Apart from exploiting naming patterns to objectify their emerging identities, Rozvi elites also simply chose to inherit and settle on the most prominent former Torwa capitals as part of placemaking. This was a simple, wise, and powerful political strategy to historically and materially transform Torwa places into Rozvi places. This strategic move has also been theoretically termed Dasein, meaning (in the context of human beings as 'there-beings') 'intense inhabitation', while others simply call it 'dwelling' [21,39,43,56,57]. Through the practice of building or merely caring for, cherishing, and even occupying dwellings already built by others, the identities of the dominant newcomers are often easily imbued into usurped places and landscapes. Henry [39] takes pains to clarify the complexity of this powerful cultural practice by arguing:

> "... we make space significant by 'dwelling', by building, by investing time in it, so that it ceases to be space. Space is transformed into place by dwelling. Yet what does dwelling entail? How do we dwell? I stress that dwelling can be nothing but human social engagement, and as much as it evokes notions of 'caring' and 'heeding', such engagement or being-in-the-world inevitably generates situations of social conflict in which tensions and contradictions of identity politics come to the fore".

Even Tilley [21] echoes similar sentiments by stipulating that domestic dwellings are the material media through which relations between self and society are both objectified and negotiated. A detailed appreciation of this logic or 'invisible bonding with place' certainly goes a long way in resolving the perennial challenge to archaeologically detecting Rozvi signatures from related Khami-phase sites in southwestern Zimbabwe. Despite having multiple forms of oral traditions and historical text affirming that Khami-phase sites, especially those in the Insiza district, are part of the Rozvi past, it has been quite difficult to articulate how the historical Rozvi actually relate to and claim ownership of these ancient monuments [1,8,10,13–15]. This is the same challenge that prompted the renewed excavation of three former Rozvi capitals, namely, Danamombe (Figure 4), Naletale (Figure 5), and Zinjanja (Figure 6), illustrated on Figure 1 above.

**Figure 4.** Site of Danamombe (Source: Authors).

**Figure 5.** Site of Naletale (Source: Authors).

**Figure 6.** Site of Zinjanja (Source: Authors).

In brief, excavations conducted around these settlements [2,8], particularly sought to retrieve well-stratified radiocarbon dates and material culture that could be used to glean information on both Torwa and Rozvi cultural strata. Although certain radiocarbon dates retrieved from the selected research sites closely coincided with the Rozvi era of governance, AD 1685–1830 (see Tables S1 and S2), it was quite difficult, if not impossible, to pinpoint other symbolic Rozvi 'finds/things' because the associated archaeological record largely remained homogenous [8,15]. However, excavation data clearly indicated that all the three research site foundations significantly predated Rozvi origins, meaning the initial builders were certainly not Rozvi by identity. Since the material record largely remained unchanged, the noted slight material culture variations are better explained in terms of shifts in trade and other design preferences. However, considering the power of dwelling as part of placemaking processes, we might as well confidently account for Rozvi ethnicity in this manner. Mere site dwelling was powerful enough for Rozvi elites to politically lay claim to all prestigious monumental architecture, thereby conflating their identities with these places. The Rozvi further consolidated their territorial influence by exploiting the power of storytelling, which significantly asserted their identities in public spaces, places, and memory.

As Palone [43] observes, since there is more than one dimension to storytelling, any failure to critically read, write, and understand all stories of the land is suicidal on our part. Through storytelling involving the use of diverse narratives, allegories, tales, folklores, legends, myths, and taboos, among others, the Rozvi systematically managed to inscribe their identities on the landscape. Hodza and Fortune [58] highlights some of the prominent Rozvi tales:

> "The Rozvi, like many other groups, say that they came from a land called Guruuswa and from a place in that land called Manyanga... Their traditions tell us that they were accompanied on their travels by a voice which they called Tovela which led them on their way, keeping them safe from dangerous places. The voice could speak from any object it chose and we are told that it spoke through grass or trees or through a little child, unable as yet to speak for itself. The voice of a presence whom some say was that of the first Rozvi ever created, the founding father of the clan. Others say he was the first person ever to be created and that after his death his spirit accompanied the different groups from Guruuswa to the countries they chose. Tovela not only protected his people on their travels but gave them food as well. When they clapped in homage and supplication to him at a tree, portions of porridge, pots of milk, and combs of honey would come out of the ground. He also gave them medicines to become

> invincible but not inaudible to their foes and to be long-lived. Tovela came to be called by other names by different groups such as Mwari...".

These narratives vividly connected their past with the broad Guruuswa landscape, the natural environment, and associated monumental settlements. Interestingly, the last Rozvi state king, called Chirisamhuru, ruled from the Khami-phase site of Manyanga, which is widely cited in numerous Rozvi oral traditions. Even today, a few spiritual Rozvi descendants still stay at Manyanga while others frequently visit it for their sacred rituals. Most importantly, several Zimbabwe Culture sites, (*madzimbahwe*) including Great Zimbabwe, sacred landscapes such as the Matobo hills, and various other mountains scattered across the Zimbabwean plateau are widely remembered in oral traditions and histories as Rozvi places [2,6]. Thus, through mere storytelling, the Rozvi managed to extensively claim much wider landscapes and places across the Zimbabwean plateau and beyond (see Figure 7 below). Hodza and Fortune [58] also highlight other oral histories that were collected from several Rozvi chiefs in the following manner:

> "...When the Rozvi reached this country, they found Munhumutapa ruling it. They attacked him and drove him away. This was the first fight in which they had ever engaged. After Munhumutapa's flight, they subdued all the chiefs who had been subject to him. They are said first to have gone north, to Pfura, then east, to Nyanga, then south to Manyika and Bikita, and finally west to Mberengwa (Belingwe). Tradition explains that the name Rozvi was given to them because they disturbed the lives of everyone, man and beast alike. All this is said to have happened under the rule of the first chief whom they remember, called Changamire...".

Another tradition collected from Mbava, a Rozvi chief well advanced in age goes as follows:

> "The Warozwi came with a great army and conquered all the country. King Chiduku of the Warozwi ruled over the country of the Mawungwe, the Manyika, the Mabocha, the Wanyashanu, the Makombe and the Mazezuru. After this the Warozwi were the over-lords of the whole country, and chose all the new chiefs. If any chief died, his successor had to be confirmed by the Warozwi. If the people chose a new chief without consulting their over-lord, it was said in common parlance that they had no chief, but the one selected or approved by the Warozwi was considered to be a true chief... When a chief died the people went and announced it to the Warozwi, and the Warozwi sent heralds with black and white calico and the skin of a sheep. When the heralds arrived, they anointed the new chief, probably the one chosen by the people. They anointed him with oil and clothed him with the black and white calico... I remembered that when Mbava was born the Portuguese had been in the country for a long time, and so the king of the Warozwi used to send his servants with elephants' tusks to buy calico for the robing of new chiefs... The place called Zimbabgwe is a holy place. Mbava says that in the old days only chosen people were allowed to enter it and they only entered once to offer sacrifice. All the common people stayed at the foot of the mount... sacrifice of three or four black hornless cattle and elephants' tusks and gold and black and white calico" [59].

Such stories asserted complex Rozvi identity construction processes by emphasizing their spiritual dexterity, power over the landscape and its people, prolonged governance, and supreme control over all chiefs through space and time. Stories highlight what communities value in the landscape, they project the intimate relationship between people and places, and above all help generations to remember their ancestral places and landscapes. Hence, Palone [43] draws our attention to the fact that placeless stories lack power and meaning, leaving readers with little space to imagine themselves. In this respect, place and landscape act as some kind of historical text or grids upon which narrative memories are anchored, they help people to remember and they foster in people a deep sense of identification with and belonging to a group [21,39]. Above all, and for purposes of this paper, stories particularly help us to find that which we cannot find through conventional methods in archaeology [18]. Put differently, stories constitute the invisible bond between people, places, and landscapes; they cement or capture the landscape in a much more robust and intimate manner. Henceforth, indigenous stories, the land, material culture, and people are always bound; any isolated approach towards any of these social and physical elements is tragically destined to fail, become frustrating, misleading, and totally confusing [21,43,54,55].

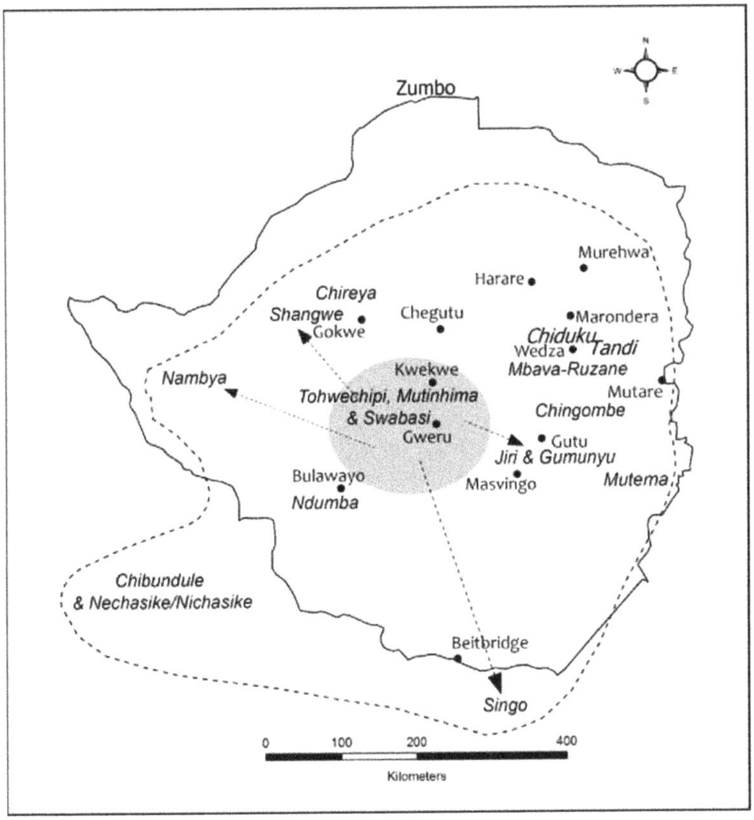


**Figure 7.** Map showing the core and broader landscapes associated with Rozvi identities.

## 7. Conclusions

As the post-colonial mindset and methodologies continue to shape archaeological thinking today, novel frameworks to free and untangle African historiography from colonial rhetoric are increasingly emerging and gaining momentum. For instance, calls to re-imagine precolonial sociopolitical landscapes in Africa using indigenous epistemic perspectives are a welcome development. While colonialism undoubtedly denigrated and perpetuated calculated racial scholarship and biases against colonized societies, it is important to also acknowledge it. It is worth noting that aligning with post-colonial or indigenous paradigm shifts is not to take away from the efforts of earlier scholars. They were children of their times, informed by the dominant theoretical and methodological environments existing then. Though disagreeing with our present sensibilities, some of the colonial perspectives explored above were indeed the prevailing academic norms less than a century ago.

Therefore, despite the initial 'aggressive tone' set out in this article, we acknowledge that not all aspects of Western colonial scholarship were mischievous, disrespectful, and hateful to African people and their histories. On the contrary, and again using archaeological science as a lens, scientific insights into the past have enabled us to examine ancient African societies that would have otherwise remained inaccessible or nearly impossible to explore beyond contemporary memory. Given the myriad biases and distortions typical of oral traditions, we cannot dare imagine how complicated it would have been for historians and archaeologists alike to comprehend diverse and conflicting narratives emanating from adamant indigenous communities seeking to legitimize their biased claims for land, identity, power, and cultural achievements among other things. With that in mind, we however remain unapologetic about the urgent need to push some of our 'prehistory archaeology' colleagues out of their false comfort zones so as to draw them much closer to indigenous communities' past realities. Material culture, even when put under advanced microscopes, is not enough to inform us about dynamic processes in the past. Hence, there is need for us to reinvest more towards exploring the invisible, spiritual, and intangible dimensions of past societies.

This paper underscores the need for archaeologists to closely engage with indigenous epistemologies and ontologies when examining processes from the past that situationally produced identities, objects, and landscapes. No matter how rigorous archaeological scientific methods become, our reading of material culture and landscapes is never complete without alternative histories that provide invaluable contextual meanings. In fact, a strict 'scientific' or material-based approach to the past largely obscures ideological meanings that relate to those 'things' we see, touch, and verify [18,21,42]. This growing realization is prompting the prioritization of metaphors as part of attempts to centralize 'people' instead of 'objects/things' as well as the 'how?' instead of the 'what?' questions [18,20,21]. Metaphors particularly open up new ways of thinking and talking about 'things' (material culture] and they help us explore those processes and practices that ultimately produced the stable entities at our disposal. More precisely, metaphors significantly enrich our accounts of the past by bringing to light those processes and historical contexts that produced fluid identities, societies, landscapes, and meanings of material culture [20]. This approach also helps us move away from the traditional inherited frameworks that encourage essentialism and a static view of the past towards social memory, identities, and agency [16,19,60]. In this context, the social and ideological dimensions of landscapes come to the fore as landscapes become 'houses' of myths, legends, folklores, narratives, tales, and local stories [17,18]. For too long, these crucial and complex indigenous discourses have been marginalized, dismissed, and ignored in archaeological knowledge production processes, yet they provide some of the deepest and best insights into the intimate relationship between people, places, and landscapes.

This submission has also advocated for the prioritization of vernacular languages and associated meanings in archaeological and heritage research endeavors [18–20,47]. Firstly, this paves way for multivocality by giving space to the once-silenced voices whose past we seek to study and understand. Secondly, indigenous community views enable us to systematically counter the colonial library by creating alternative lines of thinking. Last but not least, indigenous language proficiency always gives external researchers direct access to 'closely guarded community secrets' often embedded in complex myths, legends, proverbs, and riddles. As Shennan [61] observes, who we are affects how we perceive the world, so once we are able to unlock indigenous language barriers, we gain access into the 'Other's worldview', which naturally refreshes our traditional professional biases. Smith [23] further adds that indigenous languages are central to research inquiries because they carry a lot of complex meanings, realities, and histories of people, places, and events through cultural systems such as naming patterns. In support, Palone [43] argues that interactions between culture, language, and landscape naturally form the basis of all human stories. Since we are particularly keen on gathering and generating new stories, it is key to always bear in mind that our reading and understanding of the landscape is never complete enough without a full appreciation of the associated indigenous community perceptions of the same landscape. This is the kind of thinking that should now guide our interrogation and interpretation of landscapes, including the Rozvi archaeological past in southwestern Zimbabwe. An approach of this nature will certainly go a long way in addressing the notorious absence of fit between oral traditions and the archaeological record.

To crown it all, contested meanings such as Rozvi claims to all *madzimbabwe* (archaeological monuments) and the disparities noted between their origins and related site chronologies can now be meaningfully explained. By considering the Rozvi as a very fluid, dynamic, and imagined community, rather than a static, frozen, and bound social entity in which to root essentialized identities, we create room to informatively account for their complex archaeological past. Thus, through simply exploiting the power of naming, storytelling, and dwelling as part of their ethnic identity construction and placemaking processes, the Rozvi systematically and ideologically laid claim to vast territories and associated prestigious material culture. Their intimate connection with objects, places, and landscapes was premised on a wealth of IKS that are 'archaeologically invisible' but 'epistemologically accessible'. Therefore, in order to effectively denote Rozvi archaeological signatures, one needs to closely engage an interdisciplinary framework such as historical archaeology. Fortunately, historical archaeology has already made its mark by becoming sensitive to local needs, sensibilities, and structures. Most importantly, it recognizes that oral histories can sometimes differ from archaeological narratives, which is quite positive and realistic [17,18,62]. Tilley [21] further reinforces that while imaginations of identities constantly change, associated material symbolism may remain fixed through time. Hence, symbols of identities such as the land, objects, monuments, and landscapes, among other things, may remain fixed, while related ideas, beliefs, emotions, values, and memories constantly change. Simply put, objects, monuments, the land, and landscapes are active agents of identities, but they also double-up as ever-evolving indigenous archives, repositories of social memory, and mnemonic devices [21,43,44,55].

**Supplementary Materials:** The following supporting information can be downloaded at: https://www.mdpi.com/article/10.3390/land12081625/s1, Table S1: Detailed summary of samples submitted for radiocarbon dating and calibrated dates from the three research sites [8]; Table S2: Summary of additional samples from Danamombe submitted for radiocarbon dating and calibrated dates [8].

**Author Contributions:** Conceptualization, L.H.M. and R.K.; writing—original draft preparation, L.H.M.; writing—review and editing, R.K. All authors have read and agreed to the published version of the manuscript.

**Funding:** This research received external funding through the South African National Research Foundation.

**Acknowledgments:** Authors acknowledge our international fellowship funders Alexander von Humboldt Foundation and the British Academy and host institutions; the Institute for African Studies and Egyptology, University of Cologne; Firebird Foundation for Anthropological Research to support ethnographic research among Rozvi communities in Zimbabwe and the School of Archaeology, University of Oxford. The radiocarbon dates for this research were funded through the South African National Research Foundation (NRF) grant titled 'Great Zimbabwe in Recent History: Historical Archaeologies of South-Central Zimbabwe' (African Origins Platform Grant No. 98833), administered by the University of Pretoria.

**Conflicts of Interest:** The authors declare no conflict of interest.

## References

1. Beach, D.N. *The Shona and Zimbabwe 1900–1850*; Mambo Press: Gwelo, Zimbabwe, 1980.
2. Machiridza, L.H. Archaeology of the Rozvi: Toward a Historical Archaeology of South-Western Zimbabwe. Ph.D. Dissertation, University of Pretoria, Pretoria, South Africa, 2018.
3. Randles, W.G.L. *The Empire of Munhumutapa from the Fifteenth to the Nineteenth Century*; Mambo Press: Gwelo, Zimbabwe, 1979.
4. Huffman, T.N. *Snakes and Crocodiles: Power and Symbolism in Ancient Zimbabwe*; Witwatersrand University Press: Johannesburg, South Africa, 1996.

5. Huffman, T.N.; Piesanie, J. The Historical Archaeology of Mapungubwe: Khami, Venda and Machete. *J. Afr. Archaeol.* **2011**, *9*, 189–206. [CrossRef]
6. Machiridza, L.H. Material Culture and Dialectics of Identity and Power: Towards a Historical Archaeology of the Rozvi in South-Western Zimbabwe. Ph.D. Dissertation, University of Pretoria, Pretoria, South Africa, 2012.
7. Machiridza, L.H. Insights into the Meaning of VaNyai, Rozvi and Torwa: A Historical Archaeology Approach to Identities. In *Zimbabwean Archaeology: In the Post-Independence Era*; Manyanga, M., Katsamudanga, S., Eds.; SAPES Books: Harare, Zimbabwe, 2013; pp. 199–212.
8. Machiridza, L.H. Landscapes and Ethnicity: An Historical Archaeology of Khami-Phase Sites in Southwestern Zimbabwe. *J. Hist. Archaeol.* **2020**, *54*, 647–675. [CrossRef]
9. Mazarire, G.C. Reflections on Pre-Colonial Zimbabwe, c.850–1880s. In *Becoming Zimbabwe: A History from Pre-Colonial Period to 2008*; Raftopoulos, B., Mlambo, A.S., Eds.; Weaver Press: Harare, Zimbabwe, 2009.
10. Mudenge, S.I.G. An Identification of the Rozvi and Its Implications for the History of the Karanga. *Rhod. Hist.* **1974**, *5*, 19–32.
11. Ncube, G.T. *A History of Northwestern Zimbabwe 1850–1960*; Mond Books: Kadoma, Japan, 2004.
12. Pikirayi, I. *The Zimbabwe Culture; Origins and Decline of Southern Zambezian States*; Altamira Press: New York, NY, USA, 2001.
13. Robinson, K.R. The Archaeology of the Rozwi. In *The Zambezian Past*; Stokes, E., Brown, R., Eds.; Manchester University Press: Manchester, UK, 1966; pp. 3–27.
14. Waarden, C. The Origin of Zimbabwe Tradition Walling. *Zimbabwean Prehistory* **2011**, *29*, 54–77.
15. Koleini, F.; Machiridza, L.H.; Pikirayi, I.; Colomban, P. The Chronology of Insiza Cluster Khami-Phase Sites in Southwestern Zimbabwe: Compositional Insights from PXRF and Raman Analysis of Excavated Exotic Glass Finds. *Archeomentry* **2019**, *61*, 874–890. [CrossRef]
16. Jones, S.; Russell, L. Archaeology, Memory, Oral Tradition: An Introduction. *Int. J. Hist. Archaeol.* **2012**, *16*, 267–283. [CrossRef]
17. Lane, P.J. Possibilities for a Postcolonial Archaeology in Sub-Saharan Africa: Indigenous and Usable Pasts. *World Archaeol.* **2011**, *43*, 7–25. [CrossRef]
18. Sinamai, A. Myths as Metaphors: Understanding Narratives in Sustaining Sacred Landscapes in Zimbabwe and Australia. In *Archives, Objects, Places and Landscapes: Multidisciplinary approaches to Decolonised Zimbabwean Pasts*; Manyanga, M., Chirikure, S., Eds.; Langaa: Bamenda, Cameroon, 2017; pp. 399–419.
19. Schmidt, P.R.; Walz, J.R. Historical Archaeology in Africa: Noble Claims. *Revis. Perspect. Afr. Voices J. Afr. Hist.* **2005**, *46*, 315–319.
20. Stahl, A.B. Route Work' through Alternative Archives: Reflections on Cross-Disciplinary Practice. *S. Afr. Hist. J.* **2010**, *62*, 252–267. [CrossRef]
21. Tilley, C. Identity, Place, Landscape and Heritage. *J. Mater. Cult.* **2006**, *11*, 7–32. [CrossRef]
22. Dei, G.J.S. Rethinking the Role of Indigenous Knowledge in the Academy. *NALL Work. Pap.* **2002**, *58*, 1–25.
23. Smith, L.T. *Decolonizing Methodologies: Research and Indigenous Peoples*; Zed Books Ltd.: London, UK, 1999.
24. Machiridza, L.H.; Musindo, T. Challenges, Distortions and Prospects: A Review of Historical Archaeology in Southern Africa. In *Reference Module in Social Sciences*; Elsevier: Amsterdam, The Netherlands, 2023; pp. 1–8.
25. Semali, L.M.; Asino, T.I. Decolonizing Cultural Heritage of Indigenous People's Knowledge from Images in Global Films. *J. Decolonization Indig. Educ. Soc.* **2013**, *2*, 25–56.
26. Mawere, M. Indigenous Knowledge and Public Education in Sub-Saharan Africa. *Spectrum* **2015**, *50*, 57–71. [CrossRef]
27. Russell, L. Remembering Places Never Visited: Connections and Context in Imagined and Imaginary Landscapes. *Int. J. Hist. Archaeol.* **2012**, *16*, 401–417. [CrossRef]
28. Dei, G.J.S. Indigenous Anti-Colonial Knowledge as 'Heritage Knowledge' for Promoting Black/African Education in Diasporic Contexts. *Decolonization Indig. Educ. Soc.* **2012**, *1*, 102–119.
29. Schmidt, P.R. An Alternative to a Strictly Materialist Perspective: A Review of Historical Archaeology, Ethnoarchaeology and Symbolic Approaches in African Archaeology. *Am. Antiq.* **1983**, *48*, 62–79. [CrossRef]
30. Schmidt, P.R.; Pikirayi, I. Will Historical Archaeology Escape Its Western Prejudices to Become Relevant to Africa? *Archaeol. J. World Archaeol. Congr.* **2018**, *14*, 443–471. [CrossRef]
31. Connah, G. Historical Archaeology in Africa: An Appropriate Concept? *Afr. Archaeol. Rev.* **2007**, *24*, 35–40. [CrossRef]
32. Lucas, G. Archaeology at the Edge. An Archaeological Dialogue with Martin Hall. *Archaeol. Dialogues* **2006**, *13*, 55–67. [CrossRef]
33. Machiridza, L.H. Setting the Parameters for Reconsidering the Rozvi Archaeological Identity in South-Western Zimbabwe: A Historical Archaeology Perspective. Bachelor's Dissertation, University of Zimbabwe, Harare, Zimbabwe, 2005.
34. Robinson, K.R. *Khami Ruins*; Cambridge University Press: Cambridge, UK, 1959.
35. Kuklick, H. Contested Monuments: The Politics of Archaeology in Southern Africa. In *Colonial Situations: Essays on the Contextualization of Ethnographic Knowledge*; Stocking, G.W., Ed.; University of Wisconsin Press: Madison, WI, USA, 1991; pp. 135–169.
36. Russell, L.; McNiven, I.J. Monumental Colonialism: Megaliths and the Appropriation of Australia's Aboriginal Past. *J. Mater. Cult.* **1998**, *3*, 283–299. [CrossRef]
37. Silverberg, R. *Mound Builders of Ancient America*; University of Ohio Press: Athens, OH, USA, 1968.
38. Held, M.B.E. Decolonizing Research Paradigms in the Context of Settler Colonialism: An Unsettling, Mutual, and Collaborative Effort. *Int. J. Quant. Methods* **2019**, *18*, 1–16. [CrossRef]
39. Henry, R. *Performing Place, Practicing Memories: Aboriginal Australians, Hippies and the State*; Berghahn Books: Oxford, UK, 2012.

40. Anschuetz, K.F.; Wilshusen, R.H.; Scheick, C.L. An Archaeology of Landscapes: Perspectives and Directions. *J. Archaeol. Res.* **2001**, *9*, 157–211. [CrossRef]
41. Relph, E. A Pragmatic Sense of Place. *Environ. Archit. Phenomenol.* **2009**, *20*, 22–31.
42. Hand, J. Spatial Organization and Material Culture as Means to Study Identity. *Post Hole* **2015**, *44*, 23–30.
43. Palone, A. Dimensions of Landscape Storytelling. In *Environmental Readings: Theory*; Fleming, B., Palone, A.J.R., Eds.; Center for Sustainable Development: Austin, TX, USA, 2012; pp. 144–153.
44. Vansina, J. *Oral Traditions as History*; Heinemann: Nairobi, Kenya, 1985.
45. Huffman, T.N. Historical Archaeology of the Mapungubwe Area: Boer, Birwa, Sotho-Tswana and Machete. *S. Afr. Humanit.* **2012**, *24*, 33–59.
46. Mukwende, T. An Archaeological Study of the Zimbabwe Culture Capital of Khami, South-Western Zimbabwe. Ph.D. Dissertation, University of Cape Town, Cape Town, South Africa, 2016.
47. Chirikure, S.M.; Mukwende, M.; Moffett, A.J.; Nyamushosho, R.T.; Bandama, F.; House, M. No Big Brother Here: Heterarchy, Shona Political Succession and the Relationship between Great Zimbabwe and Khami, Southern Africa. *Camb. Archaeol. J.* **2018**, *28*, 45–66. [CrossRef]
48. Main, M.; Huffman, T. *Palaces of Stone: Uncovering Ancient Southern African Kingdoms*; Struick Travel and Heritage: Cape Town, South Africa, 2021.
49. Chirikure, S.; Manyanga, M.; Pollard, A.M.; Bandama, F.; Mahachi, G.; Pikirayi, I. Zimbabwe Culture before Mapungubwe: New Evidence from Mapela Hill, South-Western Zimbabwe. *PLoS ONE* **2014**, *9*, e111224. [CrossRef] [PubMed]
50. Pikirayi, I. *The archaeological identity of the Mutapa State: Towards an historical archaeology of northern Zimbabwe*; Studies in African Archaeology 6; Uppsala University: Uppsala, Sweden, 1995.
51. Sicard, H. The Origin of Some of the Tribes in the Belingwe Reserve. *Nativ. Aff. Dep. Annu.* **1948**, *25*, 93–104.
52. Mudenge, S.I.G. *A Political History of the Munhumutapa*; Zimbabwe Publishing House: Harare, Zimbabwe, 1988.
53. Alderman, D.H. Place, Naming and the Interpretation of Cultural Landscapes. In *The Routledge Research Companion to Heritage and Identity*; Howard, P., Graham, B., Eds.; Routledge: London, UK, 2008; pp. 195–213.
54. Berger, T. Place, Imaginary, Identity: Place Ethnography in truth or Consequences, New Mexico. Ph.D. Thesis, The University of New Mexico, Albuquerque, NM, USA, 2016.
55. Yi, I. Cartographies of the Voice: Storying the Land as Survivance in Nature American Oral Traditions. *Humanities* **2016**, *5*, 62. [CrossRef]
56. Bachelard, G. *The Poetics of Space*; Beacon Press: Boston, MA, USA, 1969.
57. Heidegger, M. *Poetry, Language, Thought*; Harper and Row: New York, NY, USA, 1971.
58. Fortune, G.; Hodza, A.C. *Shona Praise Poetry*; Oxford University Press: Oxford, UK, 1979.
59. Lloyd, E.M. Mbava and Others. *Nativ. Aff. Dep. Annu.* **1925**, 62–64.
60. Robertshaw, P. African Archaeology, Multidisciplinary Reconstructions of Africa's Recent Past, and Archaeology's Role in Future Collaborative Research. *Afr. Archaeol. Rev.* **2012**, *29*, 95–108. [CrossRef]
61. Shennan, S.J. Archaeological approaches to cultural identity. In *Archaeological Approaches to Cultural Identity*; Shennan, S.J., Ed.; Unwin Hyman Ltd.: London, UK, 1989; pp. 1–32.
62. Hershey, R.A.; McCormack, J.; Newell, G.E. Mapping Intergenerational Memories (Part I): Proving the Contemporary Truth of the Indigenous Past. *Ariz. Leg. Stud. Discuss. Pap.* **2014**, *14*, 1–27. [CrossRef]

land

MDPI

*Article*

# Older Is Not Necessarily Better: Decolonizing Ifugao History through the Archaeology of the Rice Terraces

**Stephen B. Acabado [1,*] and Marlon M. Martin [2]**

1 Department of Anthropology, University of California (UCLA), Los Angeles, CA 90095, USA
2 Save the Ifugao Terraces Movement, Kiangan 3604, Ifugao, Philippines; marlon.martin12@yahoo.com
* Correspondence: acabado@anthro.ucla.edu

**Abstract:** This study examines the intersection of archaeological data and community narratives in interpreting the Ifugao Rice Terraces in the Philippines, a UNESCO World Heritage Site. Long regarded as 2000-year-old symbols of an uncolonized cultural past, recent research challenges this view, suggesting a 16th-century origin coinciding with Spanish contact. The longstanding characterization of the Ifugao Rice Terraces as 2000-year-old monuments cemented a perception of Ifugao culture as static and unchanging, overshadowing the dynamic cultural practices that have persisted and evolved over the centuries. It is crucial to recognize that these terraces are not frozen in time but are active representations of Ifugao's living culture, which has continually adapted to social, environmental, and historical changes while maintaining its distinct identity. This paradigm shift, supported by radiocarbon dating and ethnohistorical analysis, aligns more closely with local oral histories and portrays the Ifugao not as passive inheritors of tradition but as active participants in their history. We argue for the integration of scientific data with community stories, presenting a holistic understanding of the terraces as dynamic elements of Ifugao resilience and identity. The findings advocate a move away from romanticized historical interpretations toward a narrative that respects the complexity and adaptability of Indigenous cultural landscapes.

**Keywords:** decolonizing archaeology; community engagement; Ifugao; Philippines; rice terraces

## 1. Introduction

The discipline of archaeology often finds itself at the crossroads of past and present, where the narratives of history meet the interpretations of the modern era. This intersection is fraught with challenges, particularly the tendency to romanticize the past, leading to a skewed understanding that often intertwines with national consciousness. Such romanticized views are not merely academic indulgences; they shape identities, policies, and perspectives of entire communities. A striking example of this phenomenon is the Ifugao Rice Terraces in the Philippines, a UNESCO World Heritage Site (Figure 1). These terraces are not just remarkable feats of agricultural engineering but also a palimpsest onto which various narratives of history, identity, and culture have been projected (Figure 2).

We underscore the deeply ingrained controversy centered around the ethnohistory of the rice terraces and propose a pathway to resolution that involves collaboration between archaeologists and the Ifugao community. It foregrounds the notion that the traditional archaeological narrative, heavily influenced by early 20th-century anthropologists like Roy Franklin Barton [1] and Henry Otley Beyer [2], has promulgated a view of the Ifugao and their terraces as static relics of a distant past. This perspective, entangled with nationalist pride, asserts the terraces' antiquity as a symbol of uncolonized purity and cultural identity. The terraces are reflective of cultural and historical dynamism of the Ifugao, but the 2000-year old origin argument appears to make these terraces static.

This paper aims to demystify, and in the process, decolonize, archaeological and historical narratives developed in the early 1900s by pioneer anthropologists who worked in the Cordillera region of the Philippines. In addition, the dating of the Ifugao Rice Terraces has been part of the nationalist discussions that older is better, but at the same time, relegating the Ifugao to the sidelines of history. This can be achieved by actively engaging stakeholder and descendant communities, as exemplified by our work.

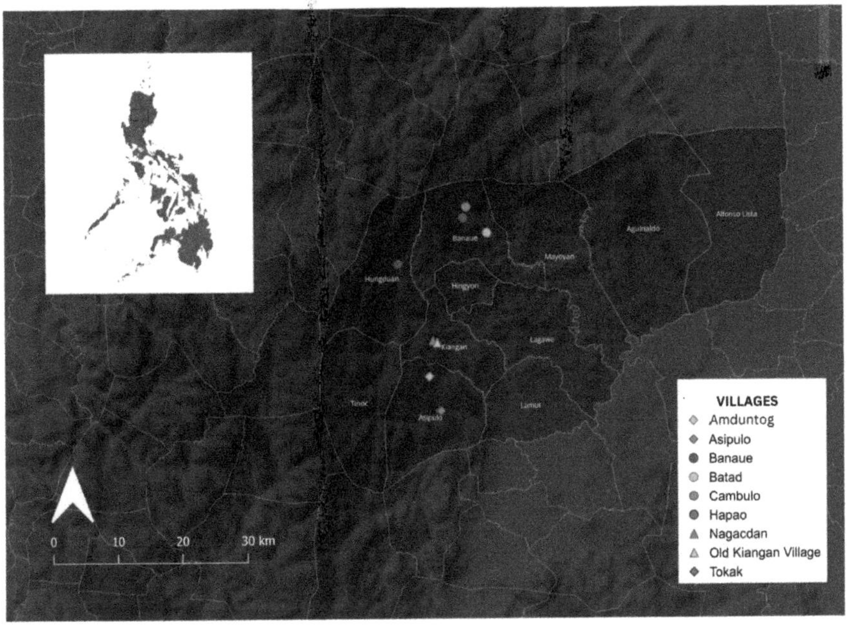


**Figure 1.** Ifugao Province, with approximate locations of major sites mentioned in the text. The municipality of Asipulo includes Tokak and Amduntog; Cambulo is adjacent to Batad; and Old Kiyyangan Village is adjacent to Nagacadan (Map by NJ Roxas and the IAP).

For decades, the dominant narrative, championed by early anthropologists such as Roy Franklin Barton [1] and Henry Otley Beyer [2], posited that the Ifugao Rice Terraces were over 2000 years old. This narrative was more than just a chronological assertion; it was a statement of cultural identity. It presented the Ifugao as custodians of an ancient, unchanging tradition, existing in isolation from the influences and tumults that shaped the rest of the archipelago. In this view, the terraces were not only an agricultural wonder but also a symbol of a pure, uncolonized past. This interpretation has been echoed in academic circles, popular media, and even in the descriptions of the terraces in UNESCO documents, reinforcing the image of the Ifugao as a timeless entity, disconnected from the currents of history.

While only two published materials purports to argue for the 2000-year origin of the terraces, this has become entrenched in the Philippine historical narratives because of the inclusion of the dating on Philippine curricula. As such, our approach challenges this long-standing narrative by actively involving the Ifugao community in the research process. By doing so, we uncover a more nuanced and historically grounded understanding of the Ifugao Rice Terraces and the people who have shaped their history over generations.

Recent archaeological findings have challenged this long-standing belief [3–7]. Utilizing a combination of radiocarbon dating, archaeobotanical analysis, and ethnohistorical research, these studies suggest a much later inception of the terraces, likely around the period of Spanish colonization in the 16th century. This revelation is not just a mere adjustment of dates; it represents a paradigm shift in how the history of the Ifugao and their iconic terraces is understood. It suggests that rather than being a relic of a distant past, the terraces are a dynamic, living testament to the Ifugao's resilience and adaptability in the face of external influences.

**Figure 2.** The Mayoyao Rice Terraces, one of the five terrace clusters listed on the UNESCO Cultural Landscapes World Heritage Sites (Photo by Bhong Tawana and the IAP).

The Implications of this new interpretation force the rethinking of dominant historical narratives. For one, it challenges the narrative of the Ifugao and their terraces as isolated and unchanging. It paints a picture of a people who are not mere passive recipients of history but active agents who have continuously adapted and evolved in response to changing circumstances. This view aligns more closely with the oral histories and narratives of the Ifugao community, which have often been sidelined in academic discourse in favor of more 'scientific' interpretations.

Community stories and oral histories are invaluable in this recontextualization, providing a complementary, as well as supplementary, narrative to the archaeological data. In Ifugao, these oral traditions serve as living records chronicling the collective memory and societal changes over generations. They provide insight into the cultural significance of the terraces, detailing the struggles and achievements of the Ifugao people that are not always visible in the archaeological record. These narratives often highlight the terraces as symbols of identity and self-determination, countering the narratives of isolation and stasis with tales of dynamism, community cohesion, and resistance to colonial pressures.

These oral traditions offer a nuanced understanding that can challenge and refine archaeological interpretations. They can affirm dates and events, suggest alternative readings of material culture, and illuminate the sociopolitical structures that facilitated the construction and maintenance of the terraces. For example, the community's recounting of the forced relocations and the strategic establishment of settlements away from Spanish incursions provides context to the changes in settlement patterns inferred from archaeological digs. Moreover, the oral histories of agricultural practices, rituals, and social organization offer a vivid picture of the day-to-day life that shaped the terraces' landscape. When oral histories are integrated into the archaeological discourse, they enrich the narrative, allowing a more holistic reconstruction of the past that acknowledges the active role of the Ifugao in shaping their history and landscape.

Our archaeological work in Ifugao represents a movement towards the decolonization of archaeology. In our work, we define decolonization as the process of re-evaluating and transforming practices to remove colonial biases and structures. It seeks to decenter Western epistemologies, prioritize Indigenous perspectives, and value local knowledge systems on par with scientific methods. This means recognizing and rectifying the historical power imbalances in the interpretation, narration, and stewardship of cultural heritage i.e., [8–10]. Decolonizing archaeology entails actively involving descendant communities in the research process, respecting their oral traditions, and acknowledging their sovereignty over their cultural narratives and material past [11–13].

### 1.1. *Changing the Narrative*

The 2000-year dating of the Ifugao Rice Terraces has played a pivotal role in shaping perceptions of Ifugao culture as static and unchanging through millennia. The assertion of such antiquity has been intertwined with the narrative of an unchanging culture, one that has ostensibly remained isolated and untouched by the broader historical and cultural dynamics that have swept through the rest of the Philippines and Southeast Asia.

The 2000-year-old construction dating assumes static culture since archaeological models strongly argue that if intensive agriculture has been documented, then we expect to see developments in other aspects of culture. It is then assumed that the Ifugao did not do anything else but plant rice for two millennia. This perspective also suggests that the Ifugao people have managed to preserve their way of life, with little to no external influence, maintaining the same techniques of rice cultivation and terrace construction passed down through generations. It romanticizes the Ifugao as custodians of an ancient tradition, living relics of a precolonial past. This narrative has been appealing because it offers a sense of timelessness and purity, resonating with nationalist sentiments that valorize Indigenous cultures as symbols of national identity and resistance to colonial forces.

#### 1.1.1. The Notion of Isolation

Closely related to the idea of an unchanging culture is the notion of isolation. The dating implies that the Ifugao Rice Terraces—and by extension, the Ifugao people—have existed in a bubble of isolation, shielded from the transformative influences of trade, colonialism, and cultural exchange that have characterized the history of the region. This perception of isolation has been used to support the argument that the terraces are a testament to Indigenous ingenuity and resilience, constructed and maintained without external influence.

However, this view oversimplifies the complex historical and cultural dynamics that have shaped the Ifugao region. It overlooks evidence of trade, interaction, and conflict with neighboring groups and colonial powers, which have undoubtedly influenced Ifugao society. Moreover, it disregards the adaptability and agency of the Ifugao people, who have navigated the challenges posed by external pressures and internal developments over centuries.

#### 1.1.2. Critique and Reevaluation

Recent archaeological findings and ethnohistorical research challenge the narrative of an unchanging, isolated culture. Studies suggesting a much later inception of the terraces, around the time of Spanish colonization in the 16th century, highlight the dynamic nature of Ifugao society. This evidence indicates that the terraces are not relics of a static past but rather the outcome of a society's adaptive strategies in response to social, environmental, and political pressures.

The reconsideration of the terraces' dating prompts a broader reflection in how cultures are perceived and represented in historical narratives. It underscores the need to move beyond simplistic notions of timelessness and isolation, recognizing the fluidity and interconnectedness of human societies. Acknowledging the terraces as a product of historical contingencies rather than a symbol of unchanging tradition allows for a more nuanced understanding of the Ifugao's cultural resilience and ingenuity.

Emerging archaeological and ethnohistorical evidence challenges notions of an unchanging and isolated culture, highlighting the dynamic and interconnected nature of Ifugao history. This reevaluation not only enriches our understanding of the Ifugao but also contributes to broader discussions about the representation and interpretation of Indigenous cultures in historical narratives.

### 1.2. Decolonizing Archaeology

In the context of the Ifugao Rice Terraces, the process of decolonizing archaeology entails a shift away from a Western-centric interpretation that idealizes and freezes the Ifugao people in a fixed moment in time. Instead, an inclusive historiography seeks to acknowledge their dynamic history characterized by resilience and adaptation. By incorporating the generational timekeeping practices of Ifugao society and showing respect for their oral histories, we can validate the lived experiences of the community that have historically been marginalized in academic discourse.

This shift not only results in a more accurate representation of the history of the terraces but also restores agency to the Ifugao people, allowing them to define their own identity and history on their own terms. Co-author Martin, an Ifugao himself, emphasizes the importance of approaching the Ifugao and their terraces from the perspective of the Ifugao people themselves. He points out that, "We, the Ifugao, have always had a deep understanding of our past and our world. However, it was the imposition and arrogance of academic scientists and the so-called educated world that led us away from our own knowledge systems." Martin, being an Ifugao, possesses a unique ability to contextualize oral historical accounts within the cultural nuances of the Ifugao community. This expertise allows him to convey insights that may be challenging for non-Ifugao individuals to discern.

For the Ifugao, the terraces are more than just an agricultural innovation; they are a living legacy, deeply intertwined with their cultural identity, spirituality, and community life. The traditional Ifugao knowledge systems, which include intricate agricultural practices, rituals, and social structures, have been maintained and passed down through generations.

The academic scientists often approached the Ifugao and their terraces through a lens of romanticism or exoticism, at times casting them as passive relics of a bygone era rather than as active, adaptive agents of their own history. This distortion stemmed from a combination of colonialist attitudes and a lack of engagement with the community's oral histories, which offer a rich, nuanced perspective on the evolution of the terraces and the people who created and maintained them.

The imposition of these external narratives has had profound implications for the Ifugao. It has led to a disconnect between the Ifugao's self-perception and the way their history is portrayed to the world. The academic community's oversight or dismissal of Indigenous knowledge systems has not only misrepresented the terraces' history but has also undermined the Ifugao's agency in their own cultural heritage.

In recent years, there has been a growing recognition of the need to decolonize archaeology and to recenter Indigenous perspectives. This involves a conscious effort to understand the Ifugao and their terraces from the viewpoint of the Ifugao themselves. It calls for a collaborative approach that respects and incorporates Ifugao oral traditions and knowledge systems into the historical narrative. Such an approach not only corrects the historical record but also empowers the Ifugao community, allowing them to reclaim their narrative and affirm their identity on the global stage.

By valuing the Ifugao's own knowledge systems and perspectives, we can encourage a more authentic and respectful understanding of their history. This shift is crucial not only for the academic integrity of historical research but also for supporting the Ifugao in preserving their cultural heritage for future generations. The Ifugao experience thus becomes a powerful example of how Indigenous knowledge can illuminate and enrich our understanding of the past, challenging us to rethink the way we engage with and represent the histories of Indigenous peoples.

Our work actively resists the exoticization of the Ifugao by reframing the terraces as active cultural landscapes rather than static monuments. The inclusion of local narratives in archaeological interpretation not only enriches our understanding of the past but also supports the Ifugao in safeguarding their intangible heritage. It is a reciprocal process where archaeological practice benefits from the depth of knowledge held within community stories, and the inclusion and synthesis of these stories gain visibility and validation in archaeological narratives.

This decolonized approach is evident in the way the Ifugao community has been integral to the archaeological process, from hypothesis formation to fieldwork and interpretation. Community members are not mere informants; they are collaborators with voices as authoritative as archaeologists. Their involvement in the Ifugao Archaeological Project has led to the creation of the Ifugao Community Heritage Galleries and the Indigenous Peoples Education Center, which are not only resources for cultural education but also embodiments of the community's ongoing engagement with their heritage.

The decolonization of archaeology is an ongoing and evolving practice, one that demands reflexivity, openness to change, and a commitment to justice and equity. In Ifugao, it is a conscious effort to dismantle the colonial legacy that has long influenced the interpretation of the terraces and to replace it with a collaborative, multidisciplinary approach that honors the sovereignty of the Ifugao over their cultural heritage. This method serves as a model for how archaeologists can and should move forward globally, advocating for a discipline that is inclusive, respectful, and supportive of the communities whose past we seek to understand.

This synergy between archaeological evidence and oral histories is crucial for a more inclusive and accurate portrayal of the past. It emphasizes the need to preserve these oral traditions as integral components of cultural heritage, just as vital as the physical structures themselves. Furthermore, it provides a model for other archaeological endeavors worldwide, advocating for a collaborative approach that respects and utilizes the knowledge embedded within local communities. As a result, such a confluence of perspectives not only deepens historical understanding but also empowers the descendant communities by centering their voices in the narratives of their own past [6,14].

### *1.3. Ethnographic Accounts and Community Stories*

We approached ethnographic accounts in Ifugao with a strong focus on cultural sensitivity and ethical considerations. We particularly underscored the respect and the amplification of the voices and perspectives of the Ifugao community while conducting research and interpreting ethnographic accounts. We understand that the Western-centric interpretation of the terraces have long romanticized and frozen the Ifugao in time, obscuring their dynamic history of resilience and adaptation. Ethical considerations guided every step of our research and actively engaged with the Ifugao community, forging partnerships that prioritized the community's interests and autonomy.

For instance, understanding the passage of time through generational reckoning and genealogical continuity is important in comprehending Ifugao history and culture. In Ifugao society, time is traditionally measured not in years, as per the Gregorian calendar, but in generations. This method of timekeeping is central to their collective memory and is exemplified in narratives concerning the origins of significant landmarks such as the Batad Rice Terraces—one of the five clusters recognized by UNESCO—the extensive terraces in Asipulo, and the community history of Tokak Village in Namal, Asipulo.

The oral histories of the Ifugao provide a glimpse into the chronology of their landscape and culture, particularly through the narratives surrounding the Batad and Asipulo terraces. Local lore, rich in detail and passed down through generations, tells the story of two brothers from Cambulo, a village near Batad. Their discovery of the Batad hillside led to the establishment of a swidden field, marking the beginnings of what would become a significant agricultural and cultural site. Subsequently, one brother brought his family to settle in the area, and together they constructed the terraces that have since become emblematic of Ifugao's rich tradition of rice cultivation. This story, believed to have taken place within the last six generations, illustrates a living history that is intrinsically linked to the land and the community's ancestral lineage.

In a similar vein, the Asipulo terraces, as recounted in the oral history of the Tokak community, were constructed within the past five generations. These narratives underscore a relatively recent and rapid development of terrace farming in these regions, contrasting with the often static and ancient depictions found in external historical accounts.

The story of the Tokak village exemplifies the impact of Spanish colonial expeditions on the Ifugao's settlement patterns. Village elders recount how their great grandparents, seeking to maintain autonomy and protect their way of life, were compelled to leave Amduntog—a village closer to colonial centers—and resettle in the more secluded village of Tokak. This strategic relocation allowed the Ifugao to evade direct colonial control and preserve their cultural practices, including rice terrace farming, which has been central to their identity and sustenance.

The return of their descendants to Amduntog, following the departure of the Spaniards, signifies not just a physical return to ancestral lands but also a cultural reclamation. It underscores the Ifugao's enduring connection to their territory and the resilience of their sociocultural systems amid colonial disruptions. These community stories, passed down through generations, are critical to understanding the nuanced history of the Ifugao. They provide context to the archaeological evidence, enriching our perception of the past and offering a more holistic view that honors the lived experiences and agency of the Ifugao in shaping their history.

The generational approach to history allows for a fluid and evolving understanding of the past, one that is directly connected to the memories and experiences of the community. Such oral traditions not only serve as historical records but also as affirmations of the community's enduring connection to their environment and their continuous adaptation to it. Through these stories, the people of Ifugao maintain a vibrant and ongoing relationship with their history, one that is characterized by resilience, familial bonds, and a deep respect for the terraced landscapes that have sustained them.

Dismissing these oral histories undermines the validity of ethnographic methods and shows disrespect to the Ifugao's worldview, as articulated through their community stories. These narratives align closely with the broader Philippine history, as recorded in both archival documents and contemporary scholarship. They are not merely stories; they encapsulate the lived experiences of the Ifugao.

Contrary to the notion that the Ifugao rice terraces are 2000 years old—an assumption unsupported by any archaeological evidence—recent archaeological and ethnographic data from five major sites including Old Kiyyangan Village, Hapao, Nagacadan, Batad, and Banaue suggest a much shorter history. In addition, Maher's decade-long work in the region (with at least seven sites dated) does not provide any support for the 2000-year origin of the terraces. The absence of evidence should not be mistaken for evidence of absence; yet the lack of archaeological support for the terraces' ancient origins cannot be ignored.

The primary objective of our research in Ifugao goes beyond merely dating the terraces. The terraces exemplify humanity's ability to negotiate environmental constraints and adapt to changing needs, as wet-rice cultivation is integral to Ifugao culture. These terraces represent not just a triumph over environmental challenges but also the aspirations and identities of the Ifugao. It is essential that collective consciousness moves in the direction of fostering an appreciation for the terraces not as static relics of the past but as living landscapes reflecting the ingenuity and resilience of their builders.

The convergence of archaeology with community narratives sheds light on the shared history of human societies and their environments. Archaeological evidence provides tangible proof of past human interactions with landscapes, while community stories offer nuanced insights into these interactions. The Ifugao Archaeological Project corroborates and enriches community narratives, offering a depth of temporal understanding that enhances our appreciation of cultural practices and ecological wisdom.

In the context of climate change, this synergy becomes a valuable resource. Insights from archaeology combined with community stories offer a comprehensive view of historical climate patterns, adaptive strategies, and resilience. This knowledge is crucial for creating climate adaptation and mitigation strategies that are both scientifically sound and culturally sensitive. Understanding how communities have historically navigated environmental challenges provides invaluable lessons for today's climate action.

The integration of archaeological findings and Indigenous narratives particularly shines in the Ifugao Rice Terraces' context. It reveals a dynamic history of sustainable adaptation to climatic challenges, with Indigenous knowledge guiding effective water and soil management techniques. This Indigenous wisdom, embedded in local stories and integrated with the landscape's history, lays a robust foundation for climate-resilient agricultural practices. Moreover, this collaboration extends to natural resource management, evident in the Ifugao's *muyung* traditional forestry system [15,16].

Archaeology paired with Indigenous narratives offers insight into historical forestry practices, sustainable resource extraction methods, and community-led conservation efforts. This holistic approach recognizes the ecological wisdom encoded in Indigenous knowledge, offering vital lessons for contemporary climate adaptation.

Embracing Indigenous knowledge in climate adaptation strategies acknowledges its role as a repository of strategies finely attuned to environmental nuances. The collaborative efforts of archaeology and Indigenous knowledge bridge historical resilience with modern scientific insights, fostering a culturally attuned, sustainable, and wisdom-rooted climate adaptation paradigm.

Finally, revising the narrative to reflect the recent origins of the terraces empowers Ifugao communities, dispelling colonial myths and promoting heritage conservation programs developed and implemented by the communities themselves. Debates over the dating of the terraces risk exoticizing the Ifugao and romanticizing the past. While dating is essential, it should be secondary to understanding the terraces' cultural context.

### 1.4. Nature of Archaeology: Hypothesis Testing

The importance of two specifically dated sites in rethinking the dating of the Ifugao Rice Terraces underscores the hypothesis-testing nature of archaeology, a discipline that relies on empirical evidence to challenge, refine, or confirm existing theories. In the context of the Ifugao Rice Terraces, the traditional narrative supported by early 20th-century anthropologists posited that these terraces were over 2000 years old, a claim that has profound implications for understanding the cultural and historical landscape of the Ifugao region. The discovery and subsequent dating of two key sites have been instrumental in challenging this long-held belief, demonstrating the dynamic and evolving practice of archaeology as a field that continuously tests hypotheses against emerging data.

Archaeology, at its core, is an empirical science that constructs narratives about the past based on material evidence. Hypotheses about human behavior, cultural practices, and historical timelines are formulated based on available data, which are then tested through further excavation, analysis, and interdisciplinary research. The process is iterative: new findings can corroborate, refine, or completely overturn existing theories, leading to a more nuanced understanding of the past.

Archaeology, often misconstrued as a quest for definitive truths about our past, is fundamentally a discipline grounded in the formulation and testing of hypotheses. Rather than seeking irrefutable facts, archaeologists engage in a dynamic process of hypothesizing, gathering evidence, and revising understandings based on new findings. This approach is illustrated in the recent archaeological work on the Ifugao Rice Terraces in the Philippines, which has led to a significant re-evaluation of their age and origins. For detail on the work of the IAP, please visit www.ifugao-archaeological-project.org.

For many years, the prevailing belief, supported by early anthropological studies, was that the Ifugao Rice Terraces were over 2000 years old. This supposition was not only an academic conjecture but also a part of the national and cultural identity of the Philippines. However, the essence of archaeological inquiry is not to cement such narratives as absolute truths but to continuously test and reevaluate them in light of new evidence and methodologies.

Acabado's research, a continuation of his graduate work [3,15–17], presented a contrasting hypothesis: the terraces were built much later, possibly as a response to Spanish colonization in the 16th century. This proposal was not an attempt to uncover a 'final truth' but rather to challenge the existing narrative based on emerging data and perspectives. The hypothesis was tested through various methods, including radiocarbon dating, archaeobotanical analysis, and ethnohistorical research, all standard tools in the archaeologist's kit for hypothesis testing.

The Ifugao Archaeological Project focused its investigations on the Old Kiyyangan Village (OKV), a site integral to Ifugao history and mythology. The scientific approach taken by the IAP was methodical and evidence based. They uncovered artifacts, such as locally produced and imported beads, a crocodile tooth, and infant burials, which provided insights into the Ifugao's interactions and adaptations over time. These findings were crucial in testing the new hypothesis about the timing and nature of terrace construction.

To test this hypothesis, the team employed a range of methods. Radiocarbon dating of charcoal samples, analysis of pollen and phytoliths, and careful stratigraphic excavation provided data points. Each method offered a piece of the temporal puzzle, contributing to a body of evidence that was then evaluated against the hypothesis.

The results of these tests were surprising. Instead of supporting the millennia-old narrative, the evidence pointed to a much more recent construction, aligning with the period of Spanish contact. This finding challenged the traditional view but was consistent with the nature of scientific inquiry, which is characterized by its openness to change when faced with new, robust evidence.

The shift in understanding about the age of the Ifugao rice terraces underscores a fundamental aspect of archaeology: it is a discipline more about asking the right questions and less about finding definitive answers. Archaeology does not deal in absolutes; rather, it embraces uncertainty and change as integral parts of understanding the past. Each new piece of evidence can redefine existing narratives, demonstrating that our understanding of history is fluid and subject to revision.

This perspective is crucial to considering the role of archaeology in society. By not positioning itself as a seeker of absolute truth, archaeology allows for multiple narratives and interpretations to coexist. This approach is particularly important in areas like the Ifugao province, where cultural identity and historical narratives are deeply intertwined. The re-evaluation of the terraces' age does not diminish their cultural and historical significance; instead, it enriches our understanding of the Ifugao people's resilience, adaptability, and agency.

Moreover, the case of the Ifugao Rice Terraces highlights the importance of integrating scientific inquiry with local knowledge and oral histories. The Ifugao community's narratives and memories played a crucial role in shaping the hypothesis and guiding the archaeological investigation. This integration underscores the importance of a holistic approach in archaeology, one that respects and incorporates various sources of knowledge.

The evolving understanding of the Ifugao rice terraces' history exemplifies the nature of archaeology as a discipline focused on testing hypotheses rather than seeking immutable truths. This case study demonstrates how archaeology is a dynamic and iterative process, one that adapts and advances as new evidence emerges. It reminds us that our grasp of the past is always provisional, and that openness to re-evaluation and revision is essential in the pursuit of knowledge. In this way, archaeology not only helps us to understand our history but also teaches us about the complexity and fluidity of the human experience.

The Ifugao case exemplifies the iterative nature of archaeological research. Hypotheses are constructed based on the best available information, then tested against empirical data. If the data align, the hypothesis is strengthened; if not, it must be revised or discarded. This process is not unique to archaeology but is the hallmark of the scientific approach across disciplines.

Through this example, we see that archaeology, like all science, is a dynamic process. It is a discipline that evolves with each new discovery, where hypotheses are continuously refined to enhance our understanding of the human past. Far from seeking immutable truths, archaeology embraces the complexity and variability of human history, offering insights that are as provisional as they are profound.

## 2. Materials and Methods

The recent archaeological work in Ifugao, Philippines, continues and expands upon Stephen Acabado's prior research [3,10,15–18], suggesting the iconic Ifugao rice terraces were constructed in response to Spanish colonization. This challenges the older belief that the terraces are millennia old and reshapes the understanding of the region's history and the Indigenous Ifugao people. The AMS determinations and spatial analyses are also supported by archaeobotanical datasets. There is a total absence of any evidence of wet-rice cultivation in the region that predates the 1600s.

The Ifugao Archaeological Project's (IAP) focus on the Old Kiyyangan Village (OKV) has been pivotal in uncovering the area's past. Fray Molano, in a Spanish document from 1801, described OKV as a large settlement—a contrast to its portrayal as isolated and unchanging. Robert Maher's [19] initial research at OKV provided important chronological data, indicating the site was much older than previously understood, with findings suggesting Ifugao ancestors settled there.

The selection of OKV for archaeological exploration was driven by the community's desire to investigate their origins. Despite its transformation into paddy fields, historical records from the American colonial period mention the Village of Otbobon, another name for OKV, suggesting a relocation by 1869 [20]. The modern town of Kiangan's proximity to the original OKV site highlights shifts in settlement patterns.

Robert Maher's [19] initial investigations into OKV, now a rice field believed to have been transformed into a paddy field just before World War II, provided important chronological data. Maher's excavation yielded two thermoluminescence (TL) dates, ranging between 1130 CE and 1230 CE, indicating a much earlier occupation of the site than previously understood (Table 1). This finding is crucial, as OKV is deeply ingrained in the cultural and mythical history of the Ifugao. According to local myths, Old Kiyyangan was the village where the first Ifugao ancestors settled and where they received divine knowledge of wet-rice cultivation. These narratives also suggest that prior to the development of the terraces, taro was the primary crop cultivated by the Ifugao. Maher's research in the region set out to provide archaeological evidence for the origins of the Ifugao and their rice terraces [21–24], which provided the initial radiocarbon dates (Table 2).

**Table 1.** Maher's TL dates in the OKV Site and adjacent Bintacan Cave.

| Site | Level Info | TL Dates | Reported by |
|---|---|---|---|
| **Bintacan Cave** | Level F | 1620 BP Alpha 476 | [24] |
| **Bintacan Cave** | Level E | 1420 BP (±20%) Alpha 480 | [24] |
| **Bintacan Cave** | Level C | 760 BP (±20%) Alpha 479 | [24] |
| **Old Kiyyangan Village** | No data presented | 820 BP Alpha 566 | [20] |
| **Old Kiyyangan Village** | No data presented | 720 BP Alpha 671 | [20] |

**Table 2.** Radiocarbon determinations from Banaue Ifugao, obtained by Maher [21,22] and Harold Conklin [25]. While some of these radiocarbon determinations show older (>500 years), these samples are not contextualized and may not be associated with the archaeological event in question. For more discussion, refer to Acabado and colleagues' discussions [3,5].

| Site/Locality | Depth (cm) | Lab # | Material | $^{14}$C BP | Cal. CE (2 σ–95%) (Recalibrated Using IntCal13) | $^{14}$C Sample Context | Reported by |
|---|---|---|---|---|---|---|---|
| **If1-Nabyun** | 91 | GX0668 | *M. sinensis* | 205 ± 100 | 1493-postbomb | Pond-field | [21] |
| **If2-Nabyun** | 91 | GX1900 | No data | 325 ± 110 | 1408-postbomb | House Platform | [21] |
| **If2-Nabyun** | 90 | GX1901 | No data | 695 ± 100 | 1052–1435 | Midden | [21] |
| **If2-Nabyun** | 90 | BX2184 | No data | 735 ± 105 | 1043–1413 | Midden | [21] |
| **If3-Banaue** | 4.4 | GX2183 | No data | 2950 ± 250 | 1867-540BCE | House Platform | [21] |
| **Gawwa, Poitan** | 5 | GX3138 | No data | 530 ± 140 | 1192–1792 | Underground chamber | [21] |
| **Gawwa, Poitan** | 5 | GaK5238 | No data | 530 ± 100 | 1273–1631 | Underground chamber | [21] |
| **Lugu** | No data | UGA2515 | No data | 395 ± 60 | 1430–1639 | Terrace embankment post | [25] |
| **If20-Banghallan** | 50 | GaK6442 | No data | 890 ± 310 | 434–1647 | Village edge | [22] |
| **If20-Banghallan** | 60 | UGA1541 | No data | 1340 ± 375 | 176BCE-1388CE | Village edge | [21] |

Between 2012 and 2016, the IAP executed twenty excavation units at OKV. The findings from these excavations have been revealing, indicating that the arrival of the Spanish in the region was a catalyst for significant changes in the village. Most notably, the shift to wet-rice cultivation seems to have occurred concurrently with the Spanish influence, supported by AMS dates. The excavations in OKV unearthed a variety of artifacts, including locally produced and imported beads, a crocodile tooth, imported metal adornments, and infant burials [26–31]. These discoveries are indicative of the active role the Ifugao community played in broader pre-colonial and colonial interactions within the Philippines, challenging the notion of their cultural and economic isolation. The initial phase yielded seven radiocarbon dates from the terraced fields of Banaue (Table 3), and subsequent excavations at the Old Kiyyangan Village (OKV) added nineteen more dates (Table 4). These were complemented by extensive analyses, including sherd residue [29], wood charcoal, and microfossil studies, to determine the advent of wet-rice agriculture in the region.

**Table 3.** AMS dates from Bocos rice terraces, Banaue, Ifugao [3].

| Lab. No. | Unit | DBS | Layer | CRA | $^{13}$C | Cal AD (BCal) | Post-AD 1585 Probability * |
|---|---|---|---|---|---|---|---|
| AA78973 | Mamag | 0.855 | II | 119 ± 38 | 25.2 | 1687–1862 | 74.6% |
| AA78974 | Mamag | 1.3 m | III | 485 ± 39 | −27.5 | 1325–1460 | 74.6% |
| AA78971 | Rasa | 0.35 m | II | 313 ± 38 | −24.4 | 1620–1800 | 98.5% |
| AA78972 | Rasa | 0.52 m | III | 164 ± 38 | −26.0 | 1527–1757 | 98.5% |
| AA78969 | Linagbu | 0.55 m | II | 180 ± 38 | −26.5 | 1736–1867 | 99.9% |
| AA78970 | Linagbu | 0.75 m | III | 131 ± 38 | −29.3 | 1663–1753 | 99.9% |
| AA78975 | Achao | 0.075 m | II | 193 ± 38 | −25.0 | 1646–1809 | N/A |

* Probability analyses (Bayesian modeling) of pre-Spanish or post-Spanish construction of the Bocos rice terraces walls in Banaue, Ifugao [3].

Excavations at OKV over several field seasons resulted in the opening of 21 trenches and 10 shovel test probes, uncovering over 20,000 artifacts, predominantly earthenware ceramics, as well as faunal, archaeobotanical samples, and human remains, mostly infant burials. Radiocarbon dating at OKV employed diverse materials—bulk soil, wood charcoal, bone collagens, and sherd residue—all showing a consistent stratigraphy. The alignment of bulk soil dates within situ radiocarbon dates, despite potential contamination risks, suggests rapid deposition within a short timeframe.

**Table 4.** Radiocarbon determinations recovered from Ifugao between 2012 and 2016 [5].

| Depth (cm) | Lab Number | Location | Material/Trench | $^{14}$C BP | Cal. BP (2 σ) | Cal. CE (2 σ) | Context |
|---|---|---|---|---|---|---|---|
| **30–40** | Beta-356307 | OKV | organic sediment/8 | 190 ± 30 | 260–200 | 1640-post-1950 | Rice field |
| **50–60** | UCIAMS-183276 | OKV | Wood charcoal (*P. insularis*)/14 | 415 ± 15 | 510–469 | 1440–1480 | Rice field |
| **55–73** | Beta-394185 | OKV | bone collagen/8 | 410 ± 30 | 530–470 | 1405–1445 | Mortuary |
| **60–70** | UCIAMS-183272 | OKV | Wood charcoal (*P. insularis*)/14 | 345 ± 15 | 477–317 | 1470–1633 | Rice field |
| **65–70** | Beta-356306 | OKV | organic sediment/8 | 620 ± 30 | 680–620 | 1280–1390 | Rice field |
| **80–90** | UCIAMS-183273 | OKV | Wood charcoal (*P. insularis*)/14 | 570 ± 15 | 634–537 | 1315–1415 | Rice field |
| **80–90** | Beta-394182 | OKV | bone collagen/8 | 600 ± 30 | 730–670 | 1265–1380 | Mortuary |
| **90–100** | CIAMS-183274 | OKV | Wood charcoal (*P. insularis*)/14 | 665 ± 15 | 669–564 | 1280–1385 | Rice field |
| **90–100** | Beta-421036 | OKV | charcoal/14 (*P. insularis*) | 660 ± 30 | 690–630 | 1280–1390 | Rice field |
| **90–100** | Beta-421037 | OKV | potsherd residue/14 | 590 ± 30 | 610–550 | 1300–1415 | Rice field |
| **90–100** | D-AMS 003446 | OKV | Organic sediment/9 | 861 ± 25 | 899–700 | 1052–1250 | Rice field |
| **100–110** | D-AMS 003447 | OKV | Organic sediment/10 (fill) | 1252 ±37 | 1279–1075 | 672–876 | Rice field |
| **100–110** | D-AMS 003448 | OKV | Organic sediment/10 (dark midden soil) | 292 ±27 | 456–291 | 1495–1660 | Rice field |
| **100–110** | Beta-356305 | OKV | organic sediment/8 | 720 ± 30 | 810–750 | 1220–1280 | Rice field |
| **110–120** | D-AMS 003445 | OKV | Organic sediment/9 | 672 ± 28 | 676–561 | 1274–1390 | Rice field |
| **110–120** | Beta-32953 | OKV | organic sediment/3 | 780 ± 30 | 741–669 | 1160–1260 | Rice field |
| **120–144** | Beta-394184 | OKV | bone collagen/9 | 800 ± 30 | 767–675 | 1045–1220 | Mortuary |
| **130–140** | Beta-329552 | OKV | organic sediment/3 | 770 ± 30 | 734–668 | 1050–1240 | Rice field |
| **150–160** | Beta-329551 | OKV | organic sediment/3 | 1000 ± 30 | 967–799 | 900–1020 | Rice field |

Paleoethnobotanical analysis indicated a potential presence of wet-rice cultivation dating back to 675 years before present (BP), but low counts of rice phytoliths suggested possible displacement within the soil column due to water seepage [32,33], casting doubt on earlier cultivation theories. Residue analysis from pottery did not indicate rice production, but instead showed evidence of taro consumption, further supporting the argument for the later emergence of wet-rice agriculture in Ifugao.

Analyses of bulk soil, charred residues, and botanical data from three excavation trenches suggest that the predominant crop in the OKV prior to 1650 CE was not wet rice, as previously believed, but taro (*Colocasia esculenta*). The absence of wet-rice cultivation before the mid-17th century was corroborated by pollen, phytolith, and starch analyses performed on sherd residues from these trenches, further dismantling the theory of an ancient origin for rice agriculture in the region. Details on the microbotanical analyses from soils obtained in the OKV site are provided by Horrocks et al. [34] report and Acabado's preliminary hypothesis [35,36].

Additionally, the stratigraphy of the site revealed a significant increase in wood charcoal within Layer 2, dated to around the 1600s (Figure 3). This finding could indicate a heightened demand for wood or point to deforestation, reflecting a change in landscape management or environmental conditions. Crucially, no rice or rice-associated weeds were detected in any soil or charred residue samples predating 1650 CE, supporting the hypothesis of a post-contact introduction of wet-rice agriculture.

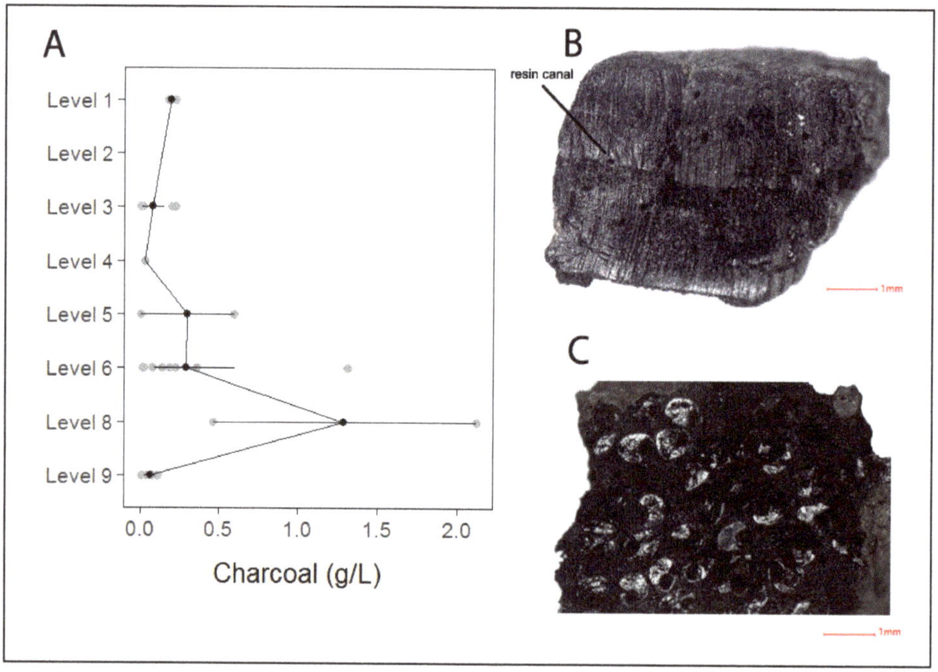


**Figure 3.** (**A**) The densities of wood charcoal (grams per liter) recovered in deposits from Trench 14, where the level (i.e., depth in 10 cm increments) is represented on the vertical axis, and the density is represented on the horizontal axis. The gray dots illustrate the values in individual samples, and the black dots are the bootstrapped mean densities of each level with 95% confidence intervals (the black line within a level). The line between levels connects the means of each level. (**B**,**C**) are the transverse (cross) sections of identified wood charcoal, where (**B**) is identified as Benguet pine with a resin canal highlighted from Trench 14, level 8, and (**C**) is a specimen from the palm family from Trench 16, level 2 [5]. Figure provided by A. Farahani.

The environmental transformation indicated by the archaeological record was also marked by a notable influx of imported goods, a surge in the remnants of animals used in rituals, and the adoption of wet-rice farming practices. These shifts signal a broader change in the Ifugao way of life, aligning with the period of increased external contact and trade [36].

This multi-evidentiary approach has been crucial in constructing a robust chronological model for the inception and expansion of the Ifugao terracing systems. Ethnographic comparisons highlight that wet-rice cultivation necessitates complex social structures. These findings support the hypothesis that the Ifugao terraces, as seen today, were constructed within a 200- to 300-year period, a rapid development indicating advanced sociopolitical organization suitable for intensive wet-rice farming.

## 3. Results

The lack of evidence for the 2000-year-old origin of the terraces across five major Ifugao sites has led to the discrediting of this long-held belief, with recent archaeological and ethnographic data suggesting a much shorter history. The significance of local time reckoning in Ifugao, which relies on generational memory rather than calendar years, has been crucial in contextualizing this narrative shift. Community stories from the Batad Rice Terraces, Asipulo terrace system, and Tokak Village provide a vivid account of the terraces' origins tied closely to the Ifugao's response to Spanish colonialism, rather than a distant, pre-colonial past.

Archaeobotanical and pottery residue data have revealed the continuation of taro as a staple carbohydrate, even after the Spanish arrival in the adjacent lowlands of the Magat Valley. The maintenance of traditional burial practices and dietary habits through faunal isotopic signatures indicates that the Ifugao retained their way of life despite the external pressures of colonialism.

These datasets collectively suggest that the Ifugao responded rapidly and adaptively to Spanish incursion, upholding their group identity and cohesion through ritual and household practices. The AMS determinations and spatial analyses of the terraces, alongside archaeobotanical datasets, point to an absence of pre-1600s wet-rice cultivation, challenging the narrative of an ancient terracing tradition.

The Ifugao terraces stand not only as a testament to humanity's ability to adapt to environmental constraints but also as a reflection of the Ifugao's identity and resistance against colonial forces. The implications of these findings are not limited to local history but contribute to broader discussions on agricultural systems' archaeology, requiring a comprehensive approach that encompasses local knowledge and generational narratives.

Spanish colonial documents recorded active irrigation systems in only two regions of the Philippines during the early years of conquest: Bicol and the Magat Valley. No other archaeological or historical investigations have documented subsistence patterns before and after the Spanish conquest. Therefore, while dating the terraces is crucial, it is secondary to the broader aim of understanding the terraces' role in Ifugao culture and heritage.

In the neighboring region of Bontoc, Bodner's [37] work has provided compelling evidence for the later inception of wet-rice cultivation, postdating 1600 CE. This assertion is grounded in the notable absence of archaeobotanical data indicative of wet rice prior to the 1600s. Supporting this perspective, an accumulation of data from various scholarly pursuits—including Maher's investigations, Bodner's own dissertation, Conklin's landscape research, and the ongoing Ifugao Archaeological Project—converge to suggest a significant shift in agricultural practices to wet-rice cultivation occurring relatively recently.

Further strengthening this argument, a genomic study of rice varieties in the Cordillera, including the Ifugao's revered tinawon rice, has shed light on the crop's lineage. According to Alam et al. [38], the highland rice varieties can trace their ancestry to Indonesian strains, suggesting a southern origin and diffusion into the highlands. This genomic lineage contradicts the theory of an ancient, Indigenous development of wet-rice cultivation in the Ifugao region, instead pointing to a more recent introduction and adaptation of these rice varieties. Together, these interdisciplinary research efforts paint a picture of dynamic agricultural evolution in the Cordillera, characterized by later adoption and continuous innovation.

This body of work underscores the need for data-driven discussions about the dating of the Ifugao terracing tradition, advocating for a narrative that recognizes the terraces as a dynamic cultural landscape shaped by historical contingencies and the Ifugao's resilient spirit.

## 4. Discussion: Archaeology and Community

Comprehensive archaeological investigations spearheaded since 2007 have significantly reshaped the narrative of the Ifugao Rice Terraces. This intensive research, however, is more than a mere quest for chronological accuracy; it is an exploration into the lived experiences of the Ifugao people, as evidenced by both material culture and enduring oral traditions. The integration of archaeological findings with local narratives has revealed a history marked not by stasis and isolation but by dynamism, resilience, and adaptation.

The ethnographic accounts and community stories of the Ifugao stand as testaments to the society's active participation in their history. These narratives, passed down through generations, serve not only as historical records but also as expressions of the Ifugao's identity and response to external pressures, particularly Spanish colonization. The recollections of the origins of landmarks such as the Batad Rice Terraces and the Asipulo terrace system underscore a temporal understanding deeply ingrained in Ifugao culture. Moreover, the tales of strategic relocations during colonial times, such as those from Amduntog to the more remote Tokak, illustrate the Ifugao's agency in preserving their way of life.

The archaeological record, bolstered by radiocarbon dating and analyses of pottery residue and paleoethnobotanical samples, has supported the contention that the terraces are of more recent origin than previously thought. These data have challenged the long-held belief that the terraces are over two millennia old, a belief unsupported by evidence from significant sites within the region. The findings suggest a rapid socioeconomic transformation in Ifugao society, coinciding with the emergence of wet-rice agriculture and the introduction of new goods and practices, indicating an era of significant change catalyzed by, but not succumbing to, colonial influence.

This cultural chronology is pivotal in understanding the terraces not as static monuments but as vibrant landscapes that encapsulate humanity's enduring adaptability. The terraces are symbols of the Ifugao's triumph over environmental and political challenges, embodying the aspirations and ingenuity of their creators. Recognizing the terraces as part of a living culture is essential for fostering respect for Ifugao heritage and for ensuring the terraces' preservation for future generations.

The role of Indigenous knowledge, particularly the *muyung* traditional forestry system, further exemplifies the Ifugao's sustainable management of natural resources [15,16]. The confluence of archaeological evidence with this Indigenous wisdom offers insights into historical practices that can inform contemporary climate adaptation strategies. Embracing this knowledge is crucial for crafting solutions that are not only effective but also respectful of cultural traditions.

Moreover, by integrating community narratives with archaeological research, a more nuanced picture of the Ifugao's past emerges. These stories provide context and color to the archaeological canvas, enriching our understanding of the terraces and the people who built and maintained them. The Ifugao Archaeological Project has shown that the terraces' recent origins do not diminish their value; rather, they highlight the Ifugao's remarkable adaptability and the terraces' ongoing significance as cultural and ecological marvels.

As global discussions on climate action and cultural conservation continue, the insights gleaned from the Ifugao terraces serve as a valuable resource. The Ifugao's historical experiences, coupled with archaeological data, provide a model for understanding how societies can navigate environmental challenges while maintaining their cultural integrity.

The narrative of the Ifugao rice terraces is being rewritten. It is a narrative that respects the intersection of archaeology and oral history, acknowledges the complex interplay between the Ifugao and their environment, and recognizes the terraces as a symbol of cultural resilience. This narrative shift has profound implications for heritage conservation, climate action, and our understanding of humanity's relationship with the land. As we move forward, it is essential to continue valuing and incorporating the voices of the Ifugao community, ensuring that their history is told with accuracy and respect, and that their wisdom is heeded in the face of contemporary challenges.

## 5. Conclusions: Changing the Narrative, Empowering Descendant Communities

The Ifugao Archaeological Project demonstrates a successful integration of community perspectives with archaeological data, a practice that has been pivotal in reshaping the narrative of the Ifugao Rice Terraces. By engaging local narratives and participation, this project has moved beyond the confines of traditional, colonial methodologies that often perpetuate a romanticized and static view of Indigenous communities. Instead, it has revealed the terraces as living emblems of the Ifugao's resistance and adaptability through centuries of sociocultural transformations. An example of this knowledge co-production is highlighted by this video link: The Old Kiyyangan Village Story.

This project has catalyzed a re-evaluation of the terraces' history, aligning archaeological findings with the oral histories of the Ifugao. This has disrupted the academic and popular portrayal of the terraces as ancient and untouched by historical events, instead highlighting the Ifugao's active role in their evolution. The Ifugao's own generational timekeeping and narratives offer an invaluable insight into the cultural significance of the terraces, detailing a history of dynamism, community cohesion, and resilience against colonial pressures [39,40].

Furthermore, the IAP's collaborative approach has empowered the Ifugao community, allowing them to reclaim their narrative and cultural heritage. The development of the Ifugao Community Heritage Galleries and the Indigenous Peoples Education Center (Figure 4) exemplifies the tangible benefits of this inclusive practice. These centers not only serve as repositories of Ifugao material culture but also as educational resources that enable the development of local history curricula, thereby preserving Indigenous knowledge for future generations.

**Figure 4.** The weaving exhibit at the Ifugao Community Heritage Galleries, which now serve as the Ifugao Indigenous Peoples Education Center.

Such practices exemplify the decolonization of archaeology, defined as the process of re-evaluating and transforming practices to remove colonial biases and structures. Decolonization in archaeology prioritizes Indigenous perspectives and knowledge systems, seeking to correct historical power imbalances in the interpretation and stewardship of cultural heritage. This approach emphasizes the need for descendant communities to be actively involved in research processes, ensuring their oral traditions and sovereignty over their cultural narratives and material past are respected and integrated into the historical narrative.

The Ifugao experience showcases how the integration of community voices with archaeological research can lead to a more accurate and respectful understanding of the past. It demonstrates that the decolonization of archaeology is not only possible but necessary for the discipline to evolve in a way that is inclusive, equitable, and just. The IAP exemplifies the paradigm of community archaeology, where local narratives and participation fundamentally reshape the interpretation and conservation of cultural heritage. This inclusive approach has illuminated the Ifugao rice terraces not merely as UNESCO-recognized aesthetic monuments but as living emblems of resistance against colonial imperialism—a legacy enduring through centuries of sociocultural transformations.

Community archaeology in Ifugao is not an isolated phenomenon. Around the world, similar movements have been gaining momentum, recognizing the value of incorporating local voices into archaeological narratives. In Australia, Indigenous archaeology has become a vital practice, with Indigenous Australian communities working alongside archaeologists to protect and interpret their ancestral lands [41–43]. This collaboration has enabled a deeper understanding of the Indigenous Australian history and culture, which spans tens of thousands of years.

Moreover, in northern Africa, community archaeology projects have engaged with local communities to explore pre-colonial history, often overshadowed by the focus on ancient monuments like the pyramids of Egypt e.g., [44,45]. These projects not only uncover rich local histories but also empower communities by involving them in the preservation and interpretation of their heritage.

The significance of community participation in archaeology is multifaceted. It democratizes the process of historical inquiry, acknowledges the enduring presence of Indigenous cultures, and ensures that heritage conservation is not just an academic exercise but a communal endeavor. By combining traditional knowledge with archaeological findings, communities gain a voice in telling their own stories—a process that validates their history and identity.

The IAP's community-based approach also has broader implications for global heritage conservation. By demonstrating the tangible benefits of inclusive practices, it sets a precedent for similar initiatives worldwide. It underscores the importance of local engagement in preserving not only the physical remnants of the past but also the intangible aspects of culture that give meaning to these relics.

In the Ifugao case, the terraces are more than agricultural feats; they are a testament to a people's ingenuity and resilience. The community's involvement in the IAP has revitalized interest among the younger generation, inspiring them to delve into their history and to value the disciplines of anthropology and archaeology. This renewed interest has the potential to create a cadre of young Ifugao who are not only aware of their cultural heritage but are also equipped to preserve and promote it.

The integration of community voices with archaeological expertise in Ifugao provides a model for how heritage sites around the world can be managed and interpreted. It shows how the past can be a source of pride and identity for contemporary communities and how archaeology can serve as a bridge between the past and the present. This collaborative model can inspire similar efforts globally, where local communities become active participants in the conservation and interpretation of their cultural heritage.

*Inclusive National Identity*

Our approach to the archaeological practice in Ifugao offers a compelling example of decolonizing methodology that aligns with Indigenous movements in the Philippines. Our work challenges the traditional colonial narratives that have often defined the history and significance of sites like the Ifugao Rice Terraces, providing a framework for understanding these landscapes that is rooted in the community's own history and knowledge systems.

We emphasize the importance of community engagement in archaeological research. This involves integrating Indigenous knowledge and oral histories with archaeological methods to construct a more holistic understanding of the past. By doing so, we not only question the colonial underpinnings of historical narratives but also elevate the role of the Ifugao people as custodians of their own history.

The decolonial perspective offered by our work engages with earlier discourses on Indigenous peoples' movements in the Philippines by validating their narratives and recognizing their agency. Instead of treating Indigenous communities as mere subjects of research, our approach involves them as active partners, thereby disrupting the power dynamics that have traditionally characterized archaeological practice.

Furthermore, our work does not shy away from the theoretical or political implications of decolonization. While it is deeply embedded in the practical aspects of archaeological work, such as site excavation and material analysis, it also confronts the theoretical frameworks that have historically marginalized Indigenous perspectives. By advocating for a participatory approach to archaeology, we implicitly critique the political structures that have perpetuated colonial attitudes in the discipline.

In terms of engaging with broader movements for Indigenous rights in the Philippines, our decolonial approach reinforces the political aims of these movements. It recognizes the sovereignty of Indigenous peoples over their cultural heritage and supports their struggle for self-determination. The emphasis on community collaboration ensures that the fruits of archaeological research benefit the Indigenous communities themselves, whether through education, tourism, or reinforcing land claims.

Our decolonial methodology in archaeological practice is not merely an academic exercise. It is a form of activism that supports the broader goals of Indigenous peoples' movements in the Philippines. By engaging with the community and placing their knowledge at the forefront, our work exemplifies how decolonization can be both a practical and a political act, addressing the call for a more inclusive and just representation of the past.

The Ifugao case has shown that community archaeology, though not perfect and not an answer to all problems, promises to minimize conflict between heritage stakeholders. The practice of community archaeology also intensifies conversations between archaeologists and descendant communities. None of this is to suggest that community archaeology solves all of the complicated problems and compromises of archaeology and of interactions with communities with their own local, regional, and national entities. The success of any heritage management program rests on the engagement of many segments of the community.

**Author Contributions:** Conceptualization, S.B.A. and M.M.M.; methodology, S.B.A.; formal analysis, S.B.A.; investigation, S.B.A. and M.M.M.; data curation, S.B.A.; writing—original draft preparation, S.B.A.; writing—review and editing, S.B.A. and M.M.M.; visualization, S.B.A.; supervision, S.B.A.; project administration, S.B.A.; funding acquisition, S.B.A. All authors have read and agreed to the published version of the manuscript.

**Funding:** Funding for this project was generously provided by the NSF-REU Site: Ifugao Archaeological Project (Award #1460665), the National Geographic Society (NGS-9069), the Hellman Fellowship, UCLA's COR and FCDA Grants, the Institute for Field Research, and the National Museum (Philippines) Grant-in-Aid of Research.

**Data Availability Statement:** There are no new data created in this article. However, landscape (GIS) and Bayesian modeling data as presented in this manuscript can be provided by email the corresponding author.

**Acknowledgments:** This paper is the culmination of a decade of dedicated research in Ifugao, with a significant focus on the Old Kiyyangan Village, Kiangan, Ifugao, in the latter half of that period. The members of the Ifugao Archaeological Project (IAP) from 2012 to 2016 were instrumental in the acquisition and organization of the data presented herein. Collaborative efforts with the Save the Ifugao Terraces Movement, Archaeological Studies Program at the University of the Philippines, and the National Museum of the Philippines were pivotal in facilitating our fieldwork endeavors. Our gratitude extends to the Butic-Baguilat family for their generosity in allowing us access to the OKV site, despite the disruptions to their agricultural activities. Our acknowledgments would be incomplete without mentioning the insightful comments and suggestions provided by Iman Nagy, which significantly enriched the content of this manuscript. We are also indebted to three anonymous reviewers who provided constructive comments. We alone are responsible for the final product and take responsibility for any errors of fact or interpretation.

**Conflicts of Interest:** The authors declare no conflicts of interest.

## References


1. Barton, R.F. Ifugao Law. *Univ. Calif. Publ. Am. Archaeol. Ethnol.* **1919**, *15*, 1–186.
2. Beyer, H.O. The Origins and History of the Philippine Rice Terraces. In *Proceedings of the Eight Pacific Science Congress, 1953*; National Research Council of the Philippines: Quezon City, Philippines, 1955.
3. Acabado, S.B. A Bayesian Approach to Dating Agricultural Terraces: A Case from the Philippines. *Antiquity* **2009**, *83*, 801–814. [CrossRef]
4. Acabado, S.B. *Antiquity, Archaeological Processes, and Highland Adaptation: The Ifugao Rice Terraces*; Ateneo de Manila University Press: Quezon City, Philippines, 2015.
5. Acabado, S.B.; Koller, J.M.; Liu, C.-H.; Lauer, A.; Farahani, A.; Barretto-Tesoro, G.; Martin, J.A.; Peterson, J.A. The Short History of the Ifugao Rice Terraces: A Local Response to the Spanish Conquest. *J. Field Archaeol.* **2019**, *44*, 195–214. [CrossRef]
6. Acabado, S.B.; Martin, M. *Indigenous Archaeology in the Philippines: Decolonizing Ifugao History*; University of Arizona Press: Tucson, AZ, USA, 2022.
7. Findley, D.M.; Acabado, S.; Amano, N.; Kay, A.U.; Hamilton, R.; Barretto-Tesoro, G.; Bankoff, G.; Kaplan, J.O.; Roberts, P.R. Land Use Change in a Pericolonial Society: Intensification and Diversification in Ifugao, Philippines Fontana, Lorenza Belinda. 2013. Evo Morales at the Crossroads: Problematizing the Relationship between the State and Indigenous Movements in Bolivia. *Iberoam.—Nord. J. Lat. Am. Caribb. Stud.* **2022**, *43*, 19–45.
8. Acabado, S.B. Current Archaeological Practice in Southeast Asia: Collaboration, Engagement, and Community Involvement in Field Research in Southeast Asia. *J. Community Archaeol. Herit.* **2020**, *7*, 158–160. [CrossRef]
9. Cheng, C.-F.J. Root Seeking and Remote Sensing with the Bunun in the Mountains of Taiwan. *Adv. Archaeol. Pract.* **2023**, *11*, 289–301. [CrossRef]
10. Heng, P.; Phon, K.; Heng, S. De-exoticizing Cambodia's archaeology through community engagement. *J. Community Archaeol. Herit.* **2020**, *7*, 198–214. [CrossRef]
11. Atalay, S. *Community-Based Archaeology: Research with, by, and for Indigenous and Local Communities*; University of California Press: Oakland, CA, USA, 2012.
12. Colwell-Chanthaphonh, C. Archaeology and Indigenous Collaboration. In *Archaeological Theory Today*, 2nd ed.; Hodder, I., Ed.; Polity: Cambridge, UK, 2012; pp. 267–292.
13. Nicholas, G.; Watkins, J.; Smith, C. Indigenous archaeologies in archaeological theory. *Am. Antiq.* **2014**, *12*, 355–370.
14. Acabado, S.B.; Martin, M.; Datar, F. Ifugao Archaeology: Collaborative and Indigenous Archaeology in the Northern Philippines. *Adv. Archaeol. Pract.* **2017**, *5*, 1–11. [CrossRef]
15. Camacho, L.D.; Gevaña, D.T.; Carandang, A.P.; Camacho, S.C. Indigenous Knowledge and Practices for the Sustainable Management of Ifugao Forests in Cordillera, Philippines. *Int. J. Biodivers. Sci. Ecosyst. Serv. Manag.* **2016**, *12*, 5–13. [CrossRef]
16. Serrano, R.C.; Cadaweng, E.A. The Ifugao Muyong: Sustaining Water, Culture and Life. In *Search of Excellence: Exemplary Forest Management in Asia and the Pacific*; Durst, P.B., Brown, C., Tacio, H.D., Ishikawa, M., Eds.; Regional Community Forestry Training Center for Asia and the Pacific, Food and Agriculture Organization of the United Nations: Rome, Italy, 2005; pp. 103–112.
17. Acabado, S.B. The Archaeology of the Ifugao Agricultural Terraces: Antiquity and Social Organization. Ph.D. Thesis, University of Hawaii, Honolulu, HI, USA, 2010.
18. Acabado, S.B. The Archaeology of Pericolonialism: Responses of the "Unconquered" to Spanish Conquest and Colonialism in Ifugao, Philippines. *Int. J. Hist. Archaeol.* **2017**, *21*, 1–26. [CrossRef]
19. Maher, R.F. Kiyyangan Village of Ifugao Province, Philippines. *Philipp. Q. Cult. Soc.* **1984**, *12*, 116–127.
20. Meimben, M. Historical Record of the Quiangan Settlement [Record Historico. Del Settlement de Quiangan], Sub-Province of Ifugao [Sub-Provincia Ifugao], Mountain Province [Provincia Montanosa]. Unpublished document, as required by Executive Order No. 2 [Segun requerido en la Orden Ejecutiva No. 2]. 26 January 1911.
21. Maher, R.F. Archaeological Investigations in Central Ifugao. *Asian Perspect.* **1973**, *16*, 39–70.
22. Maher, R.F. The great Ifugao war: A study in archaeology and oral history. *Asian Perspect.* **1975**, *18*, 64–74.
23. Maher, R. Archaeological Investigations in the Burnay District of Southeastern Ifugao, Philippines. *Asian Perspect.* **1985**, *24*, 223–236.
24. Maher, R. Excavations in Bintacan Cave, Ifugao. *Asian Perspect.* **1989**, *27*, 59–70.
25. Conklin, H. *Ethnographic Atlas of Ifugao*; Yale University Press: New Haven, CT, USA, 1980.
26. Ledesma, C.; Amano, N.; Acabado, S.B. Faunal remains from the Old Kiyyangan Village. *Natl. Mus. J. Cult. Herit.* **2015**, *1*, 21–30.
27. Bush, D. Overview of the Faunal Remains Recovered from the Old Kiyyangan Village, Kiangan, Ifugao. *Philipp. Q. Cult. Soc.* **2018**, *46*, 70–88.
28. Lauer, A.; Acabado, S.B. Infant death and burial practices in late prehistoric Kiyyangan Village, Kiangan, Ifugao. *Natl. Mus. J. Cult. Herit.* **2015**, *1*, 31–37.
29. Eusebio, M.; Ceron, J.; Acabado, S.B.; Krigbaum, J. Rice Pots or Not? Exploring Ancient Ifugao Foodways through Organic Residue Analysis and Palaeobotany. *Natl. Mus. J. Cult. Herit.* **2015**, *1*, 11–20.
30. Yakal, M.A. *Exotic Beads and Jar Burials: Social Status in the Old Kiyyangan Village, Ifugao, Philippines*; University of California: Los Angeles, CA, USA, 2017.
31. Lapeña, Q.; Acabado, S.B. Resistance through rituals: The role of Philippine "native pig" (*Sus scrofa*) in Ifugao feasting and socio-political organization. *J. Archaeol. Sci. Rep.* **2017**, *13*, 294–583. [CrossRef]

32. Fishkish, O.; Ingwersen, J.; Lamers, M.; Denysenko, D.; Streck, T. Phytolith Transport in Soil: A Field Study Using Flourescent Labelling. *Geoderma* **2010**, *157*, 27–36. [CrossRef]
33. Madella, M.; Lancelloti, C. Taphonomy and Phytoliths: A User Manual. *Quat. Int.* **2012**, *275*, 76–83. [CrossRef]
34. Horrocks, M.; Acabado, S.; Peterson, J. Plant Microfossil Results from Old Kiyyangan Village. *Asian Perspect.* **2018**, *57*, 159–176. [CrossRef]
35. Acabado, S.B. Taro Before Rice Terraces: Implications of Radiocarbon Determinations, Ethnohistoric Reconstructions, and Ethnography in Dating the Ifugao Terraces. In *Irrigated Taro (Colocassia esculenta) in the Indo-Pacific: Biological and Historical Perspectives. Senri Ethnological Studies*; Spriggs, M., Addison, D., Matthews, P.J., Eds.; National Museum of Ethnology: Osaka, Japan, 2012; Volume 78, pp. 285–305.
36. Acabado, S.B. The Ifugao Agricultural Landscapes: Complementary Systems and the Intensification Debate. *J. Southeast Asian Stud.* **2012**, *43*, 500–522. [CrossRef]
37. Bodner, C.C. On the Evolution of Agriculture in Central Bontoc. Ph.D. Thesis, University of Missouri, Columbia, MO, USA, 1986.
38. Alam, O.; Gutaker, R.M.; Wu, C.-C.; Hicks, K.A.; Bocinsky, K.; Castillo, C.C.; Acabado, S.; Fuller, D.; Guedes, J.A.D.; Hsing, Y.-I.; et al. Genome Analysis Traces Regional Dispersal of Rice in Taiwan and Southeast Asia. *Mol. Biol. Evol.* **2021**, *38*, 4832–4846. [CrossRef]
39. Acabado, S.B.; Martin, M. Decolonizing the Past, Empowering the Future: Community-led Heritage Conservation in Ifugao, Philippines. *J. Community Archaeol. Herit.* **2020**, *7*, 171–186. [CrossRef]
40. Acabado, S.B. Zones of refuge: Resisting conquest in the northern Philippine highlands through environmental practice. *J. Anthropol. Archaeol.* **2018**, *52*, 180–195. [CrossRef]
41. Greer, S. Heritage and Empowerment: Community-based Indigenous Cultural Heritage in northern Australia. *Int. J. Herit. Stud.* **2010**, *16*, 45–58. [CrossRef]
42. Prangnell, J.; Ross, A.; Coghill, B. Power Relations and Community Involvement in Landscape-based Cultural heritage Management Practice: An Australian Case Study. *Int. J. Herit. Stud.* **2013**, *16*, 140–155. [CrossRef]
43. Ross, A.; Prangnell, J.; Coghill, B. Archaeology, cultural landscapes, and Indigenous knowledge in Australian cultural heritage management legislation and practice. *Herit. Manag.* **2010**, *3*, 73–96.
44. Chirikure, S.; Pwiti, G. Community involvement in archaeology and cultural heritage management: An assessment from case studies in Southern Africa and elsewhere. *Curr. Anthropol.* **2008**, *49*, 467–485. [CrossRef]
45. Moser, S.; Glazier, D.; Phillips, J.E.; el Nemr, L.N.; Mousa, M.S.; Aeish, R.M.N.; Richardson, S.; Conner, A.; Seymour, M. Transforming Archaeology Through Practice: Strategies for Collaborative Archaeology and the Community Archaeology Project at Quseir, Egypt. In *Museums and Source Communities: A Routledge Reader*, 1st ed.; Brown, A., Peers, A., Eds.; Routledge: Abingdon, UK, 2005; pp. 208–226.

land

MDPI

*Article*

# Encoded Landscapes: A Link between Inka Wall Orientations and Andean Geomorphology

**Nicolás Palacios-Prado [1,*], Fabiola Corominas-Sustach [1], Andrés Pérez [1], Danilo Verdugo [2] and Tomaž Podobnikar [3]**

[1] Código Andino Foundation, Paine 9540000, Chile; fabiola@codigoandino.cl (F.C.-S.); andres@la-montana.cl (A.P.)
[2] Department of Geospatial and Environmental Engineering, Universidad de Santiago de Chile, Santiago 9160000, Chile; dynageo@gmail.com
[3] Faculty of Information Studies in Novo Mesto, 8000 Novo Mesto, Slovenia; tomaz.podobnikar1@guest.arnes.si
* Correspondence: nicolas@codigoandino.cl

**Abstract:** While some Inka structures and motifs align with astronomical and horizon markers, a significant portion of their constructions exhibit different patterns. We examined the potential correlation between the orientation patterns of the Inka walls and Andean geomorphology, aiming to uncover the extent to which the physical landscape guided these ancient architectural design methodologies. Using geospatial technology and specially developed peak detection and recognition software, we extensively and meticulously analyzed over 40,000 m of surveyed Inka walls and 20,000 mountain peaks across 11 distinct geographical areas. The analysis revealed a significant correlation between key wall orientations and the parallel peak alignment of the Andean Mountain Range. This suggests a purposeful encoding of landscape orientations into Inka architecture. These findings propose a novel perspective on the intricate relationship between Inka culture and the Andean highlands' topography. Furthermore, this research introduces a distinctive methodological approach to exploring the impact of natural landscapes on architectural planning, establishing a foundation for comparative studies among other ancient civilizations.

**Keywords:** Inka architecture; geospatial analysis; Andean geomorphology; mountain-peak alignment

## 1. Introduction

The Andean region, stretching from Colombia to Chile, is an emblematic testament to the indomitable spirit and adaptability of the Inka civilization. Among South America's most distinguished pre-Columbian cultures, the Inkas forged an empire that, in terms of architectural grandeur and geographical dominance, remains unparalleled. Their constructions, originating from the first half of the 13th century during the dawn of the Inka empire, were born out of a confluence of engineering knowledge, settlement planning, and a respect for nature [1,2].

The hallmark of Inka architecture is its amazing precision in stone wall masonry. This precision, coupled with their ability to sculpt structures that seamlessly merged with the mountainous topography, symbolized their advanced knowledge. Their constructions were not just functional or ornamental but were also calibrated, often aligned with astronomical and geographical elements, and showcased their holistic understanding of the cosmos [3,4].

The Inka architectural repertoire was diverse. Temples, shrines, *tambos*, *kanchas*, *kallankas*, and an extensive road network—the *qhapaq ñan*—depict a highly organized society [5]. Each of these structures, despite serving similar utilitarian purposes, exhibited a distinctive Inka blueprint. In contrast to the repetitive motifs prevalent in European architectural paradigms, Inka constructions exhibit a variety of layout configurations while maintaining distinct characteristic elements [6].

Past research narratives have primarily painted the Inka empire's expansion as a military endeavor, emphasizing conquests and territorial annexations [7]. However, a deeper view of archaeological findings and ethnohistorical accounts reveals a broader context, one where spirituality, geography, and architecture converge cohesively [8,9].

The Andes highland territory, a geological marvel spanning seven countries from Venezuela to Chile, offered more than just a backdrop to the Inka civilization. This mountain range, with its sedimentary, igneous, and metamorphic rocks, is a window into Earth's diverse geological history [10]. The Andes' fractal nature, a result of repetitive structures within the Earth's crust, offers a fascinating perspective on the potential inspirations behind Inka architectural designs [11,12]. This fractal characteristic, observable in the spatial organization of mountain peak location as well as in coastlines and valleys, underscores the recurrent patterns inherent in natural formations. Intriguingly, fractal designs and repetitive patterns were also widely utilized and depicted across various ancient civilizations.

The historical evidence suggests a deliberate intent by ancient cultures around the world to mirror the recurring natural fractal motifs in their architectural endeavors, embedding them into their architecture and meticulous urban layouts [13,14]. Moreover, numerous studies have explored the intricate relationship between geomorphology, astronomical orientation, and settlement planning, recognizing the relevance of natural landforms in the development and structure of urban spaces by ancient and modern cultures [4,15–17].

The Inkas perceived mountains not just as geological formations, but as living entities. They believed in animism, in which natural objects possessed souls [18]. This reverence was likely complemented by their advanced understanding of the physical properties of these mountains, given their extensive mining expeditions [19,20]. Their sanctification of the peaks of the Andean Mountains, especially the highest ones, was multifaceted. The snow-capped summits, melting into rivers and glaciers, were life-sustaining. Spiritually, these mountain peaks were home to the Apus, revered mountain deities believed to control elemental forces [21,22].

This study explores the relationship between highland morphology and Inka wall patterns, aiming to ascertain the extent to which the geomorphological orientations of the Andes shaped Inka architecture. With an emphasis on the azimuthal directions, this exploratory study sought to unearth the possible relationships between the orientations of the straight walls and terraces constructed by the Inka and the patterns of mountain peak alignments [23]. Our results suggest a complex relationship between the Inka civilization and its geographical setting, contributing to a more detailed comprehension of their architectural methodologies, cultural beliefs, and legacy.

## 2. Materials and Methods

### 2.1. Inka Site Selection

Inka sites spanning the diverse regions of the Inka empire, known as *Tawantinsuyu*, were carefully chosen under a well-defined set of selection criteria. The following list of selection criteria was used to choose the sites:

- Geographic Distribution across *Tawantinsuyu*: We sought to include sites from across the vast expanse of the Inka empire, spread over various South American countries.
- Reported Site Function: Sites were chosen to represent all types of Inka constructions—religious, residential, administrative, and agricultural. This diversity allows for a multifaceted view of Inka society and its various functional spaces.
- Evidence from Prior Archaeological Studies: We selected sites with confirmed Inka origins, based on previous archaeological research. This criterion lends historical accuracy and validation to the selected Inka architecture.
- State of Preservation: The preservation state of each site was considered, preferring sites with well-preserved features as they provide more reliable data and insights into original Inka architecture.
- Availability of Photogrammetric Survey Data or High-Quality Orthorectified Satellite Imagery: The presence of detailed survey data or clear satellite images is crucial

for accurate remote assessments of the sites and a detailed analysis of their walls. Good orthorectification ensures correctly aligned images representing the true surface geometry, which is necessary for correct measurements and analyses.

- Absence of Obstructive Vegetation: Sites with less vegetation cover offer unobstructed views, which are crucial for the architectural study.

In Cusco, Peru, known for its intricate and well-preserved Inka architecture, we limited our selection to four key sites to ensure that our study focused on the most representative and information-rich locations.

Additionally, we sought sites showcasing Inka architecture within diverse geological and altitudinal contexts, thus providing a deeper understanding of how they adapted their building techniques to different environments. After this selection process, we finalized a list of 20 Inka sites (Figure 1 and Table S1).

Hence, we aimed for an arbitrary, yet structured, selection representing a broader sample of all possible Inka archaeological sites to create a small representative database for the orientations and lengths of the walls and straight terraces.

### 2.2. Inka Architecture Analysis

The length of walls and azimuths (angular deviation, 1–180°) were studied and calculated using the Geographic Information System platform (ArcGIS Pro 3.0.3). The architecture of each site, specifically the walls and straight terraces, was analyzed using satellite imagery (Google Satellite, [GS]) and georeferenced orthomosaics obtained by performing photogrammetry with Agisoft Metashape Professional 1.6.1 (Agisoft LLC). Aerial photographs for generating orthomosaics were captured using DJI Mavic 2 Pro drones with a Hasselblad Sensor 1″ CMOS 20 MP camera during survey campaigns conducted by the Código Andino Foundation (CA). Drone flights were planned and executed using Map Pilot from www.mapsmadeeasy.com/map_pilot (accessed on 2 February 2023), ensuring an 80% overlap of images both along and across flight tracks. Only nadir images were acquired with constant low speed flights of 2 m/s and constant altitude of 60 m or more from the ground. Photogrammetric reports with all parameters used for each site surveyed by CA are freely available at www.codigoandino.org (accessed on 21 March 2023) and upon request. Other orthomosaics were downloaded through Open Heritage 3D (openheritage3d.org, accessed on 3 April 2023) or provided by the National Museum of Natural History of Chile (MNHN, with permission from Francisco Garrido). The references were appropriate in each case.

### 2.3. Mountain-Peak Detection and Filtering

Our in-house software, Peak Mapper 3.0, extracts mountain-peak data objectively from Digital Elevation Models (DEMs). It computes the isolated heights of peaks based on their prominence and dominance, ranging from those with low-elevation peaks in areas of gentle relief to the most prominent ones in rugged terrain. Peak Mapper conducts an automated geomorphometric analysis of the landscape to systematically extract terrain features, culminating in the precise identification of peak locations within a specified territory (Figure S1). The objectivity of this method substantiates any visual interpretations of further peak alignments, offering a reliable means of validation.

The DEMs used for the systematic search of peaks were the Global Multiresolution Terrain Elevation Data 2010 (GMTED2010), a global elevation model with a spatial resolution of 7.5 arc second (around 250 m) freely available from the earthexplorer.usgs.gov website (accessed on 7 November 2020). From these DEMs, 11 geographical zones of approximately 185,000 km$^2$ each were selected and analyzed in the Cassini map projection for our cartometric needs. Using relatively high-resolution DEMs for the large area of the entire Inka territory, an optimized algorithm was developed based on a combination of Python and PostGIS procedures as part of Peak Mapper. A complete dataset of all identified peaks across various filtering levels is available at the Código Andino Foundation GitHub repository [24].

**Figure 1.** Map of selected sites. Map of the Andes Mountain Range in South America, indicating the extent of *Tawantinsuyu* in yellow. Red circles indicate the location of the selected Inka sites. 1. Ingapirca; 2. Aypate; 3. Pariachuco; 4. Incahuasi; 5. Tambo Colorado; 6. Choquequirao; 7. Machu Picchu; 8. Ollantaytambo; 9. Tipón; 10. Raqchi; 11. Incallajta; 12. Samaipata; 13. Caspana; 14. Licancabur; 15. Shincal; 16. Punta Brava; 17. P. Incaico de la Puerta; 18. Viña del Cerro; 19. Huaca de Chena; 20. Cerro Grande de la Compañía. Blue circle indicates the location of the selected unexplored site for verification: V1. Volcán Maipo.

### 2.4. Mountain-Peak Alignments

We developed a mountain-peak-alignment score as a metric to quantify the degree of spatial alignment among multiple mountain peaks within a specified geographic area. It is calculated by an algorithm that evaluates both the azimuthal orientation within a specified search bin and the location and elevation of each peak. For this study, the algorithm spatially analyzed azimuths ranging from 1° to 180°, in steps of one degree, which segmented the analysis area into multiple parallel bins, with this study utilizing 5 km wide bins.

An alignment score is conferred to each bin, contingent on the peak count within it and the assigned elevation weights of those peaks. A bin receives an alignment score only if it contains two or more peaks; bins with fewer than two peaks are assigned a score of zero. Subsequently, the scores of all parallel bins are summed to calculate the final peak-alignment score for the entire area at each specified azimuth. This comprehensive process yields a table of peak-alignment scores ranging from 1° to 180°.

The elevation of each peak is factored into the score; peaks with higher elevations are given a greater weight, thereby exerting more influence on the score. The weight assigned to peaks increases with elevation as follows: peaks below 3500 m above sea level (masl) have a weight of zero; 3500–4000 masl, a weight of 1; 4000–4500 masl, 2; 4500–5000 masl, 3; 5000–5500 masl, 4; 5500–6000 masl, 5; 6000–6500 masl, 6; and peaks above 6500 masl have the greatest weight of 7. The peak-alignment score of a bin increases with the presence of more peaks, especially those at higher elevations. The processing for all geographical zones was performed with a Mercator map projection. The mountain-peak alignment module and code is available at the Código Andino Foundation GitHub repository [24].

### 2.5. Statistical Analysis

For the statistical analysis, the data procured through GIS and the alignment scores of peaks were analyzed using SigmaPlot 10.0 (Grafiti, Palo Alto, CA, USA). We employed histogram and column plots to depict the frequency or lengths of walls across various azimuth degrees. Regarding the peaks, we created plots that graphically represent peak-alignment scores as a function of an azimuth. Additionally, we performed correlation analyses to examine the relationship between the peak-alignment scores and azimuth orientations of walls and terraces.

## 3. Results

### 3.1. Inka Site Selection and Wall-Orientation Analysis

To explore the potential relationship between Inka architecture and Andean geomorphology, we first selected various Inka sites spread across the expanse of the *Tawantinsuyu* (Figure 1) based on a defined set of criteria (detailed in Section 2).

A brief overview of the twenty selected Inka sites, including hallmark characteristics and distinctive features inherent to the Inka architecture, is provided in Appendix A. According to archaeological and anthropological studies [25–42], all twenty sites present evidence of Inka origin or occupation.

Geospatial tools and software were employed to analyze and map the straight walls and terraces within the architectural layout of the sites. This step determines the approximate length of each wall and terrace, as well as its angular orientation (Eazimuth). Table S1 provides a list of the sites with their names, coordinates, elevation, site type, and the image source used for the layout analysis. Site V1 was selected for validation as an uncategorized, possible Inka site.

The floor plans of each Inka site were generated based on the surveyed layouts of their walls (Figure 2). Additionally, an azimuth frequency histogram (Figure 3) and wall-length bar graphs (Figure 4) were generated for each site. A color palette was designated for intervals of 10°, ranging from 1° to 90°. This color palette was then replicated for the range of 91° to 180° to enhance the visualization of azimuth pairs with a 90° difference. For each site, the two to three most predominant and preferred azimuths were highlighted

(Figure 3 and Table 1). In many instances, pairs of azimuths, approximately 90° apart, dominated the tendencies at each site, as expected from the rectangular layout of most Inka buildings. Regarding wall length, over half of the sites display a distinct azimuth based on the orientation of their longest wall (Figure 4). Numerous sites feature walls that extend beyond 100 m in length, and the longest wall is usually not part of a building enclosure.

**Table 1.** Analysis of Inka wall lengths and preferential site azimuths. V1, uncategorized site for validation.

| *Site #* | *Name* | *Number of Walls* | *Min. Length (m)* | *Max. Length (m)* | *Sum of Lengths (m)* | *Pref. Azimuth (°)* | *Longest Wall Azimuth (°)* |
|---|---|---|---|---|---|---|---|
| *1* | Ingapirca | 110 | 2.8 | 90.4 | 1486 | 12; 102; 112 | 144 |
| *2* | Aypate | 95 | 1.5 | 115 | 1285 | 59; 144 | 144 |
| *3* | Pariachuco | 32 | 3.3 | 58 | 468 | 69; 158 | 157 |
| *4* | Incahuasi | 523 | 1.4 | 116 | 8386 | 69; 157; 166 | 156 |
| *5* | Tambo Colorado | 140 | 2.8 | 139 | 2230 | 75; 166 | 135 |
| *6* | Choquequirao | 147 | 2.4 | 109.6 | 2281 | 20; 108 | 29 |
| *7* | Machu Pichu | 592 | 1.7 | 71 | 3871 | 65; 157 | 19 |
| *8* | Ollantaytambo | 132 | 3.1 | 335 | 4691 | 19; 135 | 135 |
| *9* | Tipón | 210 | 1.9 | 89.6 | 3644 | 41; 135;166 | 134 |
| *10* | Raqchi | 346 | 1.2 | 162 | 4528 | 20;111 | 173 |
| *11* | Incallajta | 136 | 1.5 | 150 | 2470 | 112; 128 | 125 |
| *12* | Samaipata | 193 | 1 | 73 | 1494 | 3; 93 | 93 |
| *13* | Caspana | 15 | 2.7 | 58 | 202 | 14; 105 | 14 |
| *14* | Licancabur | 46 | 2 | 39 | 374 | 132; 149 | 54 |
| *15* | Shinkal | 35 | 4.8 | 202 | 1368 | 3; 93 | 3 |
| *16* | Punta Brava | 49 | 1.4 | 16 | 219 | 14; 144 | 144 |
| *17* | P. Incaico de la P. | 39 | 2.6 | 34 | 408 | 3; 103 | 104 |
| *18* | Viña del Cerro | 54 | 1.4 | 58 | 562 | 60; 145 | 157 |
| *19* | Huaca de Chena | 26 | 1.7 | 60 | 251 | 8; 85 | 173 |
| *20* | La Compañía | 38 | 1.3 | 69 | 450 | 20; 112 | 21 |
| *V1* | Volcán Maipo | 48 | 2.7 | 58 | 471 | 122; 166 | 134 |

We aggregated data from all twenty sites, documenting both the wall lengths and their corresponding azimuths. This comprehensive database was then prepared for an in-depth analysis, ensuring a structured approach to understanding the patterns and insights from these measurements.

### 3.2. *Inka Wall Database Analysis*

A summary is presented in Table 1 with the parameters of the analyzed walls and terraces in each site, along with the preferred azimuths (two–three most frequent) of the site and the azimuth of the wall or terrace with the greatest length in the entire site. Certain preferred azimuths were repeated across different sites. A total of 40,671 m of walls and terraces was analyzed. Subsequently, a generalized analysis incorporated all the azimuth data obtained from the walls into a count histogram (Figure 5A), revealing a non-uniform distribution with clear count spikes regardless of site localization. To clearly visualize the data of the main and most prominent architectural features, we reselected walls and terraces greater than 5 m (Figure 5B) and greater than 40 m (Figure 5C) in length. Finally, the wall lengths were summed for each azimuth degree, and a bar graph was plotted to display the sum of the wall lengths versus azimuth (Figure 5D).

Across all the evaluated sites, the most frequent wall azimuth (±1°) was 20°, with subsequent spikes observed at 111°, 166°, 68°, 157°, and 135° (Figure 5A). When examining the most extended wall at each site, the most commonly identified azimuth (±1°) was 135°, which was repeated in four sites. In addition, 144° and 157° were present at three sites (Table 1). The longest terrace was found in Ollantaytambo with a length of 335 m and azimuth at 135°, followed by walls in Shincal and Raqchi, with lengths of 202 and 162 m and azimuths at 94° and 174°, respectively. Regardless of the site's location, walls exceeding a length of 100 m exhibit azimuths at 3°, 20°, 29°, 78°, 125°, 135°, 144°, 156°, and 174° (Figure 5B,C).

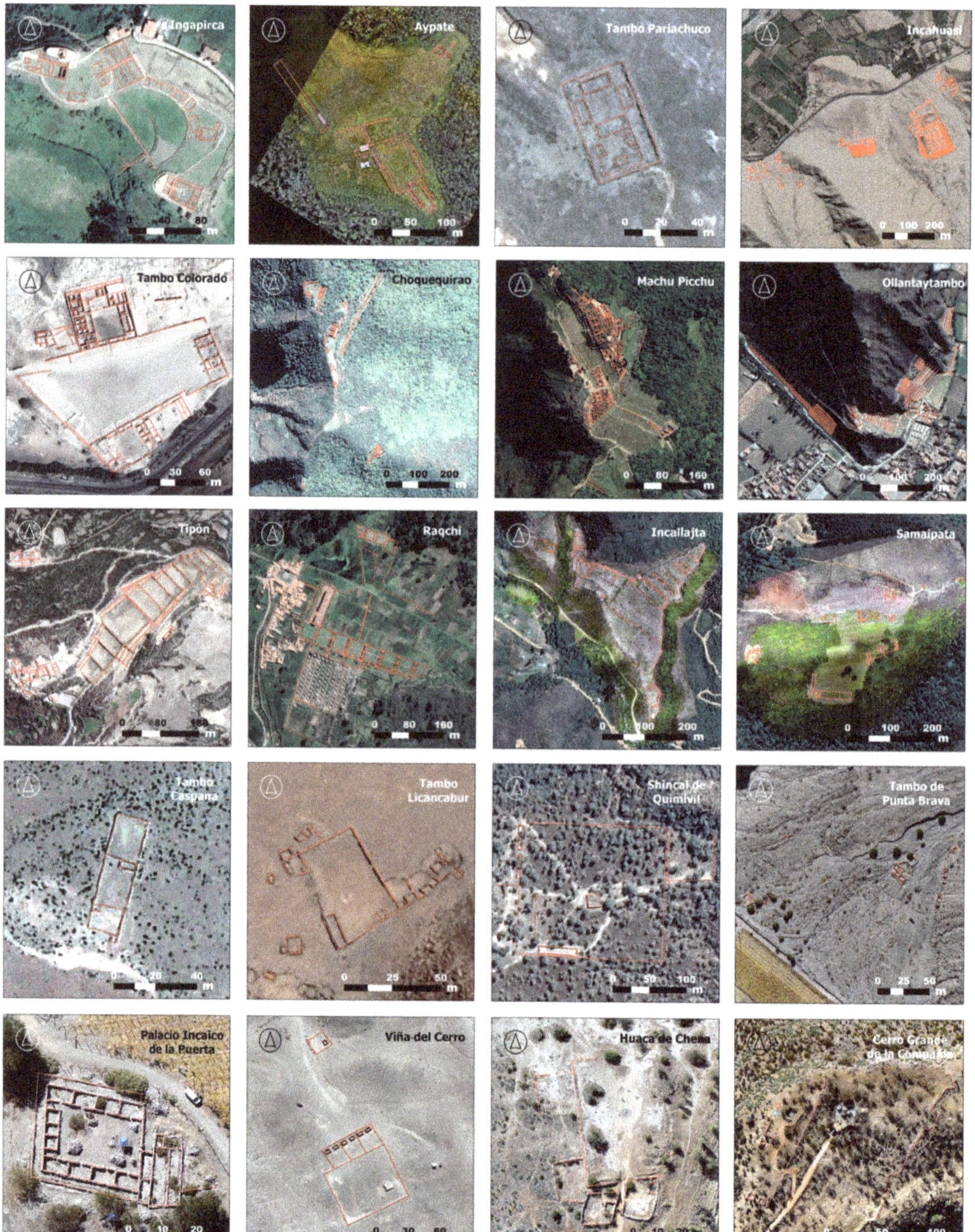


**Figure 2.** Planimetric survey of Inka sites. Maps of selected Inka sites with straight walls and terraces marked in red. High-definition maps are provided in the Supplementary Materials.

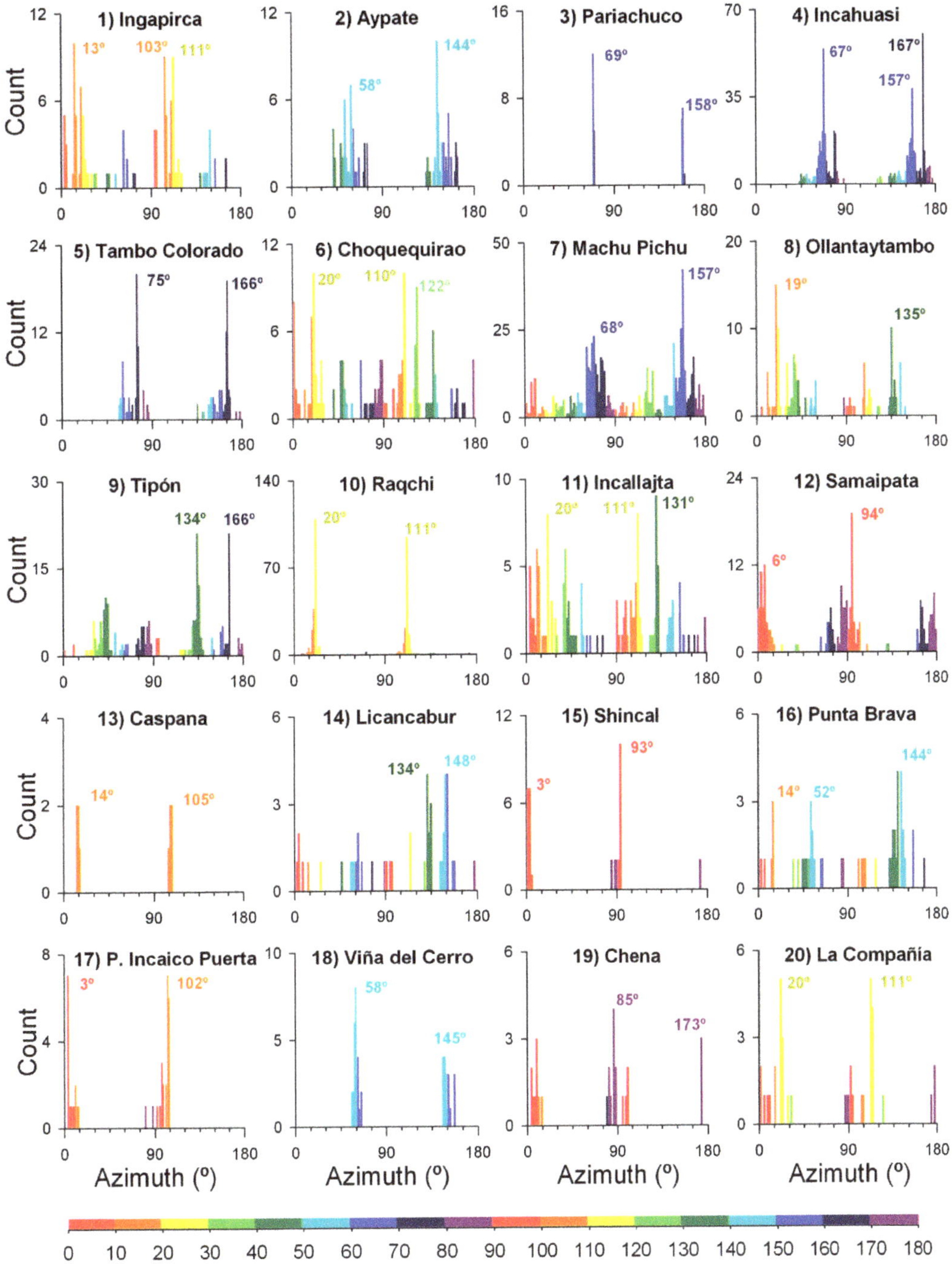


**Figure 3.** Frequency analysis of Inka wall orientation. Histograms from wall azimuths calculated for each Inka site. Numbers in degrees denote preferential azimuths. A color palette was designated for intervals of 10°, ranging from 0° to 90°, and replicated for 90° to 180°.

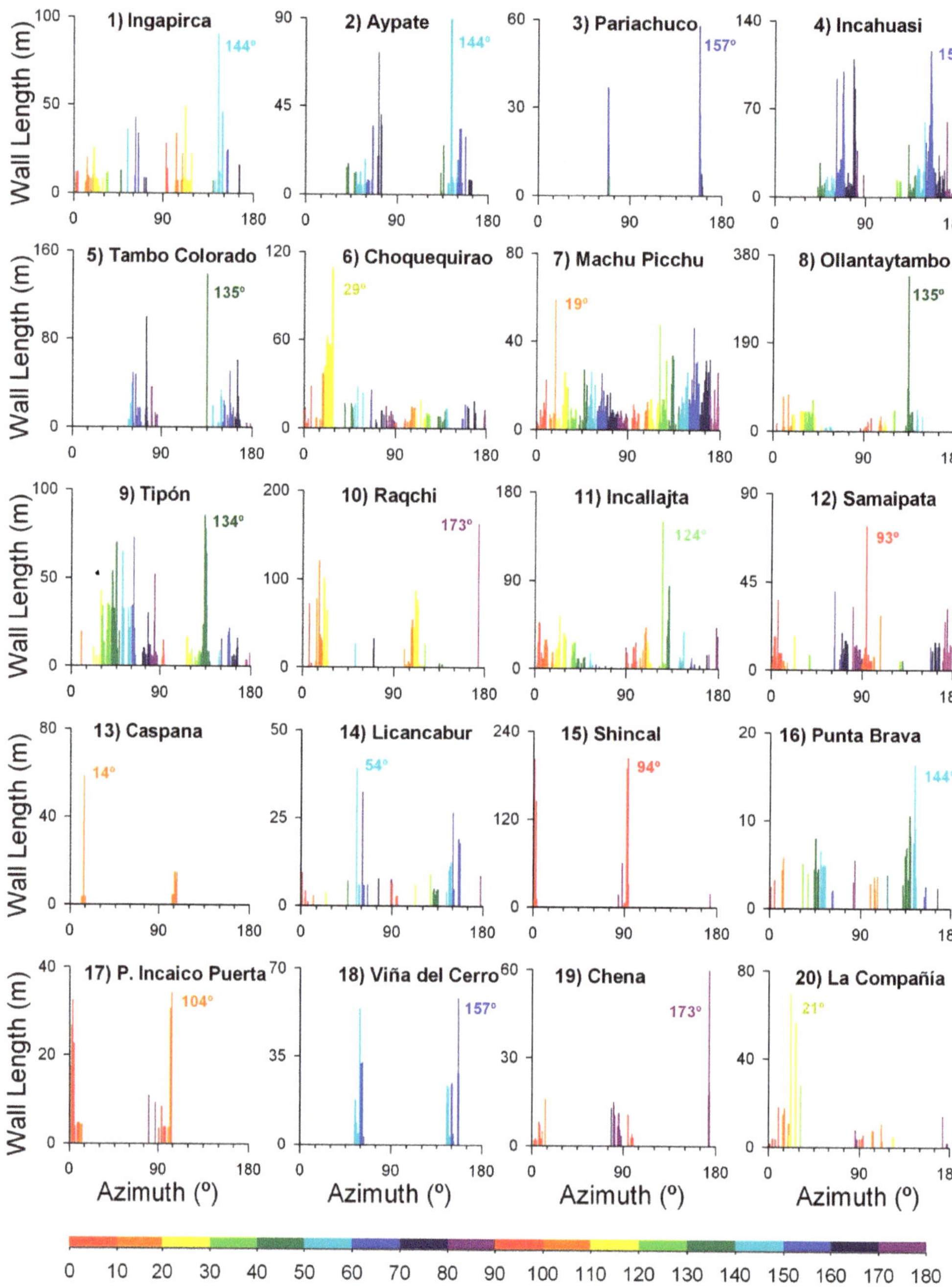


**Figure 4.** Distribution of Inka wall lengths. Bar graphs show the distribution of wall lengths by azimuths obtained for each Inka site. The azimuth of the longest wall is indicated with numbers in degrees. A color palette was designated for intervals of 10°, ranging from 0° to 90°, and replicated for 90° to 180°.

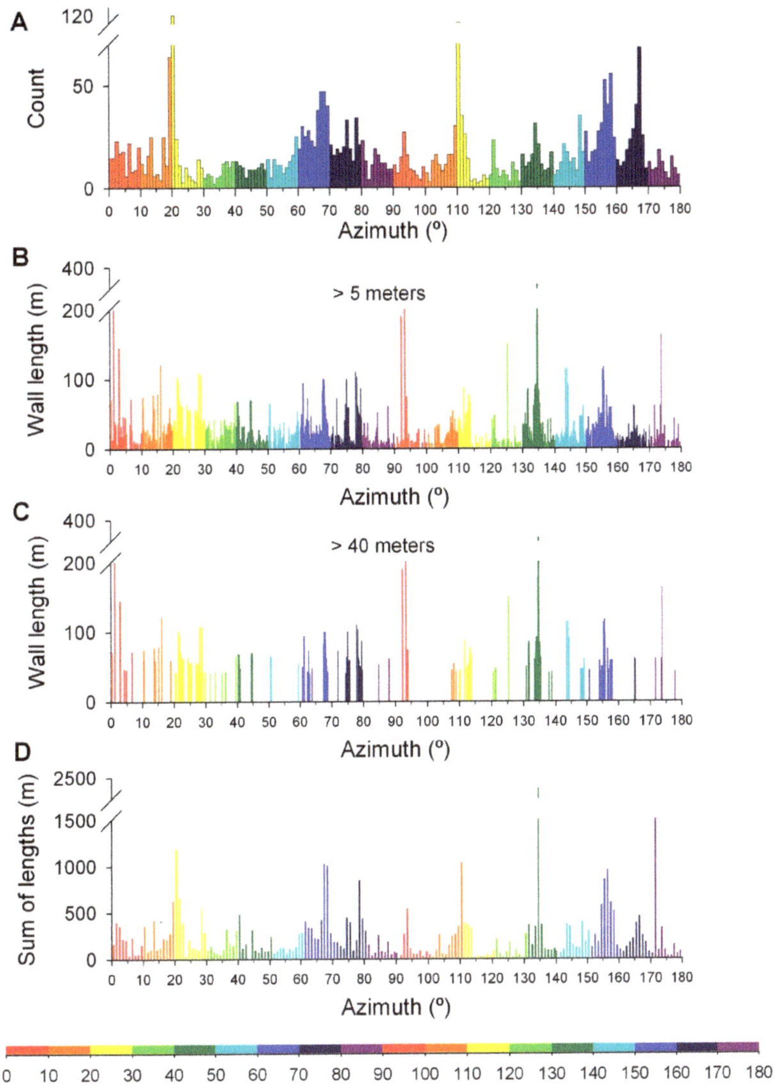


**Figure 5.** Integrated analysis of all Inka walls for all twenty sites. (**A**) Histograms of wall count vs. azimuths. (**B**,**C**) Bar graphs show the distribution of wall lengths above 5 m (**B**) and 40 m (**C**). (**D**) Bar graphs of the sum of all wall lengths grouped by azimuth from all twenty sites. A color palette was designated for intervals of 10°, ranging from 0° to 90°, and replicated for 90° to 180°.

Analyzing the cumulative lengths of all walls, the greatest summed length was oriented at the 135°, followed, in descending order, by azimuths at 20°, 111°, 68°, 156°, 78°, and 166° (Figure 5D). Undoubtedly, many of these preferred azimuths are differentiated by roughly 90°, reflecting the most frequent angle found between the rectangular enclosures of Inka buildings, e.g., 20–111°, 29–125°, 68–157°, and 78–166°.

### *3.3. Geomorphological Analysis of Andes Mountain Range*

The orientation patterns of Inka walls in relation to the surrounding Andes topography could offer compelling insights. To unveil these intricate relationships, we employed modern computational methods, specifically leveraging an algorithmic approach to dissect the geomorphological complexities of the Andes.

Utilizing our proprietary algorithm and software, Peak Mapper (Figure S1), we comprehensively mapped and selected mountain peaks across the Andes Mountain Range from Colombia to central Chile (Figures 6 and S2). The detection of prominent (topographical, visual, and even social attributes of mountain peaks according to their neighborhood) and dominant (their independence according to their neighborhood) mountain peaks and analysis of their azimuthal alignment were executed through a specially designed algorithm rooted in the DEM.

The detection algorithm identifies peaks based on both geomorphological and topographic parameters [43]. These peaks are characterized by slopes identified through grid analyses, such as differences from mean values in a circular kernel of various sizes, applying a certain threshold to select peaks in non-flat areas. We ensured a minimum distance between identified peaks, which in our analysis, ranged from 10 to 35 km. Furthermore, the relative elevation of a peak was defined as the vertical distance between the peak itself and its nearest saddle point, with a benchmark of 250 m in our study.

Adopting a multi-scale approach, we analyzed the data using three distinct spatial resolutions of the DEM: 500, 1000, and 2000 m. At a 500 m resolution, the initial calculation of local peaks yielded 1,393,498 peaks, serving as the positional reference for the peaks across all resolutions. The calculation of peak selection was segmented into three sub-algorithms (steps 1, 2, and 3; Figure S1): the first emphasizes geomorphologic parameters, the second introduces the criteria of inter-peak distance, and the third adds topographic vertical distance criteria. For each resolution, the peaks were calculated considering only the third criterion, the first and third criteria, or all three criteria, resulting in nine different sets of peaks.

After this, the nine sets of peaks underwent rasterization at a 250 m resolution. They were given equal weights, creating a grid with values ranging from 1 to 9 based on the peak frequency. The peaks were also weighted according to their absolute elevation and categorized into four segments ranging from 0 to 6880 m. This signifies that the peaks with higher elevation carried a larger weight compared to their lower counterparts, resulting in final filtering values spanning between 1 and 13. For this study, we used peak datasets filtered at peak levels 5 and 8 (F5 and F8), plus the raw dataset without filtering (F1). The total number of peaks was 44,669 for level 1 (F1) and 18 for level 13 (F13), which delineates their prominence and dominance, such that the peaks of level 1 are only prominent, and the peaks of level 13 are also remarkably dominant over the others (Figure S2).

To study the Andes' geomorphology and determine the parallel orientations of mountain peak alignments in different geographical regions, 11 areas of analysis, of approximately 185,000 km$^2$ each, were selected, covering more than 85% of the Inka Empire territory, which is approx. 2,500,000 km$^2$ (Figure 6).

A total of 20,794 (F1), 6494 (F5), and 1524 (F8) mountain peaks were included in these areas of analysis (Figure 6 and Table 2). Then, for each of these areas, mountain peaks were analyzed with our own peak-alignment algorithm to determine an alignment score for each azimuth degree and estimate the pattern orientation of the geomorphology (Figure 7). We implemented an alignment algorithm beginning with the calculation of a "circular bounding box", a technique adapted from Ritter [44], to encompass all relevant data points within a defined two-dimensional space. This bounding circle set the stage for further segmentation along the $x$-axis, in which, we divided the space into "bins" or sectors, aiding in the systematic categorization of the spatial data.

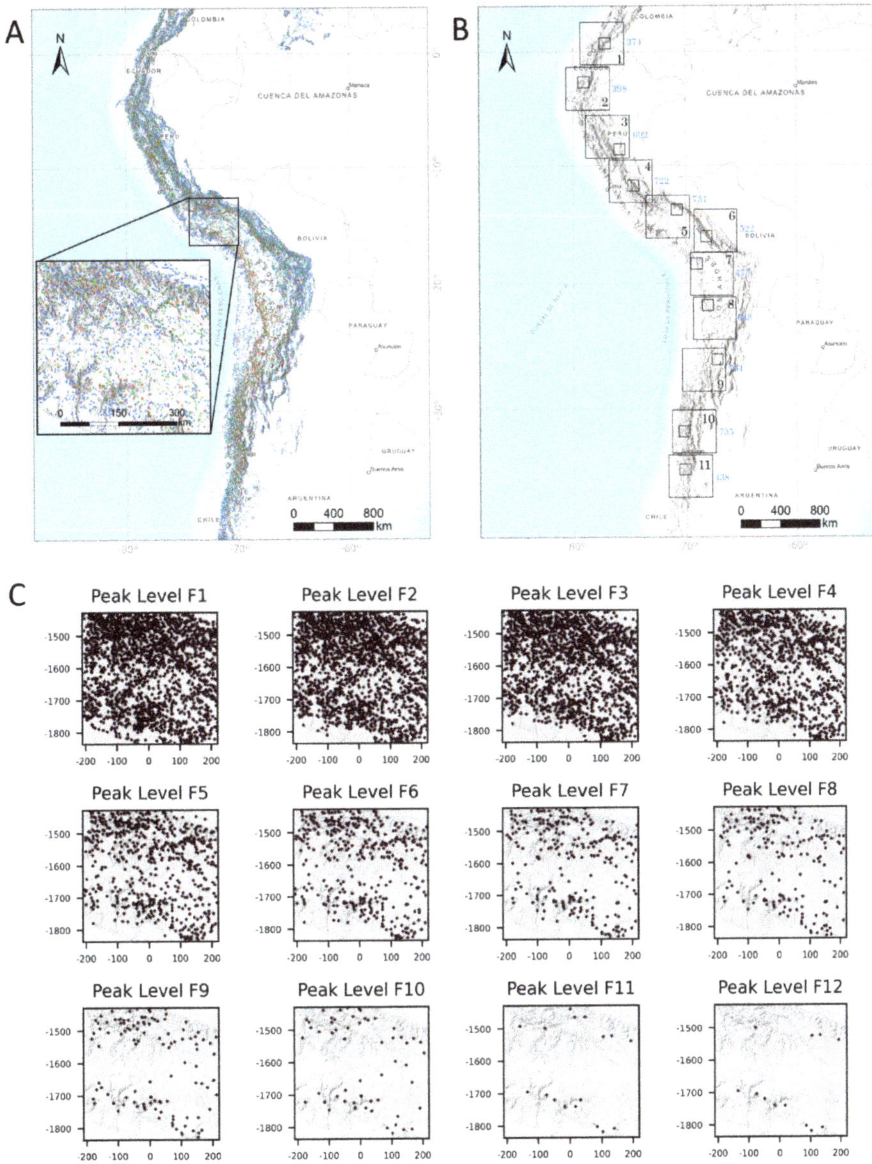


**Figure 6.** Map of detected mountain peaks and geographical zones of study. (**A**) Detected peaks are shown in blue, green, and red dots representing filtering values of levels F1, F5, and F8, respectively. The inset displays an area approximate to area #5 Cusco shown in B. (**B**) Areas of analysis are indicated by squares and numbered from 1 to 11. The total number of peaks per zone filtered at F5 is indicated in blue. Each area includes a smaller sub-area of study. (**C**) Detected mountain peaks on same area shown on inset in A at different peak filtering levels from F1–F12. Graticule numbers are in km. Peak Level 13 is not shown because, in this selected area, only one peak is left. Peak levels (1–3) are categorized as local, merely prominent; (4–10) as regional, dominant and prominent; and (11–13) global, remarkably dominant over the others. The positional resolution of these peaks is set at 250 m, consistent with that of the GMTED DEM, in the Cassini map projection.

**Table 2.** Geographical areas and sub-areas of analysis and number of peaks detected. Coordinates are in Mercator projection in meters.

| *Area #* | *Name* | *Upper Right x* | *Upper Right y* | *Lower Left y* | *Lower Left y* | *Min–Max Elev.* | *# Peaks (F1)* | *# Peaks (F5)* | *# Peaks (F8)* | *Highest Align. Azimuth* |
|---|---|---|---|---|---|---|---|---|---|---|
| 1 | Quito | −8403033 | 286740 | −8833084 | −129744 | 743–5862 | 1552 | 374 | 81 | 33° |
| s1 | Quito | −8534724 | 134229 | −8646157 | 26254 | 869–4176 | 179 | 47 | 9 | 46° |
| 2 | Cuenca | −8545303 | −153031 | −8975354 | −569524 | 720–6239 | 1716 | 398 | 77 | 13° |
| s2 | Cuenca | −8745950 | −245057 | −8857382 | −353032 | 981–4471 | 184 | 54 | 10 | 96° |
| 3 | Cajamarca | −8352630 | −622010 | −8782681 | −1038541 | 754–6525 | 2562 | 692 | 133 | 157° |
| s3 | Cajamarca | −8395975 | −897819 | −8507408 | −1005794 | 816–2499 | 147 | 22 | 1 | 158° |
| 4 | Huancayo | −8124806 | −1057264 | −8554857 | −1473856 | 757–5803 | 2470 | 722 | 124 | 142° |
| s4 | Huancayo | −8256265 | −1250819 | −8367698 | −1358794 | 943–5448 | 299 | 105 | 21 | 134° |
| 5 | Cusco | −7757108 | −1403640 | −8187159 | −1820305 | 792–6407 | 2297 | 731 | 173 | 105° |
| s5 | Cusco | −7827370 | −1489011 | −7938803 | −1596986 | 998–6275 | 195 | 89 | 17 | 121° |
| 6 | Titicaca | −7282930 | −1542982 | −7712981 | −1959688 | 794–6325 | 1895 | 522 | 93 | 134° |
| s6 | Titicaca | −7537615 | −1750833 | −7649047 | −1858808 | 1237–6275 | 235 | 97 | 24 | 136° |
| 7 | Uyuni | −7319541 | −1962979 | −7749592 | −2379788 | 1035–6524 | 1397 | 475 | 127 | 166° |
| s7 | Uyuni | −7634115 | −2014900 | −7745548 | −2122875 | 3274–6524 | 125 | 61 | 19 | 29° |
| 8 | Atacama | −6983905 | −2384989 | −7413956 | −2801892 | 1121–6184 | 1504 | 643 | 218 | 165° |
| s8 | Atacama | −7525474 | −2421368 | −7636906 | −2529343 | 3201–6082 | 142 | 100 | 36 | 136° |
| 9 | Copiapo | −7403977 | −2898592 | −7834028 | −3315652 | 894–6870 | 1924 | 764 | 222 | 31° |
| s9 | Copiapo | −7430858 | −2952249 | −7542290 | −3060224 | 3244–5964 | 113 | 48 | 14 | 14° |
| 10 | Los Andes | −7504407 | −3496853 | −7934458 | −3914109 | 676–6880 | 1978 | 735 | 168 | 174° |
| s10 | Los Andes | −7759889 | −3649963 | −7871321 | −3757938 | 1771–6635 | 260 | 114 | 35 | 165° |
| 11 | Rancagua | −7542634 | −3933173 | −7972685 | −4350576 | 613–6049 | 1499 | 438 | 108 | 19° |
| s11 | Rancagua | −7754822 | −4022352 | −7866255 | −4130327 | 1001–5228 | 258 | 94 | 28 | 14° |

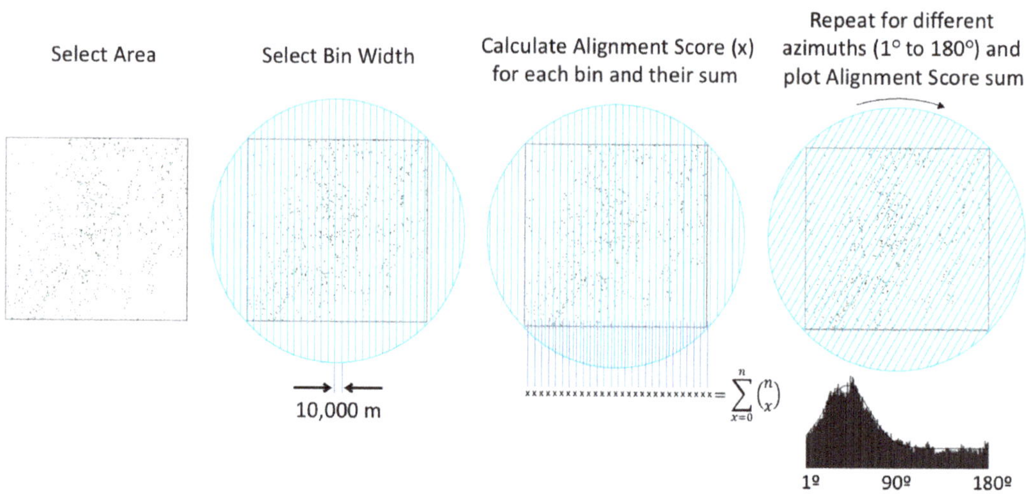


**Figure 7.** Illustration of peak-alignment score calculation. From left to right: selection of area of analysis with detected peaks (green dots); selection of bin width, usually between 1 and 10 km; calculation of alignment score for each bin by adding the number of peaks within the bin, while each peak value is weighted by elevation with a factor of 0, 1, 2, 3, 4, 5, 6, and 7 for peaks with elevation of <3500, 3500–4000, 4000–4500, 4500–5000, 5000–5500, 5500–6000, 6000–6500, and >6500 masl, respectively; repeat for each degree from 1° to 180° and plot the alignment score for each degree and find the maximum score for the area.

Upon establishing a bin width, we utilized a specific template to identify and quantify intersections between our sample peak data and the individual bins. A weight value for peak elevation was also included in the quantification. This intersection analysis was pivotal, allowing for a detailed enumeration of the samples within each sector, and preparing the data for the next phase of analysis. A score for peak alignment at each specific azimuth degree was produced and noted (Figure 7).

We then engaged in a rotational analysis, methodically turning the circular bounding box from its central axis and hence, reorienting the bins for the next peak-alignment score quantification. By maintaining consistent rotation increments of one degree, we could scrutinize the data from various iterative alignment scores from 1° to 180° (Figure 7). This methodical combination of spatial segmentation and rotational analysis allowed for an in-depth, yet efficient, exploration of parallel orientations of peak alignments over the area of analysis.

To find a good set of parameters and sharp tendencies for the peak-alignment scores, we first tested different values for peak filtering (F1, F5, and F8) and bin width (1, 5, and 10 km) in area #1 (Quito) (Figure S3). A bin width of 5 km and F5 reasonably distributed the peak-alignment score. Therefore, we used this set of parameters to completely analyze all selected areas (Figure 8). Each graph was fitted to a Gaussian or Lorentzian curve to find the maximum peak-alignment value of the area (Figure 8 and Table 2).

We further refined the scope of the investigation by designating smaller sub-areas, each encompassing roughly 12,000 km$^2$, within the broader regions under analysis. This step was undertaken to meticulously evaluate the peak-alignment values in different sub-areas (Figures S4–S6). We strategically selected these sub-areas based on their distinctive geomorphological characteristics, which are representative of their larger regions. The primary aim was to elucidate the variability in peak-alignment scores across different landscapes and to test the robustness of our analytical methods on multiple scales.

Intriguingly, our findings revealed considerable heterogeneity in the alignment patterns among these sub-regions. For instance, both area and sub-area #6 (Titicaca) exhibited a pronounced propensity for a singular preferential peak-alignment azimuth at 135° (±1°), which corresponds with the most commonly identified azimuth in the longest walls and terraces in Inka architecture (Table 1). Conversely, sub-area #3 (Cajamarca) presented several azimuths, in which, the peak-alignment scores revealed consistent spikes, recurring at intervals of approximately 10–13° (Figure S7).

### 3.4. Correlation Analysis between Inka Wall Orientations and Mountain-Peak Alignments

Considering the count frequency, preferential orientations, and longest wall azimuth, the most representative azimuths found in Inka architecture (±1°) were 4°, 14°, 20°, 29°, 41°, 68°, 78°, 93°, 111°, 122°, 135°, 144°, 156°, 166°, and 174° (Figure 5). On the other hand, the prominent geomorphological axes in the Andes obtained from the regional peak-alignment score on different geographical areas (±1°) were 14°, 20°, 31°, 105°, 134°, 142°, 157°, 166°, and 174° (Figure 8). In addition to these peak-alignment orientations from large areas, the sub-areas displayed additional azimuths at 4°, 46°, 95°, 112°, and 121° (Figure S6).

Representing these geomorphological orientations on a radial graph showcased an interesting pattern; they appeared to be separated by intervals of approximately 10° to 13° (Figure 9A). Remarkably, when the corresponding geomorphological axes were rotated by 90°, most offset orientations coincided with the azimuth found on Inka walls. This observation suggests that the most preferred Inka wall azimuth covers 180° in approximately 10° to 13° intervals, reflecting the geomorphological orientations of the Andes peak alignments.

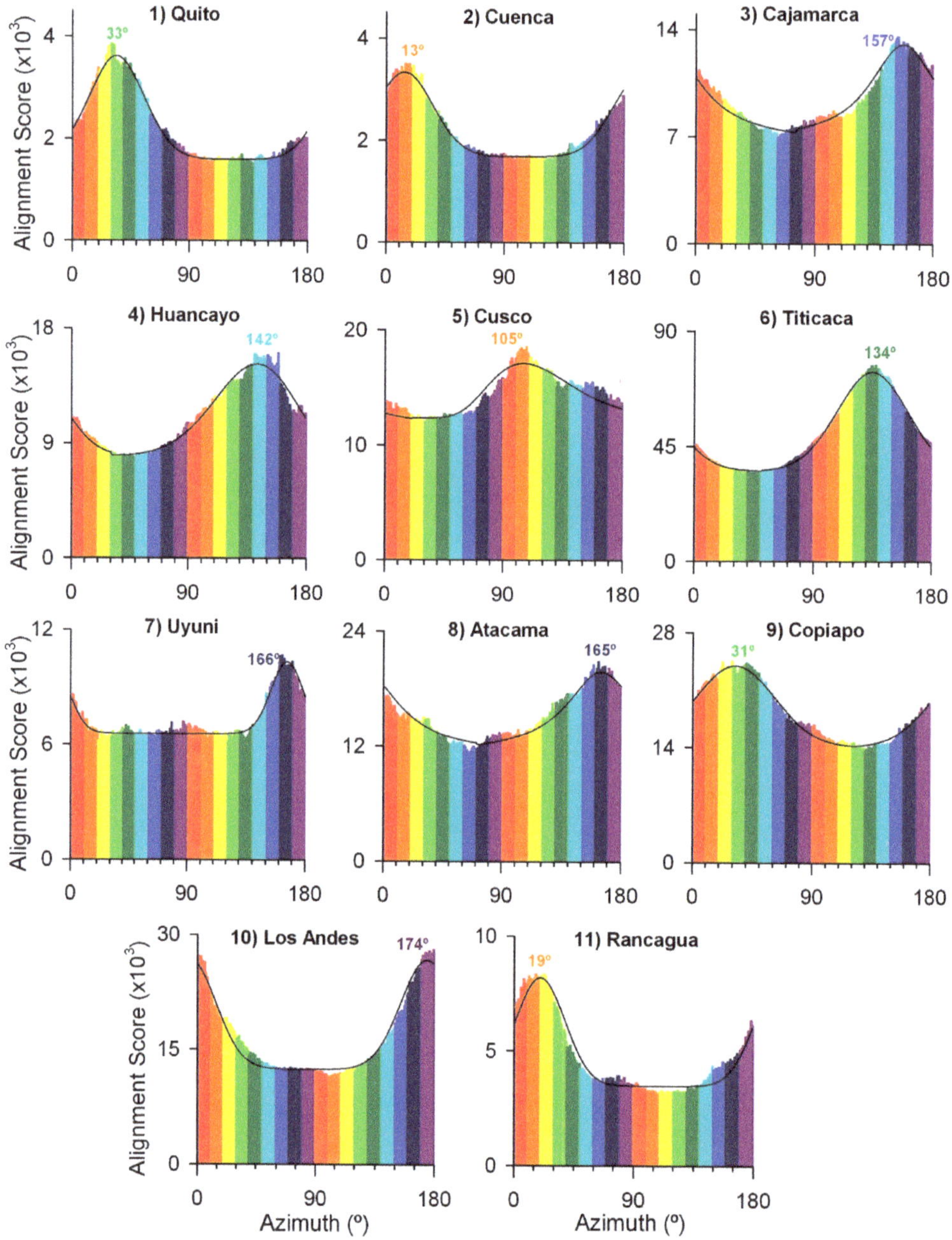


**Figure 8. Regional mountain-peak-alignment score distribution.** The black line displays a Gaussian or Lorentzian regression of four variables with the highest alignment score marked with its corresponding regional peak alignment azimuth in colored numbers. All areas were analyzed at a bin width of 5000 m and peak level of F5. A color palette was designated for intervals of 10°, ranging from 0° to 90°, and replicated for 90° to 180°.

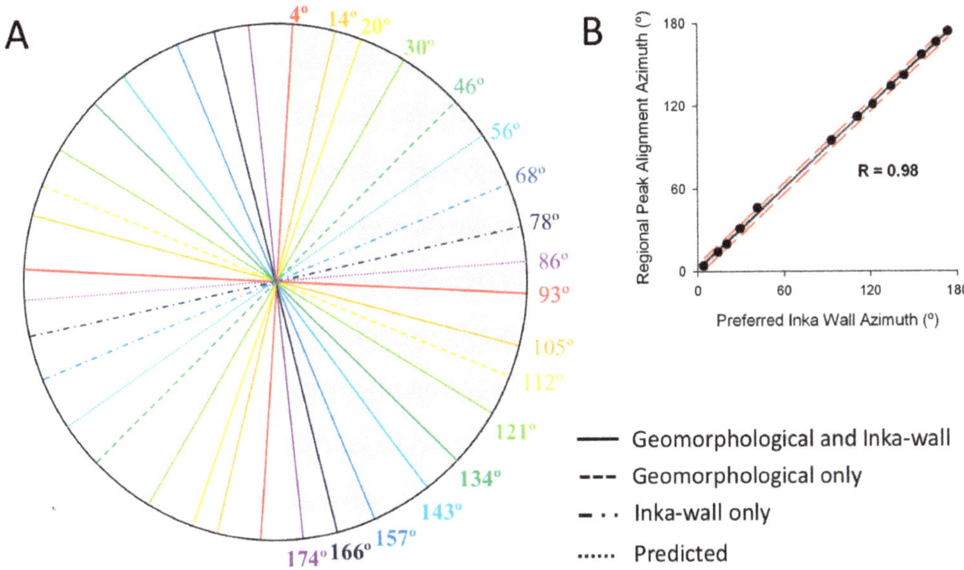


**Figure 9.** Radial distribution of preferred Inka wall and regional peak-alignment azimuths. (**A**) Representation of all prominent geomorphological peak alignments and preferred Inka wall azimuths (±1°) on a polar graph. Colors were designated for each azimuth within a 10° range from 0° to 90°, and replicated for 90° to 180°. Colored numbers show the specific azimuth for each preferred Inka wall and/or regional peak-alignment orientation. Grey half of the chart highlights the 0° to 180° range. (**B**) Correlation analysis between regional peak alignments and preferred Inka wall azimuth. Correlation coefficient (R); 95% confidence band (red).

The results obtained from both analyses were statistically compared using a correlation analysis (Figure 9B), revealing a correspondence between the azimuth of Inka walls and peak-alignment scores in the Andes Mountain Range. The achieved correlation coefficient was 0.98.

In summary, thirteen orientations were found for preferred peak alignments in the Andes Mountain Range. These orientations were also found in Inka wall azimuths (±1°). Thus, these results strongly support the notion that the Inkas integrated the orientations of Andean geomorphology into their architectural designs.

*3.5. Employing Azimuth Analysis to Determine Potential Inka Heritage*

Our research findings provide the foundation for a non-invasive method to infer the potential Inka origin of unexplored archaeological structures based on the predominant azimuth of straight walls. To test this predictive method, we analyzed the wall layout of a recently uncovered archaeological site (V1) situated on the foothills of the Maipo Volcano, near the border between Chile and Argentina (Figures 1 and 10). This significant site, nestled at an elevation of roughly 3500 masl within the Laguna Diamante National Reserve, lies within the Caldera Diamante—a prominent geomorphological feature of the Andes Mountain Range. First discovered in 2021 by Andrés Pérez, this site has been the subject of archaeological investigations since 2022.

Our wall azimuth analysis revealed that the dominant direction at this site is 165°, followed by azimuths of 122° and 76°. The longest feature is oriented at 134°, with subsequent orientations noted at 165° and 122° (Figure 10 and Table 1). Given that the wall orientations at this site coincide with several primary azimuths identified in our Inka wall analysis, there is compelling evidence to suggest that these structures are of Inka origin.

Moreover, its strategic placement suggests it served as a significant Inka shrine venerating the Maipo Volcano.

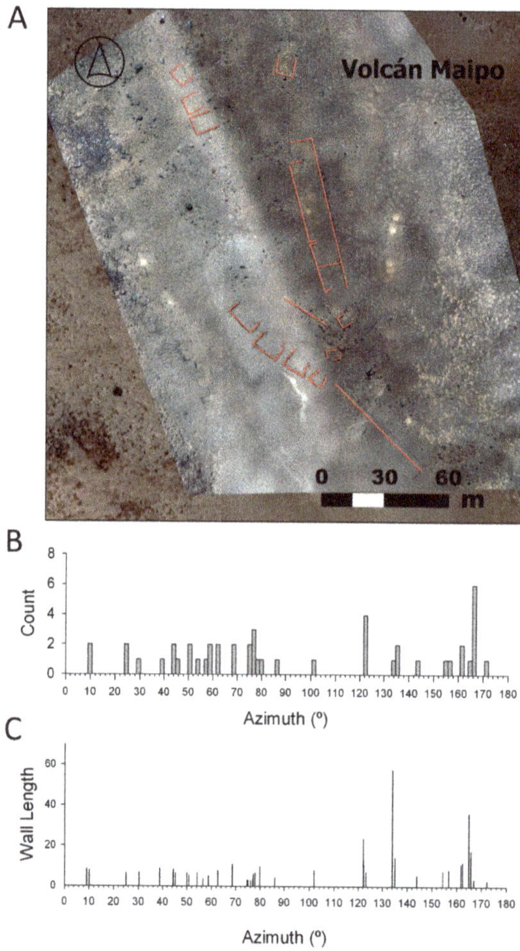


**Figure 10.** Wall azimuth analysis of possible Inka shrine at Maipo Volcano. (**A**) a map of the Volcán Maipo archaeological site and its analyzed walls in red. (**B**) count histogram for wall azimuth from A. (**C**) distribution of wall lengths with respect to their azimuth from A.

## 4. Discussion

Our investigation into the potential interplay between Inka architecture and the geomorphological patterns found in the Andean highlands has uncovered a promising avenue for expanding our comprehension of ancient building philosophies. Although the results of our exploratory data analysis are preliminary and await confirmation through cross-validation with an independent dataset, they indicate a discernible systematic relationship between the geomorphological context and the building patterns of the Inka. By delineating this relationship, we introduced a conceptual model that establishes a foundation for further inquiry, inviting rigorous examination and empirical validation in future studies.

Rock structures, from the tiniest grains to vast landscapes, showcase complex geometric patterns and, in some scenarios, specific crystal arrangements. These patterns can span from the microscale observable under a microscope, to a macroscale of entire mountain ranges. Repetitive features like rock layers, fractures, and mineral lineations visually represent some of these fractal patterns [45]. The information hidden within these geometric shapes provides valuable insight into the Earth's past and the processes that have shaped it. In fact, the anisotropic patterns we observe on a large scale, such as the parallel orientation of peak alignment, might also be mirrored on a much smaller scale [45]. This suggests that the examination of rock structures during mining could unveil repetitive lineation patterns consistent with those found in broader geomorphology.

The Inkas' rich history of mining suggest an intimate knowledge of the land's geological features, such as fault lines, mineral lineations, and mountain range orientations [19,20]. Building on the premise that the Inkas identified repetitive patterns in the natural fractures and mineral lineations of rocky landscapes, we hypothesize that they could correlate this knowledge with the shape and positioning of the surrounding geography. These insights would have not only streamlined their rock extraction processes but also informed their building practices. Within this context, rocks became more than simple construction materials for the Inkas. They represented a sacred connection, a bridge between their cosmology and the tangible, intrinsic world order around them.

The unique nature of our study is highlighted not only by its perspective on Inka architecture but also by its novel methodological approach. We have investigated the parallel orientation of peak alignments in the Andes, presenting a new avenue for analyzing the inherent fractality of continental rock in mountainous regions. While this methodology was developed within the context of this Inka research, it holds significant potential for exploring other mountain-dwelling cultures around the world. Additionally, our findings suggest that the positioning of the Andes mountain peaks does not occur randomly across the geographical area. Instead, they follow fractal patterns of repetitive parallel orientations that deviate by roughly 10 to 13 degrees. This insight provides a clue about the potential internal lattice of crystal arrangements within the continental crust from which the Andes highlands emerged.

Moreover, by analyzing wall azimuths on architectural layouts, we can employ this non-invasive approach to determine the possible origins of yet-unidentified archaeological structures. We have applied this azimuth analysis at an important uncovered archaeological site near Maipo Volcano to identify its possible Inka origin (Figure 10). It also offers an avenue to unearth potential Inka influences or origins in previously overlooked or misattributed building and urban layouts. Besides wall azimuths, this predictive method can be improved by considering the relative location of sites according to nearby mountain peaks and local geomorphological orientations. This predictive capability could be invaluable for future archaeological expeditions, guiding researchers to areas of high potential significance and saving invaluable time and resources in the process.

In discussing the locational attributes and construction orientations of Inka archaeological sites, considering the topographical characteristics influencing ancient building practices is paramount. Our research indicated that 60% of the evaluated Inka sites are situated in relatively flat terrains. This prevalence suggests that the azimuth orientations identified within these sites are less likely to be a mere consequence of constructing the walls perpendicular to the prevailing mountain slopes. Instead, these findings support the hypothesis of deliberate planning in the azimuthal orientation of the walls.

However, addressing the remaining sites found in more mountainous areas is also necessary. For these locations, the local topography, particularly slope inclinations, could have influenced construction methodologies. It remains plausible that Inka builders capitalized on the natural terrain for establishing their structures, laying walls and terraces that correspond to desired azimuthal orientations. Therefore, while our data suggest a significant pattern of purposeful design in wall orientations across numerous Inka sites, we must consider the role of geographical features in shaping these ancient practices.

This study introduces a different and innovative approach, but acknowledging its limitations is essential. To thoroughly validate the results and resulting hypothesis, detailed fieldwork is necessary. We must determine that the observed peak alignment orientations in certain regions are consistently reflected across different scales and can be identified within local rock formations. This activity requires collaborative international field research supported by in-depth geological assessments.

Regarding data collection methods, we must consider the constraints of satellite imagery and photogrammetric surveys. Factors such as the off-nadir angle and orthorectification—which adjusts the geometric properties of an image—can occasionally introduce significant errors into azimuth analyses. For instance, we meticulously chose satellite images in which straight walls were represented accurately, avoiding those appearing distorted or curved. Even minor inaccuracies can impact the reliability of the wall azimuths, which in turn can affect our broader analysis. Expanding our wall database and incorporating more sites could mitigate such errors.

The potential for these discrepancies stresses the need for diverse data collection methods. To bolster the robustness and accuracy of our predictions, on-site measurements using cutting-edge geo-positioning tools and advanced technologies are desired. Such measurements would not only validate results from remote sensing methods but also effectively bridge the gap between the different types and spatial scales of earth observations, such as satellite imagery, Lidar measurements, and tangible on-site observations.

In advancing this study, broadening the analysis beyond merely straight walls and terraces is imperative, as is incorporating a more extensive examination of various structural elements and their purposes. This might include assessing the azimuth of the main rectangular enclosure diagonals, and separately analyzing structures based on their presumed functions, such as ceremonial, residential, or administrative.

A particularly intriguing path for exploration is the orientation of Inka *ushnus*, ceremonial platforms typically characterized by rectangular or square shapes, which hold considerable cultural and religious significance. These structures are often believed to point towards the directions of the equinoxes or solstices; however, a more detailed analysis may reveal orientations corresponding to nearby summit hills [46], which could be reflected in local geomorphological peak-alignment orientations. By incorporating a more sophisticated classification of different Inka structures and examining their orientations, we can enhance the applicability of the method and also potentially discern the selection criteria for encoding different landscape orientations.

## 5. Conclusions

Our analysis of Inka wall orientations has yielded significant findings: the most prevalent azimuth is 20° (±1°), corresponding with an extensive axis orientation of the Andes Mountain Range that stretches from central Bolivia to southern Chile and Argentina. Additionally, we identified 135° (±1°) as the most frequent azimuth for the longest walls and terraces, corresponding to the predominant peak alignments found in the Titicaca basin—a region widely recognized as the birthplace of the Inka civilization. A strong correlation, with a coefficient of 0.98, between frequent wall orientations and peak alignments emphasizes the significant influence of the landscape on Inka architectural design.

This study further advances the understanding of Inka architectural practices, which traditionally focused on archaeoastronomical and horizon markers found in a subset of motifs. We propose that the Inka established a guiding principle of orientation derived from the natural geographical patterns present in the Andes highlands as a fundamental framework for organizing their architectural layouts in diverse settings across the *Tawantinsuyu* landscape. This guiding principle, reflected in the orientation of straight walls and terraces, likely extended to the layout of the Inka's urban planning and the orientation of central squares. We have noticed that this guiding principle and its main orientations appears to be encoded in the architectural layout of the walls at *Muyucmarca* in *Saqsaywaman*, an important sacred site of the Inka in Cusco.

In conclusion, our research has uncovered a statistically significant correlation between the orientations of Inka walls and the parallel orientations of peak alignments. This discovery points to a comprehensive framework for Inka architectural planning that extends beyond the confines of archaeoastronomical considerations. The link between Inka wall orientations and Andean geomorphology suggests more than a mere coincidence; it hints at purposeful encoding of ancient knowledge that identifies a natural order within a seemingly chaotic landscape. By demonstrating this connection, our study lays the groundwork for expanded exploration into the worldviews and building principles of the Inka and other ancient civilizations.

**Supplementary Materials:** The following supporting information can be downloaded at: https://www.mdpi.com/article/10.3390/land13040463/s1. Figure S1. Peak Mapper Algorithm Flowchart; Figure S2. Maps Illustrating the Distribution of Identified Mountain Peaks Across Various Levels of Filtering; Figure S3. Peak-Alignment Analysis in Area #1 Quito Using Different Bin Width and Peak Filtering Levels; Figure S4. Peak Detection on Sub-areas From #1 to #6; Figure S5. Peak Detection on Sub-areas From #7 to #11; Figure S6. Peak-Alignment Score Analysis in Sub-areas; Figure S7. Peak-Alignment Score Analysis in Sub-Areas #3 Cajamarca; Figure S8. Planimetric Survey of Ingapirca; Figure S9. Planimetric Survey of Aypate; Figure S10. Planimetric Survey of Tambo Pariachuco; Figure S11. Planimetric Survey of Incahuasi; Figure S12. Planimetric Survey of Tambo Colorado; Figure S13. Planimetric Survey of Choquequirao; Figure S14. Planimetric Survey of Machu Picchu; Figure S15. Planimetric Survey of Ollantaytambo; Figure S16. Planimetric Survey of Tipón; Figure S17. Planimetric Survey of Raqchi; Figure S18. Planimetric Survey of Incallajta; Figure S19. Planimetric Survey of Samaipata; Figure S20. Planimetric Survey of Tambo Caspana; Figure S21. Planimetric Survey of Tambo Licancabur; Figure S22. Planimetric Survey of Shincal de Quimivil; Figure S23. Planimetric Survey of Pucara de Punta de Brava; Figure S24. Planimetric Survey of Palacio Incaico de la Puerta; Figure S25. Planimetric Survey of Viña del Cerro; Figure S26. Planimetric Survey of Huaca de Chena; Figure S27. Planimetric Survey of Cerro Grande de la Compañía; Table S1. Selected Inka Sites for This Study.

**Author Contributions:** Conceptualization, N.P.-P.; methodology, N.P.-P., D.V. and T.P.; software, D.V. and T.P.; validation, N.P.-P., D.V. and T.P.; formal analysis, N.P.-P. and T.P.; investigation, N.P.-P., A.P. and F.C.-S.; resources, N.P.-P. and T.P.; data curation, N.P.-P., F.C.-S. and T.P.; writing—original draft preparation, N.P.-P.; writing—review and editing, N.P.-P., A.P., D.V., F.C.-S. and T.P.; visualization, N.P.-P., F.C.-S. and T.P.; supervision, N.P.-P. and T.P.; project administration, N.P.-P. All authors have read and agreed to the published version of the manuscript.

**Funding:** This research received no external funding.

**Data Availability Statement:** The original contributions presented in the study are included in the article/Supplementary Material, further inquiries can be directed to the corresponding author.

**Acknowledgments:** We extend our heartfelt thanks to Valentín Palacios and Gerardo Viollier for their constant support and encouragement. Our gratitude also goes to Patricia Dominguez for her support and valuable references. We appreciate the support from Victoria Roi, Felipe Combeau, Francisco Molina, Benjamin Molina, and Martín Bustamante during the expeditions to Maipo Volcano.

**Conflicts of Interest:** The authors declare no conflicts of interest.

## Appendix A

### Site Descriptions (Site #)

**Ingapirca (1)** is a significant archaeological site in Ecuador that showcases the Inka legacy in the northern region of the old empire. Built in the 15th century, it served as a ceremonial and administrative center for the Inkas [28]. The Temple of the Sun, an impressive stone structure dedicated to the sun god, stands out. Additionally, it features a system of water channels that reveals the Inkas' management of water resources. The influence of the earlier Cashaloma-Cañari culture is also reflected at the site; however, for our analysis, only the later architectural stages attributed to the Inka culture were taken into consideration.

**Aypate (2)** is a significant archaeological site in Peru that showcases the Inka legacy in the northern region of the country. This place served as an important administrative, religious, and agricultural center during the height of the Inka Empire [41]. Its elaborate stone constructions, including agricultural terraces and ceremonial structures, reflect the advanced architectural and hydraulic knowledge of the Inkas in this region.

**Pariachuco (3)** is an archaeological site located in the province of Huaylas, which has an Inka origin and represents the legacy of the Inka culture in the high terrains of the Ancash region. It was a strategic tambo used as a resting and provisioning point along the Inka routes near the Pariachuco Mountain [31]. This site exhibits a typical rectangular Inka architectural pattern in the form of a Kancha, and its location at the foot of a known Apu (sacred mountain) could reflect a ceremonial use.

**Incahuasi (4)** (Lunahuaná) is located in the Cañete region of Peru and showcases Inka architectural legacy, characterized by its monumentality and planning. It served as a residential and ceremonial complex for Inka rulers, and its engineering reflects the technical prowess of Inka builders [38]. Its strategic location indicates its role as an administrative and local control center.

**Tambo Colorado (5)** is a prominent archaeological site in Peru that stands out as one of the finest examples of Inka architecture on the southern coast of the ancient empire. It was built as a tambo, serving as a rest and provisioning point along the Inka routes, in addition to having an administrative and control function [25]. Its well-preserved adobe structures, vibrant colors, and distinctive architectural details reflect the mastery of the Inkas.

**Choquequirao (6)**, known as the "Sacred Sister" of Machu Picchu, was built in the 15th century as a llaqta or urban and ceremonial complex of the Inka culture [34]. Discovered in the 19th century, with its remote location and meticulously constructed architecture, Choquequirao has been the subject of various research and conservation projects.

**Machu Picchu (7)** is an iconic Inka archaeological site built in the mountains of Peru, representing the grandeur and mystery of Inka culture. Constructed in the 15th century and rediscovered in the 20th century, this astonishing site reveals the architectural and engineering sophistication of the Inkas [40].

**Ollantaytambo (8)** in the Sacred Valley of Urubamba is considered a living Inka city [42]. The Inka architecture in Ollantaytambo is especially notable for its impressive agricultural terraces. These terraces, known as "andenes," are stepped structures built on the mountainsides surrounding the site. They were precisely designed to maximize land utilization and enable agricultural production in a mountainous environment.

**Tipón (9)**, located near Cusco, showcases the engineering prowess of the Inka culture. This walled settlement spanning 200 hectares served as an estate for Inka nobility and stands out for its impressive irrigation system and hydraulic technology [35]. The architectural design in harmony with the natural environment makes Tipón an example of Inka architecture and hydraulic engineering.

**Raqchi (10)** (Racchi) is an Inka archaeological site located in the Cusco region of Peru, near the town of San Pedro. It is renowned for the Temple of Wiracocha, a massive rectangular structure made of adobe and stone, standing at the center of the site. This temple is dedicated to Wiracocha, the creator god in the Inka cosmovision, and is characterized by its high walls and distinctive architectural style [30]. The site also includes various other structures like living quarters, storerooms, and agricultural terraces, providing a glimpse into the daily life and administrative functions during the Inka period. Raqchi's architectural design demonstrates the Inka's advanced engineering and agricultural techniques, and it is believed to have been a significant ceremonial and administrative center positioned right in the pass of the main Inka road system.

**Incallajta (11)** is a prominent archaeological site in Bolivia known for its impressive Inka architecture. Located in the Cochabamba department, this urban and ceremonial complex represents the splendor of Inka culture in the region. The layout and distribution of buildings and spaces in Inkallajta also reveal the urban and ceremonial planning of the Inkas [32]. Central plazas, residential areas, religious enclosures, and spaces dedicated to administration and governance can be observed at Incallajta, along with one of the largest kallankas in the ancient empire.

**Samaipata (12)** in Bolivia is an archaeological site that was occupied by the Inka culture. Located in the Santa Cruz region, it stands out for its impressive carved rock and its use as a ceremonial and administrative center by the Inkas. The architecture combines Inka elements with local and other cultural influences, creating a unique fusion [37]. The structures include terraces, plazas, and a rock-carved platform known as the "Throne." The Inka occupation leaves a significant imprint on the site with buildings and enclosures of classic Inka architecture.

**Tambo Caspana (13)** is a little-known and studied Inka site near Caspana [27]. It stands out for its architectural ensembles featuring rectangular layouts and closed enclosures, which are suggestive of domestic spaces. This site could have a domestic and ceremonial function as it is located near Tatio Volcano and the great Geysers of Tatio.

**Tambo or Licancabur Inka Sanctuary (14)** was originally described by Father Le-Paige [26] and is a poorly studied site located at the foot of the Licancabur Volcano. However, this site is related to the ascents made by the Inkas to the volcano and is part of a group of structures that have been studied in the Licancabur area [47].

**El Shincal de Quimivil (15),** located in Argentina, within the southern region of the Inka Empire, was an important political, administrative, and religious center. This site stood out for its regional capital character, with distinctive elements that identified it as such. Its large plaza, flanked by imposing kallankas, and an iconic central ushnu are testimonies to its administrative and ceremonial function [36]. In addition to its significant architecture, the surrounding landscape also played an important role at El Shincal de Quimivil. Natural elements of the environment, such as hills and mountains, became landmarks and symbolic representations, similar to what has been identified in the city of Cusco.

**Pucará of Punta Brava (16)**, located near the town of Los Loros in Tierra Amarilla, Chile, is part of a fascinating archaeological complex that includes pre-Inka mining facilities and rock paintings. It is estimated that this site corresponds to the Inka dominance period in Chile [33]. Pucará de Punta Brava is composed of imposing stone walls built in the pirca style.

**The Inka Palace of La Puerta (17)**, located in the Atacama region of Chile, is an important archaeological site that was part of a key settlement in the Copiapó River basin. With primarily Inca architecture, this palace may have served as an administrative center. The strategic location of the site allowed access to resources, and its economy was diversified, including livestock and agriculture. Declared a historic monument, the Inka Palace of La Puerta continues to be the subject of archaeological research to better understand its function in pre-Hispanic societies.

**Viña del Cerro (18)** is a metallurgical center that shows the influence of the Inkas on the Diaguita culture and their mastery of copper extraction and processing techniques [19]. Discovered in 1968, it is the only known dedicated Inka smelting site in the southern Andes. The complex consists of units with specific functions, including an administrative platform, dwellings, and copper smelting furnaces. This site reveals the sophistication and technological skill of the Inkas and Diaguitas in copper metallurgy.

**Huaca of Chena (19)**, located next to Santiago de Chile, has been the subject of review and study in relation to the Inka phenomenon. Through archaeological and archaeoastronomical research, its function and symbolism have been reevaluated. The intramural plaza stands out as an engineering feat where the summit of the hill was leveled, and a retaining wall of approximately 50 m in length was constructed. The possibility has been raised that the construction of Chena had geographic and symbolic aspects [39].

**The Pucará del Cerro Grande de La Compañía (20)** is an Inka site located in the O'Higgins region of Chile. It is one of the southernmost constructions of the Inka Empire and has been the subject of various archaeological and ethnohistorical studies [29,48]. The site stands out for its strategic location and defensive characteristics, such as panoramic visibility and the presence of perimeter walls. Remains of dwellings and an intramural plaza have been identified, suggesting the presence of an active community at the site. Its archaeological significance has led to its declaration as a historic monument.

## References


1. Gasparini, G.; Margolies, L. *Inca Architecture*; Indiana University Press: Bloomington, IN, USA, 1980.
2. Hyslop, J. *Inka Settlement Planning*; University of Texas Press: Austin, TX, USA, 1990.
3. Reinhard, J.; Ceruti, C. Sacred Mountains, Ceremonial Sites, and Human Sacrifice Among the Incas. *Archaeoastronomy* **2005**, *XIX*, 1–43.
4. Pinasco Carella, A. *Pachacámac. Templos, Montañas, Astros y Agua*; Universidad Ricardo Palma: Santiago de Surco, Peru, 2019.
5. Moore, J. *Architecture & Power in the Ancient Andes. The Archeology of Public Buildings*; Cambridge University Press: Cambridge, UK, 1996.
6. Gullberg, S.R. The Cosmology of Inca Huacas. Ph.D. Thesis, James Cook University, Townsville, Australia, 2009.
7. Gonzalez Godoy, C. Comentarios Arqueológicos Sobre La Problemática Inca En Chile Central: Primera Parte. *Bol. Soc. Chil. Arqueol.* **2000**, *29*, 39–50.
8. Zuidema, R.T. The Ceque System of Cuzco: The Social Organization of the Capital of the Inca. *Int. Arch. Etnography* **1964**, *265*, 780–785.
9. Bauer, B.S. *The Sacred Lansdcape of the Inca: The Cuzco Ceque System*; University of Texas Press: Austin, TX, USA, 1998.
10. Sepúlveda, S.; Giambiagi, L.B.; Moreiras, S.M.; Pinto, L.; Tunik, M.; Hoke, G.D.; Farías, M. Geodynamic Processes in the Andes of Central Chile and Argentina: An Introduction. *Geol. Soc. Lond.* **2015**, *399*, 1–12. [CrossRef]
11. Turcotte, D.L. Fractals in Geology and Geophysics. *Pure Appl. Geophys.* **1989**, *131*, 171–196. [CrossRef]
12. Robbins, S.J. The Fractal Nature of Planetary Landforms and Implications to Geologic Mapping. *Earth Sp. Sci.* **2018**, *5*, 211–220. [CrossRef] [PubMed]
13. Brown, C.T.; Witschey, W.R.T. The Fractal Geometry of Ancient Maya Settlement. *J. Archaeol. Sci.* **2003**, *30*, 1619–1632. [CrossRef]
14. Sardar, D.; Kulkarni, S.Y. Role of Fractal Geometry in Indian Hindu Temple Architecture. *Int. J. Eng. Res. Technol.* **2015**, *4*, 532–537.
15. Vartholomaios, A. A Geospatial Analysis of the Influence of Landscape and Climate on the Location of Greek Vernacular Settlements Using GIS. *Appl. Geomat.* **2019**, *11*, 197–213. [CrossRef]
16. Aucelli, P.P.C.; Valente, E.; Di Paola, G.; Amato, V.; Cesarano, M.; Cozzolino, M.; Pappone, G.; Scorpio, V.; Rosskopf, C.M. The Influence of the Geological–Geomorphological Setting on Human Settlements and Historical Urban Development: The Case Study of Isernia (Southern Italy). *J. Maps* **2021**, *17*, 141–150. [CrossRef]
17. Rodríguez-Antón, A.; Belmonte, J.A.; González-García, A.C. Romans in the near East: The Orientation of Roman Settlements in Present-Day Jordan. *Mediterr. Archaeol. Archaeom.* **2016**, *16*, 153–160. [CrossRef]
18. Dean, C.J. *A Culture of Stone. Inka Perspectives on Rock*; Duke University Press: Durham, NC, USA, 2010; ISBN 978-0-8223-4807-8.
19. Garrido, F.; Plaza, M.T. Provincial Inca Metallurgy in Northern Chile: New Data for the Viña Del Cerro Smelting Site. *J. Archaeol. Sci. Rep.* **2020**, *33*, 102556. [CrossRef]
20. Guédron, S.; Tolu, J.; Delaere, C.; Sabatier, P.; Barre, J.; Heredia, C.; Brisset, E.; Campillo, S.; Bindler, R.; Fritz, S.C.; et al. Reconstructing Two Millennia of Copper and Silver Metallurgy in the Lake Titicaca Region (Bolivia/Peru) Using Trace Metals and Lead Isotopic Composition. *Anthropocene* **2021**, *34*, 100288. [CrossRef]
21. Vitry, C. Caminos Rituales y Montañas Sagradas. Estudio de La Vialidad Inka En El Nevado de Chañi, Argentina. *Boletín Del Mus. Chil. Arte Precolomb.* **2007**, *12*, 69–84.
22. Cruz, P.; Crubézy, E.; Gérard, P. Los Adoratorios de Altura Inkaicos. Una Mirada Desde El Cerro Cuzco, Departamento de Potosí, Bolivia. *Mem. Am.* **2013**, *21*, 93–120.

23. Palacios-Prado, N.; Corominas Sustach, F. Geomorfología Fractal de Los Andes y Su Relación Con El Trazado Urbano y Diseño Arquitectónico de La Cultura Inca. In Proceedings of the 87th Annual Meeting for the Society for American Archaeology, Chicago, IL, USA, 30 March 2022; p. 224.
24. Verdugo, D. Codigoandino/Encoded-Landscapes: Peak Alignment Release (v0.1.0). *Zenodo* **2024**. [CrossRef]
25. Engel, F. Early Sites in the Pisco Valley of Peru: Tambo Colorado. *Am. Antiq.* **1957**, *23*, 34–45. [CrossRef]
26. Le Paige, G. El Santuario Incaico Del Licancabur. *An. Arqueol. Etnol.* **1966**, *21*, 49–50.
27. Leonor Adán, A.; Mauricio Uribe, R. El Dominio Inca En La Localidad de Caspana: Un Acercamiento al Pensamiento Político Andino (Río Loa, Norte de Chile). *Estudios Atacameños* **2005**, *29*, 41–66.
28. Alcina Franch, J. Ingapirca: Arquitectura y Áreas de Asentamiento. *Rev. Española Antropol. Am.* **1978**, 127–146. Available online: https://revistas.ucm.es/index.php/REAA/article/view/REAA7878110127A/25357 (accessed on 3 April 2023).
29. Planella, M.T.; Stehberg, R.; Tagle, B.; Niemeyer, H.; Del Rio, C. La Fortaleza Indígena Del Cerro Grande de La Compañía y Su Relación Con El Proceso Expansivo Meridional Incaico. In *Proceedings of the Actas del XII Congreso Nacional de Arqueología Chilena*; Sociedad Chilena de Arqueologías: Temuco, Chile, 1991; pp. 403–421.
30. Sillar, B. Caminando a Traves Del Tiempo: Geografias Sagradas En Cacha/Raqchi, Departamento Del Cuzco (Peru). *Rev. Andin.* **2002**, *35*, 221–246.
31. Astuhuamán Gonzáles, C.W. Pariacaca: Un Oráculo Imperial Andino. In *Volume 2 of Ensayos en Ciencias Sociales*; Joffré, G.R., Ed.; Fondo Editorial de la Facultad de Ciencias Sociales, Universidad Nacional Mayor de San Marcos: Lima, Peru, 2004; pp. 15–54. ISBN 9972981274.
32. Coben, L.S. Incallajta, Performance Center of the Inkas: A Digital Reconstruction and Virtual Reality Analysis. *BAR Int. Ser.* **2006**, *1568*, 401–405.
33. Raffino, R.A.; Moralejo, R.A.; Gobbo, J.D. El Dominio Inka En La Sierra de Zapata (NOA). *Investig. Ens.* **2007**, 309–332.
34. Echevarría López, G.T.; Valencia García, Z. Choquequirao, Un Asentamiento Imperial Cusqueño Del Siglo XV En La Amazonía Andina. *Rev. Haucaypata Investig. Arqueol. Tahuantinsuyo* **2011**, *1*, 32–43.
35. Bray, T.L. Water, Ritual, and Power in the Inca Empire. *Lat. Am. Antiq.* **2013**, *24*, 164–190. [CrossRef]
36. Giovannetti, M.; Cochero, G.; Spina, J.; Corrado, G.; Valderrama, M.; Aljanati, L.; Ferraris, E. *El Shincal de Quimivil, La Capital Ceremonial Inka Del Noroeste Argentino*; Quire-Quire: La Plata, Argentina, 2013.
37. Meyers, A. *El Fuerte de Samaipata. Estudios Arqueológicos*; Biblioteca del Museo de Historia/UAGRM: Santa Cruz de la Sierra, Bolivia, 2015.
38. Chu, A. La Plaza y El Ushnu Mayor de Incahuasi, Cañete. *Cuad. Del Qhapaq Ñan. Año 3 N°3* **2015**, 92–110. Available online: https://qhapaqnan.cultura.pe/sites/default/files/mi/archivo/05%20La%20plaza%20y%20el%20ushnu%20mayor%20de%20Incahuasi-Ca%C3%B1ete-Alejandro%20Chu.pdf (accessed on 3 April 2023).
39. Stehberg, R.; Sotomayor, G.; Gatica, C. El Paisaje Ritualizado Del Pukara de Chena. In *Proceedings of the Actas del XIX Congreso Nacional de Arqueología Chilena*; Sociedad Chilena de Arqueología: Arica, Chile, 2012; pp. 141–147.
40. Bastante Abuhadba, J.M. Investigaciones Interdisciplinarias En La Llaqta de Machupicchu. *Arqueol. Soc.* **2016**, *32*, 267–276. [CrossRef]
41. Morán-Jorquera, J. Aproximaciones Al Proceso de Patrimonialización Del Territorio: Caso Del Sitio Arqueológico Aypate (2012–2016). *Espiral Rev. Geogr. Cienc. Soc.* **2020**, *2*, 107–118. [CrossRef]
42. García Bengoechea, P. Patrimonio Arqueológico, Puesta En Valor y Restauración En Ollantaytambo. *Anthropologica* **2022**, *40*, 255–285. [CrossRef]
43. Podobnikar, T. Detecting Mountain Peaks and Delineating Their Shapes Using Digital Elevation Models, Remote Sensing and Geographic Information Systems Using Autometric Methodological Procedures. *Remote Sens.* **2012**, *4*, 784–809. [CrossRef]
44. Ritter, J. An Efficient Bounding Sphere. In *Graphics Gems*; Glassner, A.S., Ed.; Academic Press Professional, Inc.: San Diego, CA, USA, 1990; Volume 1, pp. 301–303, ISBN 0-12-286166-3.
45. Kruhl, J.H. Fractal-Geometry Techniques in the Quantification of Complex Rock Structures: A Special View on Scaling Regimes, Inhomogeneity and Anisotropy. *J. Struct. Geol.* **2013**, *46*, 2–21. [CrossRef]
46. Corrado, G.; Giménez Benítez, S. The Orientation of the Ushnu of "El Shincal Del Quimivil" (Catamarca, Argentina). *Comechingonia. Rev. Arqueol.* **2019**, *22*, 251–263. [CrossRef]
47. Nielsen, A.E.; Berenguer, J.; Sanhueza, C. El Qhapaqñan Entre Atacama y Lípez. *Intersecc. Antropol.* **2006**, *7*, 217–234.
48. Planella, M.T.; Stehberg, R. Etnohistoria y Arqueologia En El Estudio de La Fortaleza Indigena de Cerro Grande de La Compañia. *Chungara* **1994**, *26*, 65–78.

land

MDPI

Article

# Living Together, Living Apart: Residential Structures in Late Bronze Age Shirenzigou, Xinjiang

**Meng Ren [1,2,3,†], Lixun Chen [1,†], Tongyuan Xi [1,2,3], Yue You [4], Duo Tian [1,2,3], Jianxin Wang [1,2,3,*], Marcella Festa [1,2,3,*] and Jian Ma [1,2,3,*]**

1 School of Cultural Heritage, Northwest University, Xi'an 710127, China
2 Key Laboratory of Cultural Heritage Research and Conservation, Ministry of Education, Xi'an 710127, China
3 China-Central Asia "The Belt and Road" Joint Laboratory on Human and Environment Research, Key Laboratory of Cultural Heritage Research and Conservation, School of Culture Heritage, Northwest University, Xi'an 710127, China
4 School of History, Capital Normal University, Beijing 100089, China
* Correspondence: moguwang@nwu.edu.cn (J.W.); merz1895@hotmail.it (M.F.); majian@nwu.edu.cn (J.M.)
† These authors contributed equally to this work.

**Abstract:** The spatial organization within ancient settlements offers valuable insights into the evolution of social complexity. This paper examines spatially and chronologically contextualized architectural structures and artifacts uncovered at the Late Bronze Age Shirenzigou site to explore the relationship between the use of space and underlying social dynamics in the Eastern Tianshan Mountains of Xinjiang (China). Central to our findings is a distinctive centripetal compound structure, consisting of a larger non-domestic building surrounded by smaller dwellings. This arrangement, along with the variety and distribution of the artifacts, reveals a complex interplay between private and communal spaces at the site, reflecting a growing complexity within the social fabric of the community. The formation of conglomerates of houses around a central communal structure which occurs across the Tianshan Mountains appears to be a strategic adaptation in response to environmental challenges and socio-political transformations across this region at the end of the second millennium BCE.

**Keywords:** Shirenzigou; spatial organization; social complexity; Eastern Tianshan Mountains; Late Bronze Age

## 1. Introduction

The study of space use and organization is crucial to our appreciation of the development of past societies [1–3]. The layout of the settlements and the location, form, and size of the buildings, sometimes (cautiously) comparable with ethnographic evidence, can provide critical understanding of the living strategies, economic activities, social interactions, and other facets of ancient communities [4–7]. Recent studies have underscored the significance of the materials and methods used in construction to gain insights into ancient lifestyles [8–10]. On an even smaller scale, examining the spatial arrangement of artifacts at archeological sites has proven crucial for pinpointing areas of specific activities, thereby aiding in the reconstruction of past behaviors and social interactions [11–13].

In the last century, Chang [14–16] introduced the concepts of 'microstructure' (i.e., the cultural and social system of a settlement) and 'macrostructure' (i.e., the broader culture and social organization) in settlement patterns in China and drew attention to the importance of anthropological approaches to address issues of social organization and complexity. His works were critical for the reevaluation of traditional theories and definitions related to social complexity in Chinese archeology, challenging established notions of states, cities, and their material implications [17,18]. Despite advancements, critiques highlight biased data from surveys favoring accessible, large, low-elevation sites, overlooking diverse environments, and a lack of control over the chronology of the findings [19–21]. While recent works in the Chifeng region, in present-day Inner Mongolia, and the high-altitude Ngawa area of Western China have broadened archeological methodologies and challenged traditional site selection and dating methodologies [22–24], research still largely relies on extensive surveys and analyses of published preliminary excavation reports, focusing on large-scale variations, with much less attention paid to small-scale studies.

The Eastern Tianshan region of Xinjiang, a crucial prehistoric cultural crossroad connecting the Hexi Corridor with the eastern Steppe and the Inner Asian Mountains [25–28], too has seen research emphasizing long-distance interactions and large-scale cultural variations. In-depth analysis of individual sites and localized social phenomena is scarce (but see [13]), which has significantly limited our understanding of the social contexts that underlay the development of these ancient societies.

The Late Bronze Age in the Eastern Tianshan Mountains (ca. 1300–800 BCE) witnessed an intensification of pastoralism [29], the expansion of crop exchange [30–32], the spread of new technologies [33,34], and the emergence of a greater social complexity, which was largely driven by the increasing diversity in pastoral mobility patterns [35,36]. Over 200 sites, dating between the second and first millennium BCE, were identified in the grasslands across both slopes of the Eastern Tianshan Range, allowing a preliminary understanding of human–environment relationships and settlement preferences [37,38]. These sites typically featured stone houses and graves, with many yielding an array of artifacts indicative of significant population growth and cultural flourishing [38,39]. The discovery of painted pottery [40–43], cast bronze items [44–48], and crops, such as wheat, barley, and millets [40,49,50], points to heightened interactions with neighboring regions to the east and west. Over 30 of these sites have been carefully surveyed revealing a unique structure, referred to as 'centripetal compound buildings' (向心式复合建筑) [51], distinctively shaped radially or as an isosceles triangle or a fan, with a central building encircled by smaller houses. Previous research has suggested that this architecture may signify a hierarchical society [52], sacred spaces [53,54], defensive strongholds [38], communication hubs [51,55] or labor division within the site [56]. However, insufficient evidence and the lack of detailed studies of the individual sites, including the architectural structures and artifacts found therein, continue to obscure the underlying function of these buildings and their potential socio-political implications on a local and regional scale.

In order to fill this gap, this article focuses on the late Bronze Age settlement of Shirenzigou, on the northern foothills of the Eastern Tianshan Mountains. Systematic excavations conducted by Northwest University and the Xinjiang Institute of Archaeology, among others, in 2006 and 2007 uncovered a distinctive centripetal compound building at the site [57]. This study compiles and examines the data from the excavation of Shirenzigou, incorporating both previously published findings and fresh insights from the excavation records, within a precise chronological and spatial framework to explore the relationship between the material sphere and the multifaceted social dynamics at the site. In doing so, it offers important insights into the complexity of the pastoral societies in the Eastern Tianshan Range in the first millennium BCE.

*The Site*

The site of Shirenzigou, also known as Dongheigou, lies on the northern side of the Tianshan Mountains, in present-day Balikun County (Figure 1). The landscape around the site includes meadows and areas covered in gravel from ancient glaciers. Towards the north, there is a large valley with wide areas of alluvial fans. The elevation of the lower river plain ranges between 1800 and 2000 m asl. While today it supports irrigated wheat farming, there is no evidence to suggest historical irrigation practices. The northern mountains, rising above 3600 m, are covered in forests and have peaks that are frozen year-round. The reliable water sources and efficient drainage system render this locale an optimal grazing area during summer. The inhabitants seem to have practiced transhumance [29], congregating in the lower lands during winter and migrating to the mountain foothills in summer, a lifestyle documented in various regions [7,58].

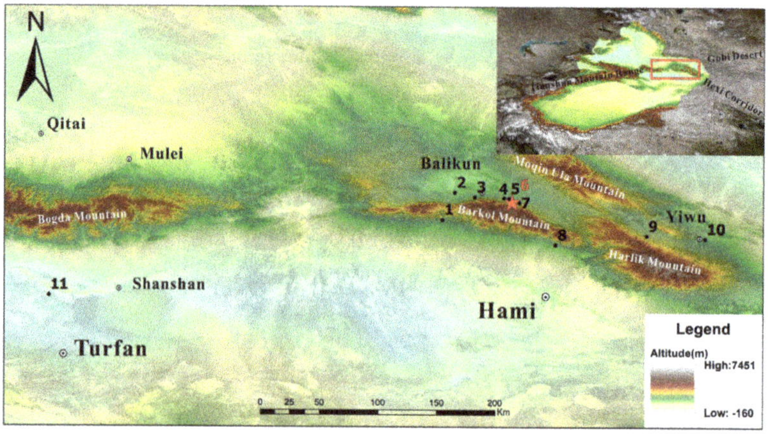


**Figure 1.** Location of Shirenzigou and other sites associated with this study. 1. Liushugou; 2. Haiziyan; 3. Yuegongtai-Xiheigou site group; 4. Xiaoheigou; 5. Xigou; 6. Shirenzigou; 7. Hongshankou No. 1; 8. Wulanbuluke; 9. Kuola northern building complex; 10. Baiqier; 11. Yanghai. The image was made using QGIS 2024 (www.qgis.org) and CorelDRAW Graphics Suite 2016 (www.corel.com).

First discovered in 1957, with initial examinations carried out in 1958 and 1981 [59], Shirenzigou underwent a thorough investigation and was mapped in the summer of 2005 by the Cultural Heritage and Archaeology Research Center of Northwest University, the Hami District Cultural Relics Bureau, and the Barkol County Cultural Management Office [60]. A large context covering roughly 8.75 km$^2$ was revealed, including mound-platforms, architectural structures, burials, and sacrificial pits. The site is believed to have been used by pastoral communities on a seasonal basis. Scattered radiocarbon dates situate the habitation of the site between 1300 BCE and 300 CE, spanning the local Bronze to Iron Ages [32,36,61].

In 2006 and 2007, a team comprising Northwest University, the Xinjiang Institute of Archaeology, and other institutions embarked on meticulous excavations of a radially structured compound building in Shirenzigou. Located in isolation on the southern section of the site, this structure comprised diverse features such as wooden and stone walls, postholes, hearths, and multifunctional pits, along with a significant collection of artifacts and bioarcheological material. Preliminary findings have been reported [57], and various studies on selected archeological evidence from the site have shed light on local livelihoods and cultural exchanges [13,29–31,35,36]. Yet, a comprehensive and detailed analysis of this unique complex within a chronological and spatial context is still pending and is the focus of the discussion in this article.

## 2. Materials and Methods

Seven excavation seasons have yielded extensive archeological insights at the site, with thorough investigations into the living spaces of F7 and F2. In the case of units F1, F5, F8, and F10, only the rear sections were excavated to preserve the slope of F7 and its overall structure. Excavations of units F3, F4, F6, and F9 were only partially conducted, limited by the scope of the excavation permit. Early findings have been documented in a series of preliminary reports and scholarly articles by the Shirenzigou research team [51,53,54,57,59]. Integrating this previously published information with new insight from the excavation record, we have compiled a detailed architectural overview of the centripetal compound structure. Through spatial analysis and statistical tools, we investigated the layout and interrelationships of various structural elements [8,62].

All artifacts and botanical and faunal remains recovered at the site were recorded (Figure 2). The artifacts were firstly divided into four large categories according to relevant activities (Food, Production, Leisure, and Others). These large divisions were further split into 16 smaller groups on the basis of the specific function of the recovered artifacts, according to published references on the topic (e.g., [63]) (Table 1). When appropriate, the evidence was further divided according to sub-type (e.g., grain processing tools were further classified into 'grinding stone plate', 'grinding stone pestles', and 'smashers') and size as follows: extra-large vessel: belly diameter $\geq$ 50 cm and height $\geq$ 50 cm; large vessel: belly diameter between 50 cm and 40 cm and height between 50 cm and 30 cm; medium vessel: belly diameter between 40 cm and 30 cm and height between 30 and 25 cm; small vessel: belly diameter $\leq$ 30 cm and height $\leq$ 25 cm. Bioarcheological material was divided in cereals grains and weed, all under the category of botanical findings and faunal remains (Table 1).

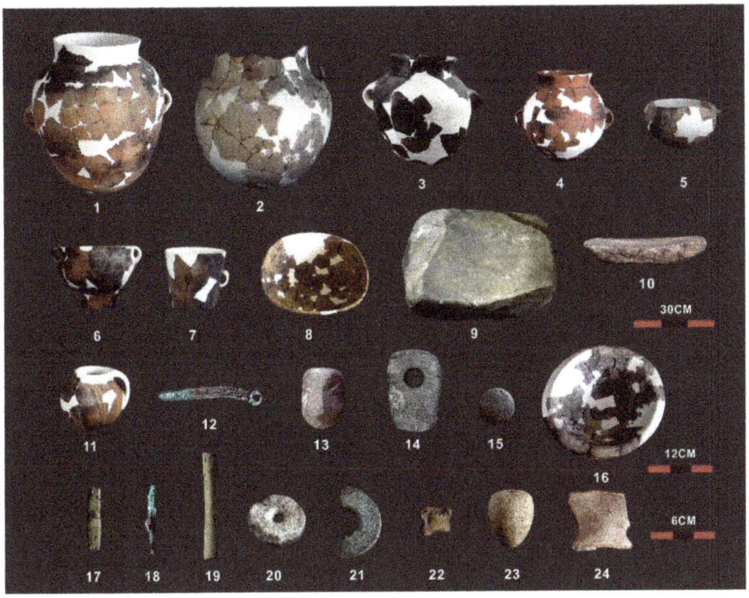


**Figure 2.** Artifacts excavated from various structures of the centripetal compound building in Shirenzigou: 1. double-handle jar (F7); 2. four-handle pot (F7); 3. double-belly pot with two handles (F7); 4. painted double-belly pot with two handles (F7); 5. double-ring flat-bottomed cauldron (F2); 6. double-handle ring-footed pot (F7); 7. single-handle cup (F7); 8. basin (F3); 9. stone grinding plate (F7); 10. stone grinding rod (F7); 11. painted single-handle pot (F4); 12. bronze knife (F7); 13. stone pestle (F1); 14. stone hoe (F1); 15. stone ball (F2); 16. crucible (F7); 17. tooth harpoon (F7); 18. bronze awl (F7); 19. bone tube (F7); 20. stone spindle whorl (F7); 21. scepter head (F7); 22. toy made out of a sheep astragalus (F3); 23. stone spinning top (F3); 24. pottery shard (F4). The image was made using CoreDRAW Graphics Suite 2016 (www.corel.com).

During the excavation, the contexts of the artifacts and bioarcheological material were recorded to the nearest cm. The number and type of artifacts were recorded to the layer or feature in which they were recovered. These features were identified as ash pits, ash heaps, post holes, stoves, and kilns. These elements were mapped in-field and the plan digitized using CorelDRAW Graphics Suite 2016 (www.corel.com). This plan was then annotated to show the arrangement of finds within the site. The analysis of artifacts and bioarcheological evidence, considering both quantity and the temporal and spatial distribution of the finds, was carried out to identify potential activity areas [11–13].

**Table 1.** Classification of the artifacts and bioarcheological material (botanical and faunal findings) from Shirenzigou used in this study.

| Category | Artifacts |
|---|---|
| Grain storage | Double-belly pots, double-handled jars, etc. |
| Food cooking | Pots, double-handled jars, etc. |
| Food serving | Basins, large single-handled cups, etc. |
| Individual dining | Small single-handled jars, bowls, cups, etc. |
| Grain processing | Grinding stone plates, grinding stone rods, pestles, stone beaters, etc. |
| Food cutting | Bronze and bone knives, etc. |
| Hunting | Stone balls, etc. |
| Construction | Stone hoes, stone axes, etc. |
| Pottery | Pottery discs, stone discs, etc. |
| Smelting | Crucibles |
| Drilling | Bone and bronze drills, etc. |
| Grinding | Whetstones, etc. |
| Textile | Spinning wheels, bone needles, etc. |
| Row material | Chalcedony, etc. |
| Toys | Sheep and deer astragalus bones (often found in groups), stone tops, etc. |
| Others | Stone scepter heads and items that do not belong to the above categories, in addition to a small number of artifacts with unclear function. |
| **Category** | **Biological remains** |
| Botanical findings | Cereals and weeds |
| Faunal remains | Animal bones, teeth, and antlers |

In order to chronologically contextualize the site and to better understand the patterns of occupation of the houses, we introduced 5 new radiocarbon dates, which were generated at the Accelerator Mass Spectrometry Dating Laboratory of Peking University. Our results were re-evaluated in light of previously published AMS C14 dating. All the dates were calibrated using the most recent calibration curves, OxCal and IntCal20 [64,65] (OxCal online version 4.4.4). Given the well-documented stratigraphic sequence of all the samples, the Bayesian modeling primarily adhered to the sequence model proposed by Ramsey [66], which enabled the integration of excavation stratigraphy to obtain the most precise dating resolution possible. Radiocarbon dates, organized according to their excavation layers, were grouped into a single phase when originating from the same stratigraphic layer.

## 3. Results

### 3.1. Layout and Buildings

The multi-space structure spanned 73 m in length from north to south and 58 m at its widest from east to west. It was composed of 10 houses, an open space (possibly a square), and several pathways (Figure 3). The largest building was F7. It was located at the highest altitude in the south, with its entrance facing east and consisted of two rooms distributed on a north–south axis. Adjacent to F7 on the east, southeast, and south sides were four small, single-room annex structures (F5, F8–F10). Further east was an open space encircled by arranged stones. To the northwest of F7, a group of four smaller units (F1–F4) were distributed from north to south, also divided into two rooms with their doors opening on the eastern walls. These buildings were placed next to each other in a sequential order, with each subsequent structure positioned slightly higher or lower than the one before it. To their east and separated by a pathway, another small house, F6, was found. F6 was divided into two rooms and had a door to the north side.

The archeological excavation revealed that this group of structures had undergone two main phases of use. Two layers of ground floors (lower and upper) were identified in F2, F3, and F7, while in F4 only one floor was found, with the late occupation occurring directly on the early ground. It is apparent that after the lower floor was abandoned, soil was added to form an upper layer of floor, while continuing to use the existing walls (Table 2). The standing walls of F7 are notably tall, exceeding 4 m in height and varying from 2.3 m to 3.6 m in thickness. In contrast, the wall heights of the smaller buildings range from 0.5 m to 0.7 m, with thicknesses between 0.7 m and 1 m [59].

**Figure 3.** Plan (**A**) and stratigraphy (**B**) of the buildings at the Shirenzigou site as per excavation records. The images were made using CorelDRAW Graphics Suite 2016 (www.corel.com) on the basis of the data recorded in the field.

**Table 2.** Excavation details and relationship between floor levels from two identified phases at Shirenzigou.

| Unit | | Lower Level | Upper Level |
|---|---|---|---|
| Core building | F7 | Excavated—ground floor | Excavated—ground floor |
| Ordinary building | F1 | Only rear part excavated | Excavated—abandoned |
| | F2 | Excavated—ground floor | Excavated—ground floor |
| | F3 | Only rear part excavated | Only rear part excavated |
| | F4 | Only rear part excavated | Only rear part excavated |
| | F6 | Not excavated | Partially excavated—ground floor |
| Others | F8 | Partially excavated—ground floor | Excavated—abandoned |
| | F5, F9, F10 | Not excavated | Excavated—abandoned/not clear |
| | Pathway | Not excavated | Not excavated |
| | Open space (square) | Not excavated | Not excavated |

In the first phase, the smaller structures F1–F4 and F6 were semi-subterranean, while F7 and its annex F8 were built at ground-level. The status of units F9 and F10, whether they were also ground-level constructions, remains unclear due to limited excavation data. In the later phase, the floor level of the smaller houses was raised slightly, approximately 0.45 m to 0.5 m, while the floor of F7 was elevated by 2 m, effectively turning it into a platform structure encircled by retaining walls. A sloping pathway leading to F7's original entrance was built, linking this central structure with the rest of the settlement. In the late occupation phase, this path began 9 m below F7's highest point, highlighting a significant elevation difference between the smaller structures and the central building. The walls of F7 pressed down F1 and the annexes F5, F8, and F10, which suggests that these structures had already been abandoned by the later period. The buildings retained abundant relics and artifacts in their original positions, supporting a rapid abandonment with minimal disturbance. Both layers of ground floor above F7 showed collapsed accumulations of burned wooden structure roofs, indicating that likely the abandonment was due to fire [51,53,54,57] (Figure 3).

The indoor layout of all the structures is consistent across the lower and upper phases, with a division into two spaces, a spacious front room and a rear room, which contained most of the artifacts [57]. The smaller dwellings (F1–F4, F6) show little variation in size, maintaining a consistent ratio of indoor floor space to the overall footprint area at 1:2 (Figure 4). The interior space of F7 does not stand out when compared to the smaller units, being comparable to that of F3 and F4. Yet, considering the walls, surrounding slopes, and additional architectural elements, F7's total area significantly exceeds that of the smaller structures in both periods (3 to 5 times larger in the early phase; 1.3 to 2 times larger in the later phase) (Figure 4; Table S1).

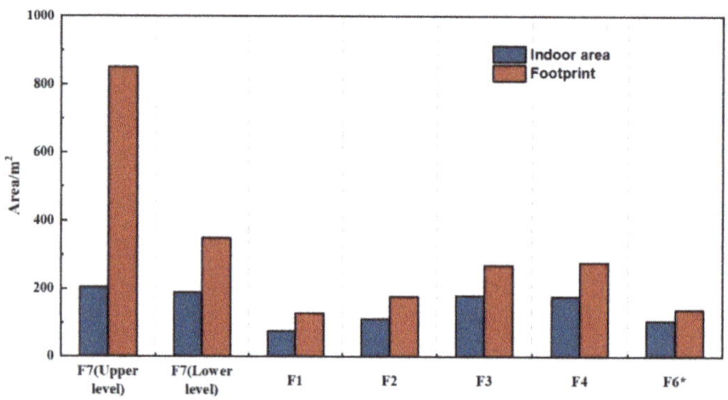


**Figure 4.** Ratio of the interior floor space to the total footprint area for each building in Shirenzigou (* walls not fully exposed during excavations). The figure was made using Origin 2024 (www.originlab.com).

### 3.2. Features and Artifacts

In all buildings, features were concentrated around the hearth, with grouped arrangements of large grinding stones, ash pits, and ash mounds distributed around it. Clustered collections of artifacts, including kitchenware and dining utensils, production and processing tools, and toys were found around the grinding stones, inside the ash pits, and under the foundations of the walls (Figure 5). Many of the features and artifacts showed signs of frequent use: for instance, the stones around the hearth were blackened and polished, some even cracked from the heat [57], and some pottery showed wear or repair marks. This suggests that these buildings were used for a relatively long period on a fairly regular basis.

In structures F1–F4 and F6, the amount and volume of various types of features and artifacts were roughly proportional to the interior floor space (Table 3; Figures 5 and 6A). The densities of pottery shards and other artifacts, including ceramic, stone, and metal items, as well as their distribution, type, size, ratio, and combination exhibited little variation across the smaller houses (Figures 5 and 6B,C; Tables S2 and S3), indicating their similar nature. Yet, significant disparities were observed when compared to the larger structure, F7 (Figure 6; Tables S2 and S3). The stove and ash pit areas were considerably larger compared to those in the smaller dwellings (Figure 6A; Table S2). The artifact count in F7 was markedly higher, with cooking and dining utensils found in quantities ranging from 7 to 23 times those found in the other buildings. The dimensions of cooking implements, grain storage, and serving vessels in F7 were also substantially larger, with diameters and heights exceeding 0.5 m and some reaching up to 0.7 m in diameter and 0.6 m in height (Figure 2). Noteworthy were 11 and 9 large stone grinding plates found on the lower and upper floors of F7, respectively, along with 19 storage devices on each floor of the building. Unit F7 also contained unique relics and artifacts not found in other houses, including ceramic kilns, bronze ware, painted large pottery vessels, jars with double-looped ring feet, oversized single-handled goblets, crucibles, and stone scepter heads, among others (Figure 2; Table 3). These observations highlight the potential special function of F7 as a central architectural feature within the settlement.

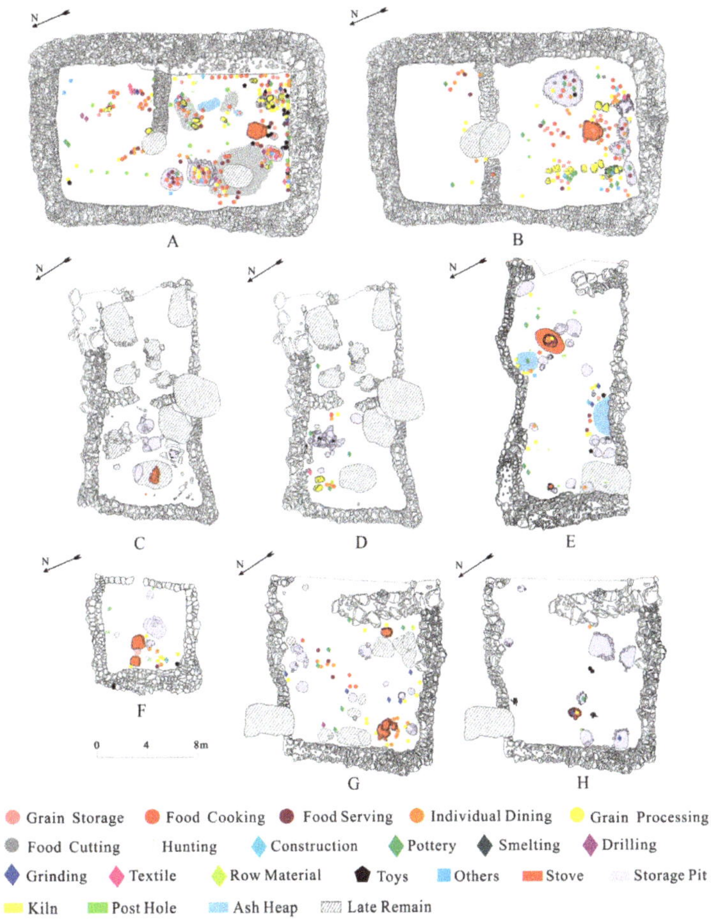


**Figure 5.** Plan of the structures uncovered in Shirenzigou divided according to stratigraphy with features and artifacts marked on each layer: (**A**). lower floor level of F7; (**B**). upper floor level of F7; (**C**). lower floor level of F2; (**D**). upper floor level of F2; (**E**). upper floor level of F4 (only one floor was present in F4); (**F**). lower floor level of F1; (**G**). lower floor level of F3; (**H**). upper floor level of F3. The figure was made using CorelDRAW Graphics Suite 2016 (www.corel.com).

**Table 3.** Types and quantities of artifacts excavated from different structures and floor levels in Shirenzigou (* partially excavated).

| Category | Function | Relics | F7 | | F1 | F2 | | F3 | | F4 | | F6 |
|---|---|---|---|---|---|---|---|---|---|---|---|---|
| | | | Early | Late | Early | Early * | Late | Early | Early * | Early * | Late | Early * |
| Food | Grain storage | | 19 | 19 | | | | | | | 1 | 4 |
| | Food cooking | Extra-large cooker | 1 | 2 | | | | | | | 0 | |
| | | Large cooker | 7 | 3 | | | | 3 | | | 1 | |
| | | Medium cooker | 16 | 8 | | | | 1 | | | 1 | |
| | | Small cooker | 21 | 6 | 1 | 1 | 3 | | | | 1 | |
| | Food serving | | 20 | 10 | 3 | | | 3 | | | 4 | |
| | Individual dining | | 19 | 16 | 3 | | 3 | 11 | 1 | | 4 | |
| | Grain processing | Grinding stone plate | 11 | 9 | | | 2 | 0 | 0 | | 1 | |
| | | Grinding stone pestle | 24 | 13 | | | 1 | 1 | 1 | | 3 | |
| | | Smasher | 21 | 10 | 5 | | 1 | 11 | 0 | | 9 | |
| | Food cutting | | 1 | 3 | | | | 0 | 1 | | 0 | |
| Production | Hunting | | 1 | 0 | 1 | | 2 | 1 | 1 | | 1 | |
| | Construction | | 3 | 2 | 1 | | | 0 | 0 | | 1 | |
| | Pottery | | 6 | 17 | | | 3 | 6 | 1 | | 8 | |
| | Smelting | | 1 | 0 | | | | 0 | 0 | | 0 | |
| | Drilling | | 8 | 0 | | | | 1 | 0 | | 0 | |
| | Grinding | | 4 | 0 | 1 | | | 10 | 2 | | 2 | |
| | Textile | | 7 | 1 | | | 1 | 0 | 0 | | 1 | |
| | Row material | | 0 | 0 | | | 1 | 3 | 0 | | 0 | |
| Leisure | Toy | | 14 | 1 | 1 | | | 1 | 2 | | 2 | |
| | Others | | 7 | 3 | | | | | 0 | | 0 | |
| | Total | | 211 | 123 | 16 | 1 | 17 | 52 | 9 | 0 | 40 | 4 |

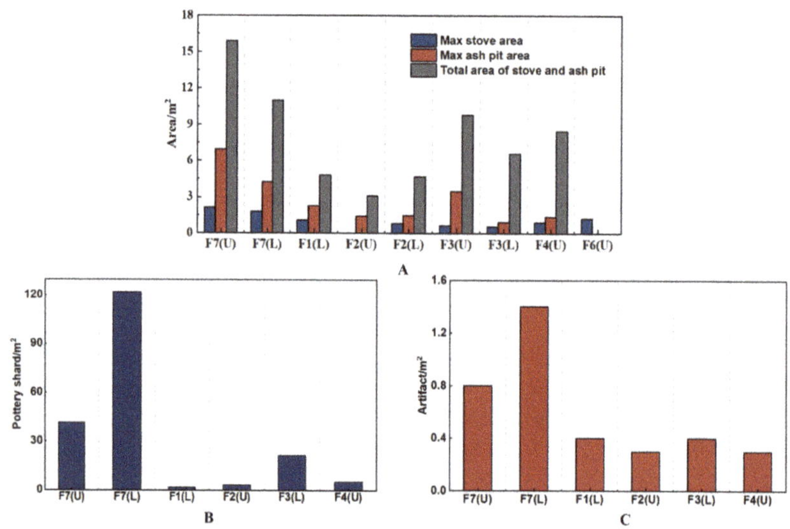


**Figure 6.** Relationship between structures and artifacts excavated in Shirenzigou: (**A**). density of the features in each structure; (**B**). density of ceramics in the lower and upper floor of each structure; (**C**). density of artifacts (pottery, stone and metal items) in the lower and upper floor of each structure. Because it was only partially excavated, F6 was excluded from the counts. U = upper level; L = lower level. The figures were made using Origin 2024 (www.originlab.com).

During the first phase, botanical and faunal remains in F7 far surpassed those found in the other structures. Apart from a single cereal grain in F1, no plant remains were discovered in the smaller houses. In the subsequent phase, the quantity of plant and animal remains decreased in F7, with botanical findings being more prevalent in F3, hinting at a possible subtle shift in the dynamic between the two buildings (Table 4). Beneath both levels of F7, there was a notable concentration of large animal bone deposits, barley seeds, and pottery showing fire marks within ash pits. Directly under F7's lower floor, pits H24 and H25 each contained seven relatively intact sheep skeletons bearing artificial marks [29]; pit H29 held a significant number of plant seeds, with around 4127 naked barley seeds extracted from about 40 L of soil [30]; pit H30 was rich in pottery fragments and animal bones [29]. Below the upper level, pit H19 was found to contain 681 animal bones and a large cache of barley seeds, estimated to total approximately 448,000 grains [30].

**Table 4.** Types and quantities of bioarcheological findings (plants seeds and animal bones, teeth, and antlers) excavated from different structures and floor levels in Shirenzigou.

| Type of Bio-Remain | | Location | | | | | | | | |
|---|---|---|---|---|---|---|---|---|---|---|
| | | Lower Level | | | | Upper Level | | | | |
| | | F7 | F1 | F2 | F3 | F7 | F2 | F3 | F4 | F6 |
| Botanical remains | Cereals | 9425 | 1 | | | 53 | | | | |
| | Weed | 348 | | | | 83 | | | | |
| Faunal remains | | 607 | 90 | 112 | 326 | 110 | 106 | 614 | 251 | 12 |

### 3.3. Chronology

Our radiocarbon dates were consistent with previous results, supporting a long-term occupation of the site between 1200 BCE and 900 BCE (Figure 7), with an early phase around 1200–1000 BCE and a later phase around 1000–900 BCE. More details on the radiocarbon dates can be found in Table S3.

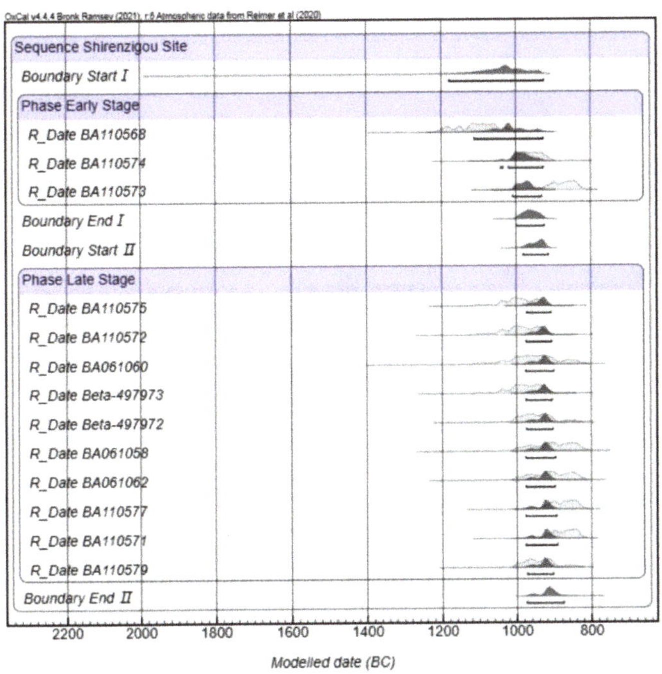


**Figure 7.** Calibrated radiocarbon dates for Shirenzigou. Accelerator Mass Spectrometry Dating Laboratory of Peking University and Beta Lab, calibrated using OxCal and IntCal20 [64,65] (OxCal online version 4.4.4).

## 4. Discussion

### 4.1. The Shirenzigou Community

The findings at Shirenzigou, marked by an abundance of caprine skeletal elements and plant remains, indicate that the late Bronze Age inhabitants led a pastoral lifestyle complemented by farming and hunting [13,31,57,67], a practice still observable in the area today [29,67].

The arrangement of the settlement, including the positioning, forms, and dimensions of the buildings, along with the organization of their internal areas, suggests a certain degree of intentional spatial division. Although it is challenging to determine the degree of the potential site planning at Shirenzigou, the alignment of buildings along pathways and the establishment of an open space in front of the larger building lend further credence to the notion that there was a deliberate conceptual framework guiding the construction of the site [9,68,69]. The arrangement of the site with devoted activities' areas, which is discussed below, provides additional support for this structured approach.

Units F1–F4 and F6 exhibited consistent architectural layout and spatial organization. Their parallel arrangement, lacking direct connection between them, suggests a deliberate effort to maintain a degree of autonomy [5,70]. Their dual-suite design further underscores this emphasis on privacy [8]. The larger front room likely fulfilled various roles, such as hosting guests, providing limited storage, and possibly serving as a temporary waste disposal area. The rear room, organized around the stove, was designated for cooking and probably sleeping. Near ash pits and walls, areas were seemingly assigned for food preparation, storage, and elementary crafts like weaving and pottery making. The significant number of basic cooking and dining utensils, greatly outnumbering other artifacts, along with animal bones bearing signs of processing, highlights the domestic nature of these spaces and their primary function as dwellings for small households. Comparable buildings have been found in nearby pastoralist Bronze Age settlements at Liushugou (F1 and F2; [71,72]) and Wulanbuluke (F5; [73]). Today similar structures are employed by small groups engaged in seasonal pastoralism for their routine living and dining needs across the Eastern Tianshan Mountains [7,29,67] (Figure 8). The archeological evidence and our observations of modern pastoralist households on-site suggest that these ancient spaces could potentially accommodate up to 4–6 people.

**Figure 8.** Dwelling of pastoralists in Xigou, Shirenzi village, Barkol. The photo was taken by Ren Meng in August 2015.

The variety of cultural material excavated from units F1–F4 and F6, including artifacts for cooking, production, and leisure, suggests a diverse community engagement in domestic life, potentially involving men, women, children, the elderly, and other identities. The consistency in the distribution and types of remains in the smaller houses indicates minimal wealth differentiation among the occupants. The coeval funerary contexts of Liushugou [71] and Baiqier [74], located on the northern slopes of the Eastern Tianshan Mountains in Barkol and Yiwu, respectively, suggest a similar scenario: burials were grouped in clusters of similar scales and included comparable grave goods, providing no evidence for social stratification.

Building F7 stands out due to its larger size and strategic placement. It was not only the tallest structure in the settlement but was also raised to a higher elevation in the second construction phase. A pathway ascended from the ordinary dwellings to F7, climbing 9 m and winding through terraced houses to reach this central structure, thereby emphasizing its prominence within the settlement. F8, with its small, single-room layout, likely functioned as F7′s storage space for various cooking devices [57]. F5 and F10 were single-room spaces sharing their walls with F7. Their doors, when present, faced various directions. For this reason, it is suggested they were designed as annexes to support the main activities of the central building. The status of F9 is unclear, due to the lack of excavation data. The presence of these annexes indicates a complex centered around F7.

F7 stands also out for the greater scale and sophistication of its indoor facilities and artifacts. Notable is the inclusion of 19 unique grain storage devices, absent in standard residential units, which hints at a cohesive community structure in Shirenzigou, at least partially, reliant on shared food resources. Communal approaches to storage are usually more evident in strategic arrangements for 'livestock storage'—utilizing fenced areas, repurposed structures, and pens—which are fairly well documented for both historical and contemporary pastoral communities throughout the Eastern and Western Tianshan Mountains [7,29,75]. The identification of grain storage containers in unit F7 not only corroborates previous arguments about small-scale agriculture practices at the site [31,61,75], but also further challenges the notion of this building serving merely as dwelling. Instead, it repositions F7 as a potential communal hub, central to the sustenance and social interaction of the Shirenzigou residents.

F7 exhibited a clear abundance of artifacts, outnumbering those in other houses and surpassing what would be expected for ordinary household needs. For example, the presence of large stone grinding plates—11 on the lower floor and 9 on the upper floor—far exceeds the 1–2 plates typically found in ordinary houses in Shirenzigou and elsewhere [57,76]. This lends further credence to the notion that this building was dedicated to collective activities, potentially including foundation-laying events, food-sharing ceremonies, and feasts. The accumulation of large quantities of animal bones, seeds, and pottery deposits beneath both floors of F7 [29,30] corroborates this argument. Similar prehistoric sites with ceremonial activities, characterized by oversized cooking facilities and numerous animal and plant remains, have been identified across the Tianshan Mountain [77–79]. Notably, the late Bronze Age centripetal compound structures at Lanzhouwanzi [80], and Haiziyan [81,82], in Barkol County, featured these elements prominently in their core houses, suggesting they served as community gathering points. Ethnographic research on current pastoralist groups in Shirenzigou by [29,67] documents seasonal collective ceremonies involving food distribution and sharing within the community. One of these studies estimates that the seven sheep found beneath F7 (in H24 or H25) could have been consumed by about 70 people in a single event [29]. The existence of a plaza to the east of F7 reinforces the interpretation of this space as being designed for communal gatherings. Distinctive artifacts such as large painted pottery vessels, bronze items, and scepter heads, predominantly discovered in F7, mirror the types of objects typically found in late prehistoric high-status burial sites across the Eastern Tianshan Mountains, like Yanghai M21 and Xigou M1 [83,84], further hinting at F7′s special role in the Shirenzigou settlement.

The architectural layout of the Shirenzigou centripetal compound building, characterized by strategic placement, varied forms and sizes of buildings around a central structure (F7), suggests a communal and possibly deliberate organization of space. Coupled with the extent, volume, and arrangement of relics and artifacts, this setup indicates different degrees of engagement in site activities, ranging from everyday tasks likely conducted at the household level in regular dwellings to broader participation in communal events associated with F7.

### 4.2. Centripetal Compound Buildings in the Eastern Tianshan Mountains: The Interplay between Environment and Social Change

The distinctive configuration of the centripetal compound building in Shirenzigou suggests an increasing complexity in the community's social organization. Similar late Bronze Age structures identified across the Tianshan Mountains, such as in Hongshankou [85], Lanzhouwanzi [80], and Haiziyan [81,82] (Figure 9), indicate a broader adoption of this architectural style, reflecting substantial social ramifications at a regional scale.

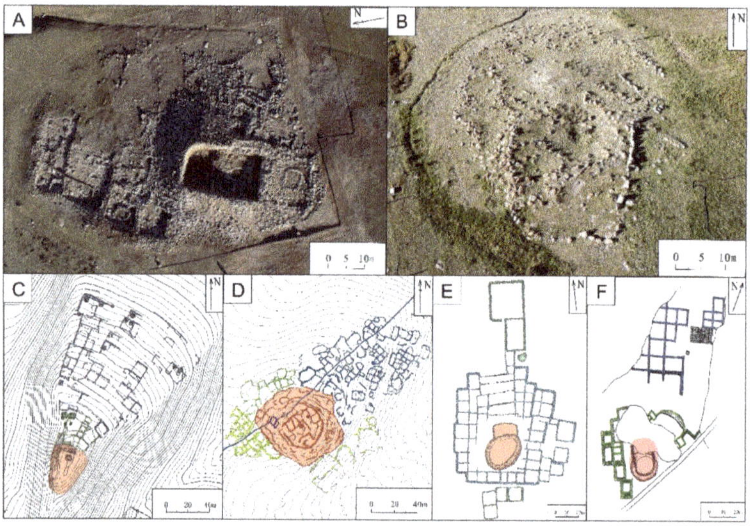


**Figure 9.** Centripetal compound structures identified in the Tianshan Mountains: (**A**). Shirenzigou [57]; (**B**). Xiaoheigou [86]; (**C**). Hongshankou [85,87]; (**D**). Kuola North Building Group [88]; (**E**). Yuegongtai-Xiheigou (Nijiaebo Group) [80,89]; (**F**). Yuegongtai-Xiheigou (Shuangzhe'ebo Group) [80,89].

[37] proposes that the selection of settlement locations and the site architectural arrangement during the late prehistory of Xinjiang were strategically based on climatic conditions, implying that these historical communities might have intentionally chosen their sites to mitigate the challenges of an increasingly hostile climate. The climate of Xinjiang during the late Holocene closely resembled its current state, being significantly more arid than in eastern China [90–92]. There is broad consensus that aridity in Northwest China has increased since 2000 BCE, leading to the expansion of deserts and the shrinkage of lake surface [93–96]. It has been argued that climatic deterioration in Xinjiang from the second millennium BCE was a crucial factor in settlement relocation, with a shift from lowland basins to the more sheltered mountain piedmonts and valleys [37]. This shift is evidenced by the decline and disappearance of the 'Xiaohe civilization' in the Lop Nur region around 1500 BCE [97]. Surveys across the Eastern Tianshan region have revealed various early burials, yet only one or two houses dating to 1500–1300 BCE have been identified in the Liushugou area [71]. While it is possible that the more obvious funerary contexts were

documented, whereas less visible residential sites might have been overlooked [98], the current archeological evidence suggests sparse occupation during the Neolithic and Early Bronze Age, with main human activities located in the lowlands of the Hami Basin [99,100]. The identification of numerous house conglomerates on the mountains' piedmonts and valleys of the Eastern Tianshan Range, dating to the Late Bronze Age, indicates a significant increase in the inhabitation of this area in the late second millennium BCE. Archaeological evidence shows that during this period, the Tianshan Beilu cultural groups, who had settled in the Hami oasis since around 2000 BCE [99,100], expanded northwards to the grasslands in front of the mountains on both the southern and northern slopes, forming the Nanwan and Shirenzigou cultures [101].

Climate deterioration has often spurred people to congregate around available (albeit increasingly diminishing) resources [102,103]. A growing number of studies have shown historical examples of such phenomena in Northwest China [104,105]. The steady water supply and good drainage in the Shirenzigou area make it an optimal summer pasture, allowing large herds to be grazed there. Additionally, until the middle of the last century, the area around the site was extensively covered with green forests and grass, making it suitable for hunting wild fauna [29]. It is probable that, in response to increasing aridity, ancient communities would have clustered around these key resources. Most of the centripetal compound buildings in the Eastern Tianshan Mountains are located in similar settings, suggesting they could have been structural adaptations aimed at strengthening communal bonds among the increasing number of occupants of these new key locations.

Between the second and first millennium BCE, the mountainous regions of Central Asia and Xinjiang saw significant transformations, including a growing demography, the emergence of different pastoral models, an increased degree of mobility, a greater dependence on herding—primarily of caprines and, to a lesser extent, cattle—low-investment agricultural practices, and the development of new metallurgical technologies [31,33,36,106,107]. These new socio-economic patterns would have prompted changes in the use of the landscape, as well as a re-negotiation of relationships within and between communities [28,108]. The pasture at Shirenzigou is expansive and allows prolonged grazing, therefore it could have been used simultaneously by multiple people, as it happens today. Research conducted in Central Asia and Xinjiang documents that winter and summer camps were central gathering spots for herders to share pastures, while engaging in social and political interactions face-to-face [7,98,108]. In the Neolithic period in Western Asia, the emergence of clustered neighborhoods, special buildings, and feasting practices were linked to growing cooperation in crop production and shared beliefs and identities, in response to enhanced aridity, aiming to ensure social cohesion to allow survival of increasingly populated communities [8,69,109,110]. In China, the historical Yi Li (仪礼or Liji 礼记Book of Rites) describes a social structure extant in Northern China in the first millennium BCE, characterized by family-based clans where people '异居而同财, 有余则归之宗, 不足则资之宗' (live separately but share wealth; the surplus goes to the clan, and in times of need, the clan supports) [111], emphasizing a blend of separate living with shared communal wealth and support to reinforce economic production and social cohesion. The development of centripetal compound buildings in the Eastern Tianshan Mountains could represent a regional adaptive strategy in response to emerging socio-political shifts. In Shirenzigou, this strategy would have particularly focused on reinforcing social cohesion among individuals and groups and securing resource availability by allocating individual tasks in ordinary houses and collective efforts in unit F7.

The Shirenzigou centripetal compound building, positioned in a valley surrounded by mountains and rivers with mountains at its back and facing steep slopes, is visible from a significant distance. During our surveys, we could see F7 from the highway in the lowlands north of the Tianshan Range, at a distance of at least 8 km. This strategic positioning likely added symbolic meaning to the structure, perhaps even serving as a social marker recognizable by nomadic pastoral groups from afar. This is in line with previous archeological and geomorphological research across Xinjiang, which has shown that during the Bronze and Iron Ages, special buildings devoted to collective ceremonies and rituals were strategically placed in locations like mountain peaks, open valleys, and occasionally alluvial fans, chosen for their ecological, climatic, environmental, and socio-political and symbolic significance [112–114]. Centripetal structures, which have been identified across the Eastern Tianshan region, were located in similar settings. For instance, the Yuegongtai-Xiheigou Group, in Barkol County, is sited on open slopes [52,80,89], while the Hongshankou Site No. 1, also in Barkol County, was atop a hill on an open slope [85]. The largest such structure identified to date—the Kuola Site North Building Group, in Yiwu County—lies on a piedmont slope with an extensive core house's footprint of 70 m in diameter and a notable height of 6 m, emphasizing its prominence within the grasslands [88,115] (Figures 1 and 9). The prominence of these structures could suggest they were a manifestation of the power of an emerging elite [116]. The relationship between the ordinary dwellings and the main unit F7 in Shirenzigou, however, seem to indicate that the centripetal compound building was part of a comprehensive strategy to create a stable social landscape. This meant that pastoralists in the Eastern Tianshan Mountains could have fostered greater social and political integration among people (and groups of people) based on cooperation and the regularity of face to face interaction during seasonal communal ceremonies.

In addition, the strategically positioned centripetal compound buildings could have served as measures to manage and control the movement of people, as well as regulating access to pastures and the settlement. However, this function may have been secondary. A Bronze Age structure, located less than two kilometers from the centripetal compound building, was identified in a higher position (approximately 20 m higher). This building would have had a broader view of the landscape [117]. Nevertheless, a more thorough analysis needs to be conducted on-site to better understand its nature and potential function.

In the Late Bronze Age, a significant number of non-ordinary dwellings and non-residential structures, alongside ordinary houses, appeared in the Tianshan Mountains, reflecting a growing regional socio-political complexity [10,13,98,118]. From this perspective, the emergence of the centripetal compound buildings across the Tianshan region may reflect a broader social strategy of integration of, but also divisions between, pastoralist groups in response to socio-political shifts between the second and first millennium BCE.

## 5. Conclusions

The late Bronze Age site of Shirenzigou was characterized by the emergence of a distinctive structure known as the centripetal compound building. This configuration, featuring smaller, independent dwellings encircling a larger, centrally located communal structure, illustrates a nuanced balance between private living spaces and public, communal areas. The arrangement points to an evolving social complexity within the pastoral communities of the Eastern Tianshan region. The proliferation of such communal-centric dwellings across the region may reflect a broader social strategy aimed at enhancing communal bonds among pastoral groups, possibly in reaction to environmental pressures and socio-political shifts in the late second millennium BCE.

Yet, several aspects remain unclear. Investigations at Shirenzigou have uncovered a systematic spatial division, with uniformly sized living quarters suggesting an organized setup potentially linked to household units. The question of whether these units were family-based remains open. While ethnographic studies suggest the predominance of nuclear families [29], genetic analyses from various prehistoric Eurasian sites reveal complex dynamics of household compositions, hinting at both genetic ties and social relationships as key in the structuring of ancient communities [119–121]. Moreover, while this study has demonstrated the unique role of unit F7, the specifics of its accessibility—whether it was exclusive to certain individuals or open to the wider community—still need to be clarified, raising further questions about the nature of social relationships within and across households. Despite significant research effort [29,31,36,75], further examination of production activities at the site are essential to uncover the complex processes and interactions that influenced daily life in Shirenzigou over time.

On a broader scale, the development and diffusion of centripetal compound buildings across the Eastern Tianshan Mountains likely had significant regional social implications. While our study provides important insights into the social landscape at the end of the second millennium BCE, a deeper understanding of these regional transformations—and their implications for the emergence of complex, potentially even proto-urban and urban societies in Xinjiang—requires more comprehensive excavations and analyses at both macro and micro levels

**Supplementary Materials:** The following supporting information can be downloaded at: https://www.mdpi.com/article/10.3390/land13050576/s1, Table S1: Data relevant to the ratio of interior floor space to the total footprint area in the Shirenzigou structures; Table S2: Data relevant to the density of the features, ceramics, and other artifacts (pottery, stone and metal items) in the Shirenzigou structures; Table S3: Calibrated radiocarbon dates for Shirenzigou from the Accelerator Mass Spectrometry Dating Laboratory of Peking University and calibrated using OxCal and IntCal20 [64,65]. Reference [122] is cited in Supplementary Materials.

**Author Contributions:** Conceptualization: M.R.; methodology: M.R. and M.F.; validation: M.R. and L.C.; formal analysis: M.R. and L.C.; resources: J.W., M.R., J.M. and T.X.; data curation: L.C., Y.Y. and D.T.; original draft preparation: M.R. and L.C.; review and editing: M.F., J.M., M.R. and L.C.; visualization: M.R. and L.C.; supervision: J.M.; funding acquisition: J.W., J.M., M.F.; M.R. and L.C. equally contributed to this work and act as first authors; J.W., M.F. and J.M. act as corresponding authors. All authors have read and agreed to the published version of the manuscript.

**Funding:** Financial support was provided by the China National Social Science Foundation (No. 14ZDB052) and the China National Foreign Young Talents Program (No. QN203152300001).

**Data Availability Statement:** The original contributions presented in the study are included in the article/Supplementary Materials, further inquiries can be directed to the corresponding authors.

**Acknowledgments:** Ran Wanli, Zhang Feng, Mo Zhanxiong, Xi Lin, Huang Shan, Shinohara Norio, Miyawaki Shiro, and Kanasugi Taishiare are gratefully acknowledged for their support during the excavations. Thanks are due to Ma Zhikun for providing some of the novel information about the site. We are grateful to the three anonymous reviewers and the editors for their insightful comments and suggestions on this manuscript.

**Conflicts of Interest:** The authors declare no conflicts of interest.

## References

1. Kent, S. *Domestic Architecture and the Use of Space: An Interdisciplinary Cross-Cultural Study*; Cambridge University Press: New York, NY, USA; Cambridge, UK, 1990.
2. Kent, S. Partitioning Space: Cross-Cultural Factors Influencing Domestic Spatial Segmentation. *Environ. Behav.* **1991**, *23*, 438–473. [CrossRef]
3. Willey, G. The Virú Valley Settlement Pattern Study. In *Archaeological Research in Retrospect*; Willey, G., Ed.; Winthrop Publishers: Cambridge, MA, USA, 1974; pp. 149–178.
4. Flannery, K.V. The Origins of the Village as a Settlement Type in Mesoamerica and the Near East. In *Man, Settlement and Urbanism*; Ucko, P.J., Tringham, R., Dimbleby, G.W., Eds.; Duckworth: London, UK, 1972; pp. 23–53.
5. Flannery, K.V. The origin of the village revisited: From nuclear to extended households. *Am. Antiq.* **2002**, *67*, 417–425. [CrossRef]
6. Garfinkel, Y.; Miller, M. Sha'ar Hagolan Vol. 1. Neolithic Art in Context. *J. Am. Orient. Soc.* **2003**, *123*, 872–873. [CrossRef]
7. Jia, P.; Caspari, G.; Betts, A.; Mohamadi, B.; Balz, T.; Cong, D.; Shen, H.; Meng, Q. Seasonal movements of Bronze Age transhumant pastoralists in western Xinjiang. *PLoS ONE* **2020**, *15*, e0240739. [CrossRef] [PubMed]
8. Duru, G.; Ozbasaran, M.; Yelozer, S.; Uzdurum, M.; Kuijt, I. Space making and home making in the world's first villages: Reconsidering the circular to rectangular architectural transition in the Central Anatolian Neolithic. *J. Anthropol. Archaeol.* **2021**, *64*, 101357. [CrossRef]
9. Banning, E.B.; Chazan, M. *Domesticating Space: Construction, Community and Cosmology in the Late Prehistoric Near East*; Ex Oriente: Berlin, Germany, 2006.
10. Jia, X.; Ge, L. Xinjiang Wenquan faxian yichu guimo pangda de Qingtong Shidai zaoqi yizhi 新疆温泉发现遗存规模庞大的青铜时代早期遗址 (Large-scale early bronze age sites found in Wenquan, Xinjiang). *Zhongguo Wenwu Bao* **2016**, *8*, 55.
11. Carr, C. The nature of organization of intra-site archaeological records and spatial analytic approaches to their investigation. *Adv. Archeol. Method Theory* **1984**, *7*, 103–222. [CrossRef]

12. Carrer, F. Interpreting intra-site spatial patterns in seasonal contexts: An ethnoarchaeological case study from the Western Alps. *J. Archaeol. Method Theory* **2017**, *24*, 303–327. [CrossRef] [PubMed]
13. Wang, Y.; Monteith, F.; Xi, T.; Ren, M.; Li, D.; Hu, S.; Wang, J.; Festa, M.; Ma, J. New evidence for regional pastoral practice and social complexity in the Eastern Tianshan Mountains in the first millennium BCE. *Sci. Rep.* **2023**, *13*, 4338. [CrossRef]
14. Chang, K.C. *Settlement Archaeology*; National Press: Paolo Alto, CA, USA, 1968.
15. Chang, K.C. Tan Juluo Xingtai Kaogu 谈聚落形态考古 (Discussion on Archaeology of Settlement Form). In *Kaoguxue Zhuanti Liujiang*; Chang, K.C., Ed.; Wenwu Chubanshe: Beijing, China, 1986.
16. Chang, K.C. Ancient China and Its Anthropological Significance. In *Archaeological Thought in America*; Cambridge University Press: New York, NY, USA; Cambridge, UK, 1989; pp. 155–166.
17. Flad, R. Urbanism as Technology in Early China. *Archaeol. Res. Asia* **2018**, *14*, 121–134. [CrossRef]
18. Liu, L.; Chen, X. *State Formation in Early China*; Duckworth: London, UK, 2003.
19. Jaffe, Y.Y.; Castellano, L.; Shelach-Lavi, G.; Campbell, R.B. Mismatches of scale in the application of paleoclimatic research to Chinese archaeology. *Quat. Res.* **2021**, *99*, 14–33. [CrossRef]
20. Liu, L.; Chen, X. Settlement Archaeology and the Study of Social Complexity in China. *Rev. Archaeol.* **2001**, *22*, 4–22.
21. Hein, A. Theory and methods of settlement archaeology—The Chinese contribution. *World Archaeol.* **2023**, *54*, 667–679. [CrossRef]
22. Shelach, G. *Leadership Strategies, Economic Activity, and Interregional Interaction: Social Complexity in Northeast China*; Plenum: New York, NY, USA, 1999.
23. Chifeng International Collaborative Archaeological Research Project. *Regional Archaeology in Eastern Inner Mongolia: A Methodological Exploration*; Kexue Chubanshe: Beijing, China, 2003.
24. Guedes, J.; Hein, A. Landscapes of Prehistoric Northwestern Sichuan: From Early Agriculture to Pastoralist Lifestyles. *J. Field Archaeol.* **2018**, *43*, 121–135. [CrossRef]
25. Høisæter, T.L. Polities and nomads: The emergence of the Silk Road exchange in the Tarim Basin region during late prehistory (2000–400 BCE). *Bull. Sch. Orient. Afr. Stud.* **2017**, *80*, 339–363. [CrossRef]
26. Zhu, J.; Ma, J.; Zhang, F.; Cui, Y.; Festa, M.; Xi, T.; Ren, M.; Wang, Y.; Li, B.; Huang, F. The Baigetuobie cemetery: New discovery and human genetic features of Andronovo community's diffusion to the Eastern Tianshan Mountains (1800–1500 BC). *Holocene* **2021**, *31*, 217–229. [CrossRef]
27. Ning, C.; Wang, C.; Gao, S.; Yang, Y.; Zhang, X.; Wu, X.; Zhang, F.; Nie, Z.; Tang, Y.; Robbeets, M.; et al. Ancient genomes reveal Yamnaya-related ancestry and a potential source of Indo-European speakers in Iron Age Tianshan. *Curr. Biol.* **2019**, *29*, 2526–2532. [CrossRef] [PubMed]
28. Frachetti, M. *Pastoralist Landscapes and Social Interaction in Bronze Age Eurasia*; University of California Press: London, UK, 2008.
29. You, Y. *Xinjiang Dongheigouyizhi Chutu Dongwuguge Yanjiu* 新疆东黑沟遗址出土动物骨骼研究 *(Study on Animal Bones Unearthed from Dongheigou Site in Xinjiang)*; Chinese Academy of Social Sciences (CASS): Beijing, China, 2012.
30. Tian, D. *Gongyuanqian Yiqianji Dongtianshandiqu de Zhiwukaoguxueyanjiu—Yi Shirenzigouyizhiqun Weizhongxin* 公元前一千纪东天山地区的植物考古学研究——以石人子沟遗址群为中心 *(Study on Plant Archaeology in the East Tianshan Mountains in the Millennium BC—Focusing on Shirengzigou Site Group)*; Northwest University: Xi'an, China, 2018.
31. Tian, D.; Sun, M.; Ritchey, M.M.; Xi, T.; Ren, M.; Ma, J.; Wang, J.; Zhao, Z.; Ling, X.; Liu, X. Varying cultivation strategies in eastern Tianshan corresponded to growing pastoral lifeways between 1300 BCE and 300 CE. *Front. Ecol. Evol.* **2022**, *10*, 966366. [CrossRef]
32. Ma, Z.; Liu, S.; Ren, M.; Xun, X.; Xi, T.; Wang, J.; Ma, J. Xinjiang dongtianshandiqu balikun shirenzigouyizhi chuliangkeng fenxi 新疆东天山地区巴里坤石人子沟遗址储粮坑分析 (Analysis of Grain Storage Pit at Shirengzigou Site in Balikun, East Tianshan, Xinjiang). *Disiji Yanjiu* **2021**, *41*, 214–223.
33. Mei, J. *Copper and Bronze Metallurgy in Late Prehistoric Xinjiang: Its Cultural Context and Relationship with Neighbouring Regions*; BAR international Series; Archaeopress: Oxford, UK, 2000. [CrossRef]
34. Hanks, B.; Linduff, K. *Social Complexity in Prehistoric Eurasia: Monuments, Metals and Mobility*; Cambridge University Press: New York, NY, USA; Cambridge, UK, 2009.
35. Ma, J. The rise of nomads on the Barkol Steppe and their cultural interflow with the Altai Region during the 1st millennium BCE. In *Eurasian Studies II*; Yu, T., Li, J., Eds.; Asia Publishing Nexus: Sydney, Australia, 2014; pp. 29–37.
36. Li, Y.; Zhang, C.; Taylor, W.; Chen, L.; Flad, R.; Boivin, N.; Liu, H.; You, Y.; Wang, J.; Ren, M.; et al. Early evidence for mounted horseback riding in northwest China. *Proc. Natl. Acad. Sci. USA* **2020**, *117*, 29569–29576. [CrossRef]
37. Tan, B.; Wang, H.; Wang, X.; Yi, S.; Zhou, J.; Ma, C.; Dai, X. The study of early human settlement preference and settlement prediction in Xinjiang, China. *Sci. Rep.* **2022**, *12*, 5072. [CrossRef]
38. Xi, T. *Qingtong Shidai Zhi Zaoqi Tieqi Shidai Dongtianshan Diqu Juluo Yizhi Yanjiu* 青铜时代至早期铁器时代东天山地区聚落遗址研究 *(Study on Settlement Sites in East Tianshan Mountains from Bronze Age to Early Iron Age)*; Northwest University: Xi'an, China, 2014.
39. Guo, W. *Xinjiang Shiqian Wanqi Shehui Kaoguxue Yanjiu.* 新疆史前晚期社会的考古学研究 *(Archaeological Research on the Late Prehistoric Society in Xinjiang)*; Shanghai Guji Chubanshe: Shanghai, China, 2012.
40. Li, H. Cong kaogufaxian kan gongyuanqian erqiannian dongxiwenhua pengzhuang he jiaoliu 从考古发现看公元前二千年东西文化的碰撞和交流 (The collision and communication between eastern and western cultures in 2000 BC from archaeological discoveries). *Xinjiang Wenwu* **1999**, *1*, 60–62.
41. Han, J. "The Painted Pottery Road" and Early Sino-Western Cultural Exchanges. *Anabasis Stud. Class. Orient.* **2012**, *3*, 35–37.

42. Liu, X. Shiqian caitaozhilu: Zhongguo wenhua xilaishuo zhi zhongjie 史前彩陶之路: "中国文化西来说"之终结 (Prehistoric painted pottery road: The end of "China culture coming from the west"). *Zhongguo Shehui Kexuebao* **2012**, *5*, 1–3.
43. Han, J. Zailun Sichouzhiluqian de Caitaozhilu 再论丝绸之路前的彩陶之路 (Re-discussion on the road of painted pottery before the Silk Road). *Wenbo Xuekan* **2018**, *1*, 29–31.
44. Mei, J.; Liu, G.; Chang, X. Xinjiang Dongbudiqu chutu Zaoqi Tongqi de chubu fenxi he yanjiu 新疆东部地区出土早期铜器的初步分析和研究 (Preliminary analysis and research on early bronzes unearthed in eastern Xinjiang). *Xiyu Yanjiu* **2002**, *2*, 5–8.
45. Mei, J.; Gao, B. Saiyimabinuo xianxiang he Zhongguo xibeidiqude zaoqi qingtongwenhua 塞伊玛图比诺现象和中国西北地区的早期青铜文化 (Saima Tubino Phenomenon and Early Bronze Culture in Northwest China). *Xinjiang Wenwu* **2003**, *1*, 50–54.
46. Qian, W. *Xinjiang Hamidiqu Shiqianshiqi Tongqi Jiqiyu Linjindiqude Guanxi* 新疆哈密地区史前时期铜器及其与邻近地区文化的关系 *(Prehistoric Bronzes in Hami, Xinjiang and Their Relationship with Neighboring Cultures)*; Zhishichanquan Chubanshe: Beijing, China, 2006.
47. Li, H. *Dongfengxijian: Xibei Shiqianwenhua Zhi Jincheng* 东风西渐: 西北史前文化之进程 *(East Wind Spreading to the West: The Process of Prehistoric Culture in Northwest China)*; Wenwu Chubanshe: Beijing, China, 2009; pp. 276–293.
48. Chen, K.; Mei, J.; Qian, W. Sichouzhilu yu zaoqi tongtiejishu de jiaoliu 丝绸之路与早期铜铁技术的交流 (The Silk Road and the Exchange of Early Copper and Iron Technology). *Xiyu Yanjiu* **2018**, *2*, 129–133. [CrossRef]
49. Dong, G.; Yang, Y.; Han, J.; Wang, H.; Chen, F. Nongzuowu chuanbo shijiaoxia de ouyadalu shiqian dongxifang wenhua jiaoliu农作物传播视角下的欧亚大陆史前东西方文化交流 (Cultural exchanges between East and West in Eurasia from the perspective of crop communication). *Zhongguokexue Diqiukexue* **2017**, *47*, 535–536.
50. An, C.; Wang, W.; Duan, F.; Huang, W.; Chen, F. Yazhou zhongbu ganhanqu sichouzhilu yanxian huanjing yanbian yu dongxifang wenhua jiaoliu亚洲中部干旱区丝绸之路沿线环境演化与东西方文化交流 (Environmental Evolution along the Silk Road in Arid Area of Central Asia and cultural exchanges with the East and the West). *Dili Xuebao* **2017**, *72*, 880–886.
51. Ren, M. *Gongyuanqian Yiqianji Dongtianshan Diqu Kaoguxue Wenhuayicun Yanjiu* 公元前一千纪东天山地区考古学文化遗存研究 *(Study on Archaeological Cultural Remains in the East Tianshan Mountains in the Millennium BC)*; Northwest University: Xi'an, China, 2012.
52. Liu, R. Xinjiang Balikun Yuegongtai-xiheigou yizhiqun chubu renshi 新疆巴里坤岳公台—西黑沟遗址群初步认识 (A preliminary understanding of Yuegongtai-Xihegou site group in Balikun, Xinjiang). *Xibeidaxue Xuebao* **2009**, *39*, 49.
53. Zhang, F. *Xinjiang Dongheigouyizhi Shizhugaotai, Juzhi Yanjiu* 新疆东黑沟遗址石筑高台, 居址研究 *(Study on the Stone Platform and Residence of Dongheigou Site in Xinjiang)*; Northwest University: Xi'an, China, 2008.
54. Ren, M. *Cong Heigouliangmudi, Dongheigouyizhi Kan Xihanqianqi Dongtianshandiqu Xiongnu Wenhua* 从黑沟梁墓地, 东黑沟遗址看西汉前期东天山地区匈奴文化 *(The Xiongnu Culture in the Eastern Tianshan Mountains in the Early Western Han Dynasty as Seen from Heigouliang Cemetery and Donghegou Site)*; Northwest University: Xi'an, China, 2008.
55. Han, B. *Dilixinxixitong Zai Gudai Daxing Youmuwenhua de Yingyong Yu Yanjiu—Yi hongshankou-Shirenzigouyizhiqun Weili* 地理信息系统在古代大型游牧文化遗址的应用与研—以红山口-石人子沟遗址群为例 *(Application and Research of Geographic Information System in Ancient Large-Scale Nomadic Cultural Sites—Taking Hongshankou-Shirengzigou Site Group as an Example)*; Northwest University: Xi'an, China, 2012.
56. Bao, S.; Jiang, Y. Gongyuanqian yiqianji dongtianshandiqu youmurenqun de tuzaixingwei 公元前一千纪东天山地区游牧人群的屠宰行为 (Slaughtering behavior of nomadic people in the eastern Tianshan mountains in the first Millennium BC). *Nongye Kaogu* **2021**, *6*, 57–62.
57. Wang, J.; Zhang, F.; Ren, M.; Yahefu, J.; Yu, J. Xinjiang balikun dongheigouyizhi 2006–2007 nian fajuejianbao 新疆巴里坤县东黑沟遗址2006～2007年发掘简报 (Brief report on excavation of Dongheigou site in Barkun County, Xinjiang in 2006~200). *Kaogu* **2009**, *1*, 3–27.
58. Milne, S. Landscape learning and lithic technology: Seasonal Mobility, enculturation and tool apprenticeship among the Early Palaeo-Eskimos. In *Structured Worlds: The Archaeology of Hunter-Gather Thought and Action*; Cannon, A., Ed.; Equinox: Sheffield, UK, 2011; pp. 95–115.
59. Wang, J.; Zhang, F.; Ren, M.; Yahefu, J.; Yu, J. *Shirenzigouyizhi 2006–2007 Niandu Fajuebaogao*石人子沟遗址2006～2007年度发掘报告 *(Excavation Report of Shirenzigou Site in 2006~2007)*; Kexue Chubanshe: Beijing, China, *in press*.
60. Chen, X.; Ren, M.; Wang, J.; Yahefu, J. Xinjiang balikun dongheigouyizhi diaocha 新疆巴里坤东黑沟遗址调查 (Investigation on Dongheigou Site in Barkol, Xinjiang). *Kaogu Yu Wenwu* **2006**, *5*, 16–26.
61. Tian, D.; Ma, J.; Ren, M.; Xi, T.; Wang, J.; Zhao, Z. Xinjiang Diqu De Zaoqi Damai Shengchan: Laizi Tianshan Beilu Shirenzigou Yizhi De Zhiwu Yicun Zhengju 新疆地区的早期大麦生产: 来自天山北麓石人子沟遗址的植物遗存证据 (Early Barley Production in Xinjiang: Evidence from Plant Remains of Shirengzigou Site at the Northern Foot of Tianshan Mountain). *Zhongguo Nongshi* **2021**, *40*, 44–55.
62. Paliou, E.; Lieberwirth, U.; Polla, S. *Spatial Analysis and Social Spaces: Interdisciplinary Approaches to the Interpretation of Prehistoric and Historic Built Environments*; Topoi, 18; De Gruyter: Berlin, Germany; Boston, MA, USA, 2014.
63. Xu, S. *Zhongguo Gudai Taoci Cidian* 中国古代陶瓷辞典 *(The Dictionary of Ancient Ceramics in China)*; Zhongguo Wenshi Chubanshe: Beijing, China, 2013.
64. Bronk, R.C. Bayesian analysis of radiocarbon dates. *Radiocarbon* **2009**, *51*, 337–360. [CrossRef]

65. Reimer, P.J.; Austin, W.; Bard, E.; Bayliss, A.; Blackwell, P.G.; Ramsey, C.B.; Butzin, M.; Cheng, H.; Edwards, R.L.; Friedrich, M.; et al. The IntCal20 Northern Hemisphere radiocarbon age calibration curve (0–55 cal kBP). *Radiocarbon* **2020**, *62*, 725–757. [CrossRef]
66. Ramsey, C.B. Radiocarbon calibration and analysis of stratigraphy: The OxCal program. *Radiocarbon* **1995**, *37*, 425–430. [CrossRef]
67. You, Y.; Zhong, H.; Yu, Y. Xinjiang Barkol Shirenzigou yizhi shengye kaogu de minzuxue diaocha yu yanjiu 新疆巴里坤县石人子沟遗址生业考古的民族学调查与研究 (Ethnology Investigation and Research on the Archaeology of Shirengzigou Site in Balikun County, Xinjiang). *Nanfang Wenwu* **2016**, *2*, 116–122.
68. Garfinkel, Y. The Social Organization at Neolithic Sha'ar Hagolan: The Nuclear Family, the Extended Family and the Community. In *Domesticating Space. Construction, Community, and Cosmology in the Late Prehistoric Near East*; Benning, T., Chazan, M., Eds.; Studies in Early Near Eastern Production, Subsistence and Environment 3; Ex Oriente: Berlin, Germany, 2006; pp. 103–111.
69. Bogaard, A.M. Neolithic "Cooperatives": Assessing Supra-Household Cooperation in Crop Production at Çatalhoyük and Beyond. In *Neolithic Corporate Identities*; Benz, M., Gebel, H.G.K., Watkins, T., Eds.; Studies in Early Near Eastern Production, Subsistence, and Environment 20; Ex Oriente: Berlin, Germany, 2017; pp. 1–9.
70. Byrd, B.F. Public and Private, Domestic and Corporate: The Emergence of the Southwest Asian Village. *Am. Antiq.* **1994**, *59*, 639–666. [CrossRef]
71. Wang, Y.; Zhang, J. Xinjiang Hamishi liushugouyizhi he mudi de kaogufajue 新疆哈密市柳树沟遗址和墓地的考古发掘 (Archaeological Excavation of Liushugou Site and Cemetery in Hami City, Xinjiang). *Xiyu Yanjiu* **2015**, *2*, 124–126. [CrossRef]
72. Wang, Y. *Liushugouyizhi Yanjiu 柳树沟遗址研究 (Study on Liushugou Site)*; Northwest University: Xi'an, China, 2015.
73. Wang, Y.; Aihemaiti, Y.; Hao, Y.; Zhang, F.; Wang, Y.; Zhang, J.; Chen, X.; Liu, Y. Xinjiang Hamishi wulanbulukeyizhi kaogu fajuejianbao 新疆哈密市乌兰布鲁克遗址考古发掘简报 (Brief report on archaeological excavation of Wulanbuluke site in Hami, Xinjiang). *Tulufanxue Yanjiu* **2022**, *2*, 45–56. [CrossRef]
74. Xinjiang Institute of Cultural Relics, Archaeology School of Cultural Heritage of Northwest University, Hami Municipal Cultural Heritage Administration and Yiwu County Cultural Heritage Administration Xinjiang. *Xinjiang Baiqier Mudi—2004~2005 Niandu Fajue Baogao 新疆拜其尔墓地—2004～2005年度发掘报告 (Bechir Cemetery: Excavation Report 2004–2005)*; Wenwu Chubanshe: Beijing, China, 2020.
75. Ma, Z.; Liu, S.; Ren, M.; Ma, J.; Xi, T.; Wang, J.; Wan, Z.; Tian, D.; Ren, W. Xinjiang Dongtian Diqu Shirenzigouyizhi nongmu huodong qingkuang: Tanhua zhongzi、zhiguiti he baofen zhengju 新疆东天山地区石人子沟遗址农牧活动情况：炭化种子, 植硅体和孢粉证据 (Agricultural and pastoral activities of Shirenzigou site in eastern Tianshan area of Xinjiang: Evidence from carbonized seeds, phytoliths and sporopollen). *Wenti Gushengwu Xuebao* **2023**, *40*, 181–193. [CrossRef]
76. Ren, M.; Ma, J.; Xi, T.; Tian, Y.; Jiang, X. Haiziyanyizhi 2017 nian fajue jianbao 新疆巴里坤海子沿遗址2017年发掘简报 (Brief Report on Excavation of Haiziyan Site in Balikun, Xinjiang in 2017). *Wenwu* **2020**, *12*, 21–26. [CrossRef]
77. Wu, X. Xinjiang Yutianxian liushui qingtongshidaimudi kaogufajue yu wenhua qingkuang jianshu新疆于田县流水青铜时代墓地考古发掘与文化情况简述 (Brief introduction of archaeological excavation and cultural situation of bronze age cemetery in Yutian County, Xinjiang). *Xinjiang Yishu* **2020**, *2*, 20–24.
78. Xinjiang Institute of Cultural Relics and Archaeology. Nilekexian wutulanmudi kaogufajue baogao 尼勒克县乌吐兰墓地考古发掘报告 (Archaeological Excavation Report of Wutulan Cemetery in Nileke County). *Xinjiang Wenwu* **2014**, *1*, 36–56.
79. Xinjiang Institute of Cultural Relics and Archaeology. 2014 nian Nilekexian Wutulanmudi kaogufajue baogao 2014 年尼勒克县乌吐兰墓地考古发掘报告 (Archaeological Excavation Report of Wutulan Cemetery in Nileke County in 2014). *Xinjiang Wenwu* **2015**, *2*, 4–17.
80. Wang, J.; Liu, R.; Ding, Y.; Ya, H.; Yu, J. Xinjiang Balikun Yuegongtai-xiheigou yizhiqun diaocha 新疆巴里坤岳公台—西黑沟遗址群调查 (Investigation on Yuegongtai-Xihegou site group in Balikun, Xinjiang). *Kaogu Yu Wenwu* **2005**, *2*, 3–12.
81. Ren, M.; Ma, J.; Xi, T.; Wang, J.; Li, W.; Tian, Y.; Ai, H.; Jiang, X. Xinjiang Balikun Haiziyanyizhi kaogufajue shouhuo yu sikao 新疆巴里坤海子沿遗址考古发掘收获与思考 (Harvest and Thinking of Archaeological Excavation of Haiziyan Site in Balikun, Xinjiang). *Xiyu Yanjiu* **2021**, *4*, 141–148. [CrossRef]
82. Ren, M.; Ma, J.; Xi, T.; Wang, J.; Li, W.; Tian, Y.; Ai, H.; Jiang, X. 2017、2018 niandu Xinjiang Balikun Haiziyanyizhi kaogu fajue 二〇一七, 二〇一九年度新疆巴里坤海子沿遗址考古发掘 (Archaeological Excavation of Haiziyan Site in Balikun, Xinjiang in 2017 and 2019). *Wenwu Tiandi* **2021**, *7*, 42–47.
83. Li, X.; Lv, E.; Zhang, Y. Xinjiang Shanshan Yanghaimudi fajuejianbao 新疆鄯善洋海墓地发掘简报 (Excavation on the Yanghai Cemetery in Shangshan (Piqan) County, Xinjiang). *Acta Archaeol. Sin.* **2011**, *1*, 99–166.
84. Ma, J.; Cheng, X.; Tang, Y.; Wang, Y.; Chen, X.; Wang, J.; Tian, Y.; Jiang, X. Xinjiang Hami Balikun Xigouyizhi 1haomu fajue jianbao 新疆哈密巴里坤西沟遗址1 号墓发掘简报 (Brief Report on Excavation of No.1 Tomb in Xigou Site of Balikun, Hami, Xinjiang). *Wenwu* **2016**, *5*, 15–31. [CrossRef]
85. Ma, J.; Wang, Y. 2012~2013 Niandu Xinjiang Balikun Hongshankouyizhi Fajuejianbao 2012～2013年度新疆巴里坤红山口遗址发掘简报 (Brief report on excavation of Hongshankou site in Barkun, Xinjiang in 2012–2013). *Wenwu in press*.
86. Ma, J.; Mou, J.; Xi, T.; Ren, M.; Tian, Y.; Aihemaiti, Y.; Ma, Y.; Wang, Y.; Liu, Q.; Fu, Y. 2018 nian Xinjiang Balikun Xiaoheigouyizhi diaocha jianbao 2018 年新疆巴里坤小黑沟遗址调查简报 (Brief Report on the Investigation of Xiaoheigou Site in Balikun, Xinjiang in 2018). *Xibu Kaogu* **2022**, *1*, 99–116.
87. Ma, J.; Xi, T.; Ren, M.; Wang, J. Xinjiang Balikun Hongshankouyizhi 2008 nian diaocha jianbao 新疆巴里坤红山口遗址2008年调查简报 (A Brief Report on the Investigation of Hongshankou Site in Balikun, Xinjiang in 2008). *Wenwu* **2014**, *7*, 17–30. [CrossRef]

88. Tang, Y.; Fu, Y.; Xi, T.; Rou, Z.; Ma, J.; Ren, M. Xinjiang Yiwu Kuolayizhi 20172018 年 diaocha jianbao 新疆伊吾阔腊遗址2017～2018年调查简报 (A Brief Report on the Investigation of Yiwu Kuola Site in Xinjiang from 2017 to 2018). *Wenwu* **2020**, *8*, 14–28. [CrossRef]
89. Ding, Y. *Yuegongtai-Xiheigou Yizhiqun ji Xiangguanwenti Yanjiu* 岳公台—西黑沟遗址群及相关问题研究 *(Research on Yuegongtai-Xihegou Site Group and Related Issues)*; Northwest University: Xi'an, China, 2003.
90. Chen, F.; Yu, Z.; Yang, M.; Ito, E.; Wang, S.; Madsen, D.; Huang, X.; Zhao, Y.; Sat, T.; Birks, H.; et al. Holocene moisture evolution in arid central Asia and its out-of-phase relationship with Asian monsoon history. *Quat. Sci. Rev.* **2008**, *27*, 351–364. [CrossRef]
91. Wünnemann, B.; Chen, F.; Riedel, F.; Zhang, C.; Mischke, S.; Chen, G.; Demske, D.; Ming, J. Holocene lake deposits of Bosten Lake, southern Xinjiang, China. *Chin. Sci. Bull.* **2003**, *48*, 1429–1432. [CrossRef]
92. Wünnemann, B.; Mischke, S.; Chen, F. A Holocene sedimentary record from Bosten Lake, China. Palaeogeography, Palaeoclimatology. *Palaeoecology* **2006**, *234*, 223–238. [CrossRef]
93. Zhong, W.; Shu, Q. Palaeoclimatic and Palaeohydrologic Oscillations Since about 12.0 ka B.P. at Bosien Lake, Southern Xinjiang. *Oceanol. Limnol. Sin.* **2001**, *32*, 213–220.
94. Huang, X.; Chen, C.; Jia, W.; An, C.; Zhou, A.; Zhang, J.; Jin, M.; Xia, D.; Chen, F.; Grimm, E. Vegetation and climate history reconstructed from an alpine lake in central Tienshan Mountains since 8.5 ka BP. *Palaeogeogr. Palaeoclimatol. Palaeoecol.* **2015**, *432*, 36–48. [CrossRef]
95. Huang, X.; Xiang, L.; Lei, G.; Sun, M.; Qiu, M.; Storozum, M.; Huang, C.; Munkhbayar, C.; Demberel, O.; Zhang, J.; et al. Sedimentary Pediastrum record of middle–late Holocene temperature change and its impacts on early human culture in the desert-oasis area of northwestern China. *Quat. Sci. Rev.* **2021**, *265*, 107054. [CrossRef]
96. Xie, H.; Liang, J.; Vachula, R.; Russell, J.; Chen, S.; Guo, M.; Wang, X.; Huang, X.; Chen, F. Changes in the hydrodynamic intensity of Bosten Lake and its impact on early human settlement in the northeastern Tarim Basin, Arid Central Asia. *Palaeogeogr. Palaeoclimatol. Palaeoecol.* **2021**, *576*, 110499. [CrossRef]
97. Zhang, Y.; Mo, D.; Hu, K.; Bao, W.; Li, W.; Abuduresule, I.; Storzum, M.; Kidder, T. Holocene environmental changes around Xiaohe Cemetery and its effects on human occupation, Xinjiang, China. *J. Geogr. Sci.* **2017**, *27*, 752–768. [CrossRef]
98. Jia, P.; Betts, A.; Dupuy, P.; Cong, D.; Jia, X. Bronze Age hill forts: New evidence for defensive sites in the western Tian Shan, China. *Archaeol. Res. Asia* **2018**, *15*, 70–81. [CrossRef]
99. Lv, E.; Chang, X.; Wang, B. Xinjiang Qingtongshidai Kaogu Wenhua Qianlun 新疆青铜时代考古文化浅论 (Brief Discussion on Archaeological Culture of Bronze Age in Xinjiang). In *Subingqi yu dangdai Zhongguo kaoguxue*; Su, B., Ed.; Kexue Chubanshe, 2001; pp. 179–184.
100. Han, J. *Xinjiang de Qingtongshidai he Zaoqitieqishidai Wenhua* 新疆的青铜时代和早期铁器时代文化 *(Bronze Age and Early Iron Age Culture in Xinjiang)*; Wenwu Chubanshe: Beijing, China, 2007.
101. Shao, H. *Xinjiang Shiqianshiqi Wenhua Geju de Yanbian Jiqi yu Zhoulin Diqu Wenhua de Guanxi* 新疆史前时期文化格局的演进及其与周邻地区文化的关系 *(The Evolution of Prehistoric Cultural Pattern in Xinjiang and Its Relationship with the Cultures in Neighboring Areas)*; Jilin University: Jilin, China, 2007.
102. Fang, J.; Liu, G. Relationship between climatic change and the nomadic southward migrations in eastern Asia during historical times. *Clim. Change* **1992**, *22*, 151–169. [CrossRef]
103. Brooks, N. Cultural responses to aridity in the Middle Holocene and increased social complexity. *Quat. Int.* **2006**, *151*, 29–49. [CrossRef]
104. Liu, F.; Li, H.; Gui, Y.; Yang, Y.; Lee, H.; Ding, D.; Hou, Y.; Dong, G. Chronology and Plant Utilization from the Earliest Walled Settlement in the Hexi Corridor, Northwestern China. *Radiocarbon* **2019**, *61*, 971–989. [CrossRef]
105. Dang, Z.; Chen, Z.; Zhang, X.; Francesca, M.; Xu, Y.; Li, C.; Zhu, J.; Ma, J.; Festa, M. Chronology of the Early Iron Age settlement of Koyuk Shahri in Xinjiang, China. *Radiocarbon* 2024, *in press*.
106. Wang, L.; Wang, Y.; Li, W.; Spate, M.; Reheman, K.; Sun, Q.; Wang, B.; Xu, H.; Zhang, Z.; Zhang, G.; et al. Inner Asian agro-pastoralism as optimal adaptation strategy of Wupu inhabitants (3000–2400 cal BP) in Xinjiang, China. *Holocene* **2021**, *31*, 203–216. [CrossRef]
107. Frachetti, M.; Benecke, N. From sheep to (some) horses: 4500 years of herd structure at the pastoralist settlement of Begash (south-eastern Kazakhstan). *Antiquity* **2009**, *83*, 1023–1037. [CrossRef]
108. Frachetti, M.D. Multiregional emergence of mobile pastoralism and nonuniform institutional complexity across Eurasia. *Curr. Anthropol.* **2012**, *53*, 2–38. [CrossRef]
109. Wright, K. Domestication and inequality? Households, corporate groups and food processing tools at Neolithic Çatalhöyük. *J. Anthropol. Archaeol.* **2014**, *33*, 1–33. [CrossRef]
110. Bogaard, A.; Charles, M.; Twiss, K.C.; Fairbairn, A.; Yalman, N.; Filipovic, D.; Demirergi, G.A.; Ertuğ, F.; Russell, N.; Henecke, J. Private pantries and celebrated surplus: Storing and sharing food at Neolithic Çatalhöyük. *Antiquity* **2009**, *83*, 649–668. [CrossRef]
111. Ding, D. *"Yi Li·Sangfu" Kao Lun* 《仪礼·丧服》考论 *(Textual Research on "The Yi-Li: Mourning")*; Shehui Kexue Wenxian Chubanshe: Beijing, China, 2013.
112. Yin, C.; Zhang, J.; Yu, X. Mountain valleys, alluvial fans and oases: Geomorphologic perspectives of the mixed agropastoral economy in Xinjiang (3000–200 BC). *Front. Earth Sci.* **2023**, *11*, 1109905. [CrossRef]
113. Chan, A.; Cong, D. Results of Field Research on Ancient Stonework in the River Valleys of Bortala and Ili in Western Tian Shan (Xinjiang, China). *Asian Perspect.* **2020**, *59*, 385–420. [CrossRef]

114. Caspari, G. Quantifying Ritual Funerary Activity of the Late Prehistoric Southern Kanas Region (Xinjiang, China). *Asian Perspect.* **2020**, *59*, 421–452. [CrossRef]
115. Fu, Y. *Kuolayizhi Diaocha Yanjiu* 阔腊遗址调查研究 *(Investigation and Study on Kuola Site)*; Northwest University: Xi'an, China, 2019. [CrossRef]
116. Houle, J.L. Socially integrative facilities and the emergence of social complexity on the Mongolian Steppe. In *Social Complexity in Prehistoric Eurasia, Monuments, Metals and Mobility*; Hanks, B.K., Linduff, K.M., Eds.; Cambridge University Press: Cambridge, UK, 2009; pp. 358–377.
117. Yang, Y. *Xinjiang Balikun Shirenzigou Yizhi Juluo Xingtai Yanjiu* 新疆巴里坤石人子沟遗址聚落形态研究 *(The Study on the Settlement Patterns of the Shirenzigou Site in Barlow, Xinjiang)*; Northwest University: Xi'an, China, 2023.
118. Wang, J.; Ma, J.; Xi, T.; Zhao, H.; Han, B.; Ye, Q.; Zhao, J.; Chen, A.; Wang, T.; Zhang, K.; et al. 2009 nian Xinjiang Balikun Shirenzigouyizhi F2 fajue baogao 2009年新疆巴里坤石人子沟遗址F2发掘报告 (Excavation report on F2 of Shirenzigou Site, Barkol, Xinjiang in 2009). *Archaeol. Cult. Relics* **2014**, *5*, 25–36.
119. Yaka, R.; Mapelli, I.; Kaptan, D.; Dogu, A.; Chyleński, M.; Erdal, Ö.D.; Koptekin, D.; Vural, K.B.; Bayliss, A.; Mazzucato, C.; et al. Variable kinship patterns in Neolithic Anatolia revealed by ancient genomes. *Curr. Biol.* **2021**, *31*, 2455–2468. [CrossRef] [PubMed]
120. Ning, C.; Zhang, F.; Cao, Y.; Qin, L.; Hudson, M.; Gao, S.; Ma, P.; Li, W.; Zhu, S.; Li, C.; et al. Ancient genome analyses shed light on kinship organization and mating practice of Late Neolithic society in China. *iScience* **2021**, *24*, 103352. [CrossRef]
121. Zhang, F.; Liu, Y.; Ning, C.; Zhang, J.; Ma, P.; Zhang, R.; Yun, Z.; Duan, C.; Cai, D. Haibing Yuan Ancient genomes reveal the origin and kinship burial patterns of human remains during the 11th to 13th centuries in northern China. *Int. J. Osteoarchaeol.* **2023**, *34*, e3274. [CrossRef]
122. Chen, J.; Mei, J.; Wang, J. Xinjiang Balikun Dongheigou yizhi chutu Tieqi Yanjiu 新疆巴里坤东黑沟遗址出土铁器研究 (Study on the Iron Objects unearthed from Dongheigou Site in Balikun, Xinjiang). *Wenwu* **2013**, *10*, 77–84.

land

MDPI

Article

# Revisiting Gubbio: Settlement Patterns and Ritual from the Middle Palaeolithic to the Roman Era

**Marianna Negro [1,*], Nicholas Whitehead [2], Caroline Malone [3] and Simon Stoddart [1,*]**

1 Department of Archaeology, University of Cambridge, Cambridge CB2 3DZ, UK
2 Independent Researcher, Cambridge CB3 9EE, UK; nwquahoag@gmail.com
3 Archaeology, Geography and Palaeoecology, School of Natural & Built Environment, Queen's University Belfast, Belfast BT9 3AZ, UK; c.malone@qub.ac.uk
* Correspondence: mesn2@cam.ac.uk (M.N.); ss16@cam.ac.uk (S.S.)

**Abstract:** The *Gubbio Revisited* project, initiated to reinterpret the archaeological evidence collected during the 1980s Gubbio Project, primarily by a conversion from a paper to a digital record, has revealed significant insights into the evolving settlement patterns and religious expression in the Gubbio valley in Central Italy. This reanalysis of the survey evidence underscores the rhythms of settlement and ritual practice from the Neolithic through the Bronze and Iron Ages, into Roman times. Key excavations in the 1980s at Monte Ingino, Monte Ansciano, San Marco Romano, and San Marco Neolitico added details not only of settlement activity but also of embedded ritual, evidenced by material culture including pottery, faunal remains, and votive offerings. The foundation myth of indigenous religious practices, even amidst Roman influence, is documented through the Iguvine Tables alongside the introduction of new cults, showcasing a blend of local and imperial religiosity, a common feature in the Roman world. This research enriches the understanding of Gubbio's historical and cultural landscape, emphasizing the demographic rhythms of the valley alongside the integral role of ritual in its societal evolution.

**Keywords:** longue durée; radiocarbon; ritual; Italy; Umbria; Gubbio; settlement; kernel density

## 1. The Prehistory of the Valley

The valley of Gubbio is located (Figure 1) in the northeast part of the modern region of Umbria (Central Italy), bridging the Marche to the East and Tuscany to the West. The region is thus a zone of transition, nestling in an upland valley, created out of the intermontane tectonic orogeny of the Apennine foothills. Its geographical location has led to a sharing of political and cultural influence with both these flanking regions. In some ways, the cultural links are mapped onto the geography since the northeastern watershed of the valley drains both into the Adriatic and the Tyrrhenian seas. The intermontane valley itself runs from the northwest to the southeast, linking the uppermost reaches of the Tiber Valley to the Chiascio catchment which itself runs into the Tiber to the west and, the approximately north–south valley later occupied by the Via Flaminia to the east. The valley is occupied by extensive alluvial fans, running off the limestone escarpment to the northeast, and the higher central ground thus created has led to the curious drainage of the valley in two directions, both northwest and southeast, even if both flow ultimately into the Tyrrhenian sea. The southwestern flank of the valley is occupied by ancient relict terraces, and the lower more incised valleys, most marked to the southeast, are occupied by more recent Holocene deposits.

Gubbio Revisited is a project, centered around the doctoral dissertation of Marianna Negro, which has the goal of reanalyzing and reinterpreting the archaeological evidence collected during the Gubbio Project, led by Simon Stoddart and Caroline Malone with the University of Cambridge in the 1980s and published in 1994 in *Territory, Time and State*. This publication focused primarily on the excavations of the upper slopes of Monte Ingino and Monte Ansciano, as well as the Roman farmhouse of San Marco Romano and the Neolithic site of San Marco Neolitico, located close to one another on the valley floor. The project also included an intensive survey of the valley whose results were broadly presented in 1994 but are now about to be published in greater detail. Reanalyzing the archaeological remains has both verified the long-term settlement trends and shed new light on the role and importance of the sacred in the Gubbio valley, beyond the pre-Roman sanctuaries of Monte Ingino and Monte Ansciano.

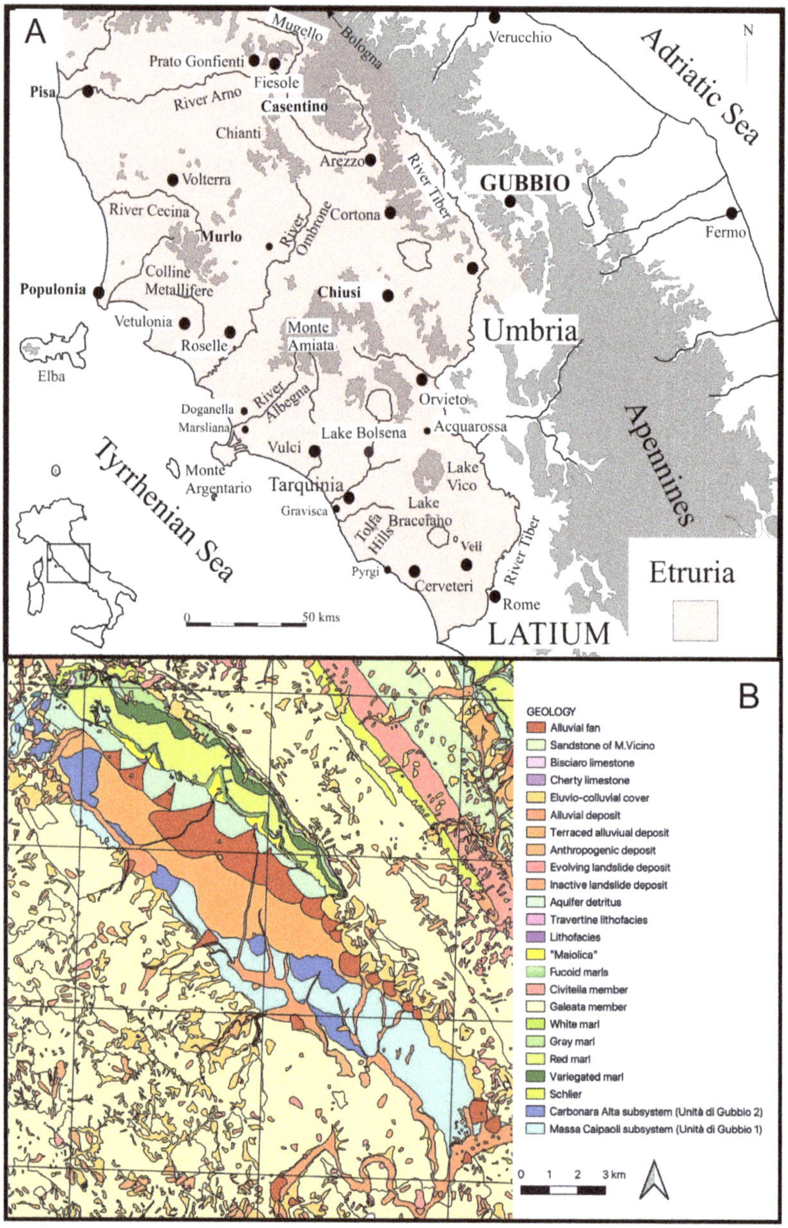


**Figure 1.** (**A**) The location of Gubbio in the wider region of Central Italy; (**B**) The essential geology of the Gubbio basin.

Any presentation of survey evidence needs to examine the underlying assumptions in the survival and collection of data. The original survey in the 1980s took the natural confines of the well-defined Gubbio valley as the region under study. The evidence from Figure 1B demonstrates that not all the valley was equally accessible from a geomorphological perspective, given the processes of erosion and alluviation over time, a fact that was much more severe for the Palaeolithic distributions compared with the Roman. The evidence from Figure 2 adds the fact that not all parts of the valley have been surveyed, most usually in response to difficult land use and limited human resources. Furthermore, fields were generally only surveyed once, and it is well known that conditions can change rapidly between seasons. A further matter to be considered is that a generous interpretation has been applied to the chronology of this surface material, a generosity that we think appropriate given the broad trends we are seeking to establish. Primacy has been given to the survey records made in the field, even though more accurate designations were made of individual artifacts in the 1994 publication. There is a considerable literature on issues related to surface survey recovery and interpretation (e.g., [1–3]) and since a broad-brush approach has been developed in this article, these complexities need to be considered when interpreting the overall trends presented. The term "site" has been employed generically, and simply indicates the location where artifacts were discovered.

### 1.1. Middle Palaeolithic

The first human exploitation of the valley can be dated back to the Middle Palaeolithic (210,000–35,000 years ago) although it is difficult to judge the intensity of occupation given the long timescale involved. Furthermore, the preservation of Middle Palaeolithic deposits was substantially determined by geological action, or more precisely the preservation of the previously mentioned, distinctively red, relict terraces located above Ponte d'Assi directly south across the valley from the modern city (Figure 1B). The pattern of Palaeolithic material in the Gubbio valley is very similar to the discoveries made in the upper Tiber valley [4], the only other area of Umbria where there has been systematic recovery of material over many years, in this case largely by the local archaeological group on the border between Umbria and Tuscany near Anghiari. In so far as it is possible to distinguish surface material, concentrations 4 and 6 in Figure 3 were largely Middle Palaeolithic in date, located on the terraces of similar geological date or earlier, whereas concentrations 1, 2 and 3 in Figure 3 were generally from the Upper Palaeolithic (35,000–10,000 years ago) period placed on an alluvial fan of a later geological date. At the end of the Palaeolithic (10,000–6000 years ago) (Figure 4), the distribution shows less evidence of geomorphological impact, with a wider distribution of more diffuse activity across the valley, even if with a strong element of continuity from the previous period in concentration 2 and to a lesser extent in concentration 5.

### 1.2. Neolithic (and Chalcolithic)

It is only with the Neolithic (6000–3000 BC) (Figure 5) that evidence for more substantial, less mobile settlement emerged, with a major focus on the mid contour of the alluvial fans in the center of the valley. This focus may be partly an ecological position in the landscape [5], suited to the collection of mixed resources, but may also be partly a geomorphological artifact, at a crucial change in slope of the alluvial fan where the balance of deposition and erosion led to a maximum revelation of the archaeological layers beneath. Taken at face value, there is concentrated evidence for Neolithic settlement to the west of Gubbio, under the slopes of Monte Semonte (concentration 1), on an alluvial fan descending down to Ponte d'Assi (concentration 2) and finally at San Marco to the East (concentration 3), where the site was excavated. San Marco Neolitico is a fundamental site for understanding the Gubbio, and indeed Umbrian, Neolithic, especially through the analysis of the faunal and floral remains, which permit the reconstruction of resource exploitation at the site, and its environmental context [5,6]. The excavated material also illustrates the character of lithic production and the use of pottery [6,7]. In particular, the lithic production is especially characterized by collections of specialized bladelet cores of local chert [6]. As far as the pottery is concerned there can be found affinities in style and decoration in the Italian regions of both Marche and Lazio, underlining the role of Neolithic Gubbio "a bridge between two worlds" [7]. The use of pottery is one of the most interesting points, since almost intact vessels, now displayed in the National Archaeological Museum of Perugia, were found in the ditch which formed a major feature of the site, suggesting a deliberate act of deposition. If this deliberate agency is correct, then this can be considered ritual action embedded within social practice, a phenomenon that is common in many prehistoric societies, seeking, in this case, to mark with culture a part of the landscape both by deposition and by construction (of a earth cut feature in this case) [8].

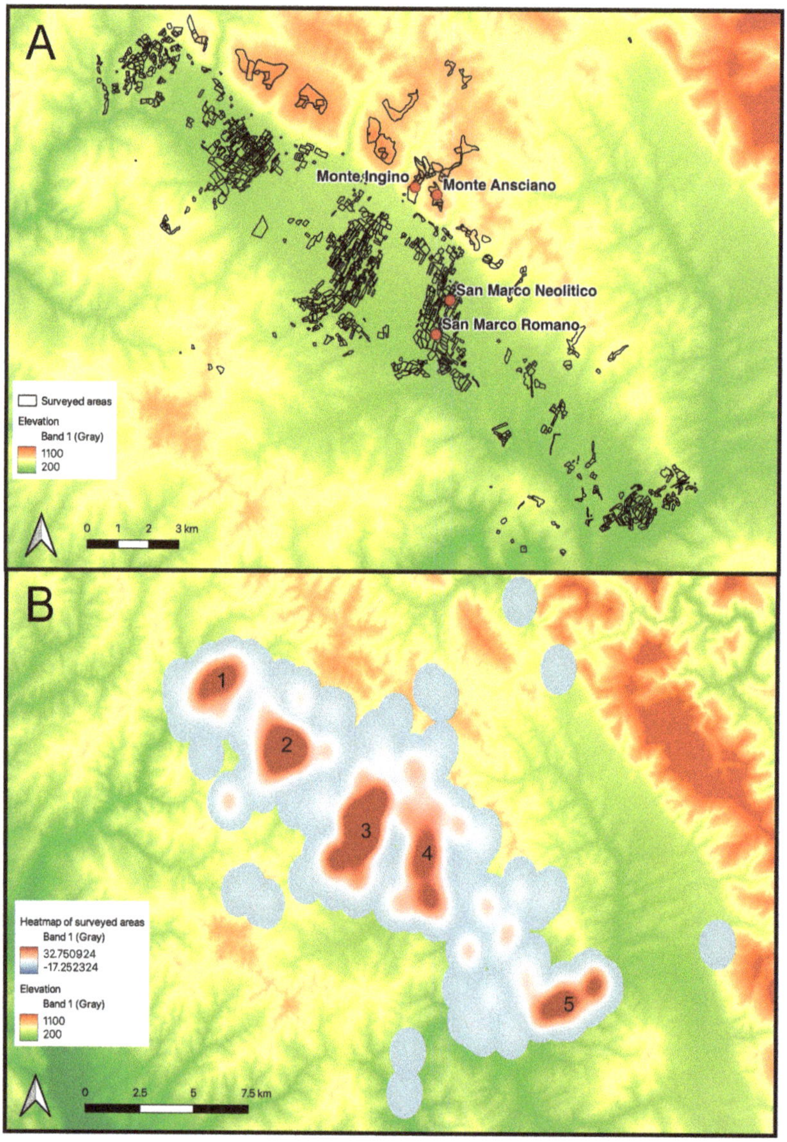


**Figure 2.** The areas surveyed in the valley of Gubbio. (**A**) The fields surveyed; (**B**) A heat map representation of the fields surveyed to permit a better comparison with the density plots that follow. Essentially five transects (1–5) were placed across the valley.

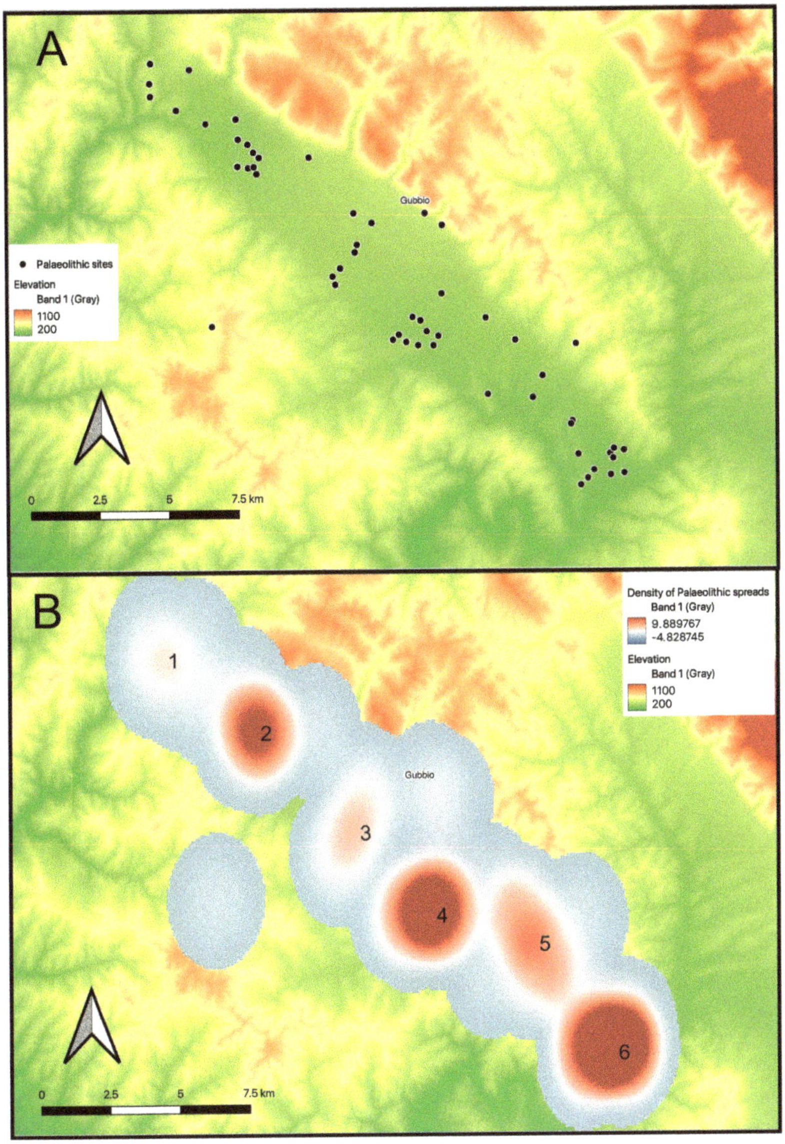


**Figure 3.** (**A**) Distribution of Palaeolithic sites in the valley; (**B**) Heatmap of Palaeolithic activity based on the spreads of lithic material recorded during the Gubbio Survey, using kernel densities estimation, with the same range as Figure 4. Numbers 1–6 refer to clusters of sites cited in the text.

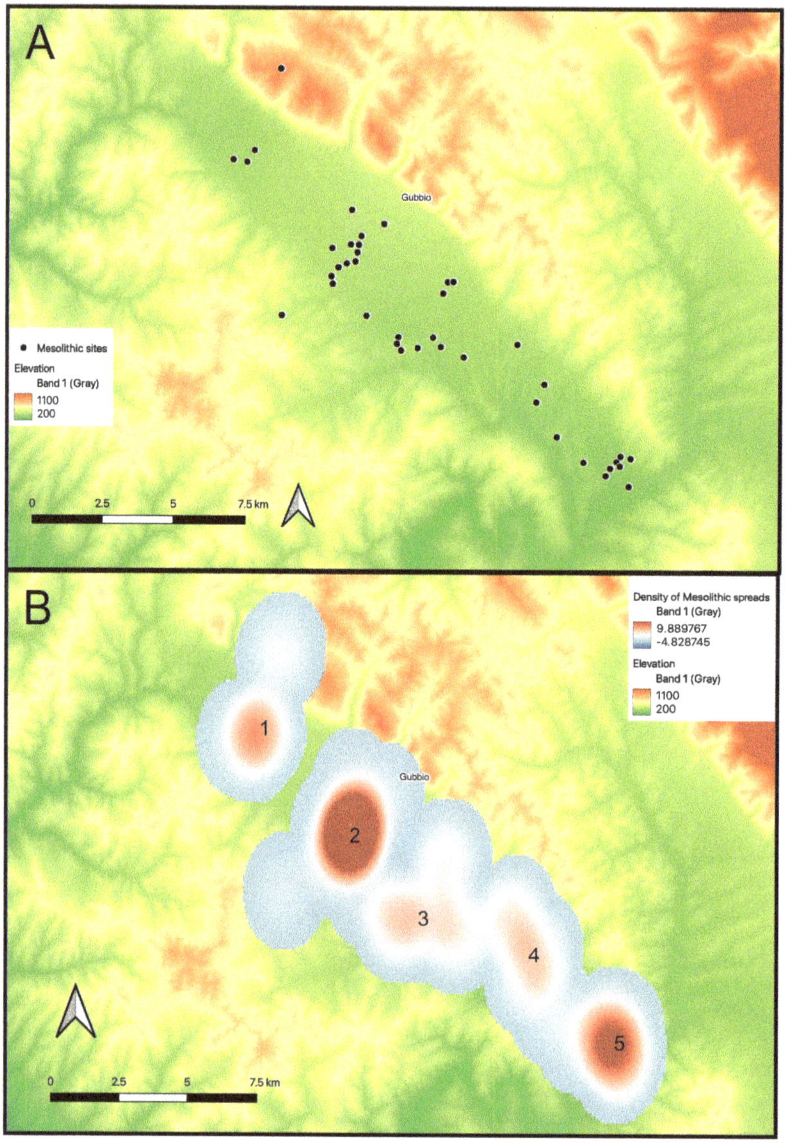


**Figure 4.** (**A**) Distribution of Mesolithic (Epi-palaeolithic) sites in the valley; (**B**) Heatmap of Mesolithic (Epi-palaeolithic) activity based on the spreads of lithic material recorded during the Gubbio Survey, using kernel densities estimation with the same range as Figure 3. Numbers 1–5 refer to clusters of sites cited in the text.

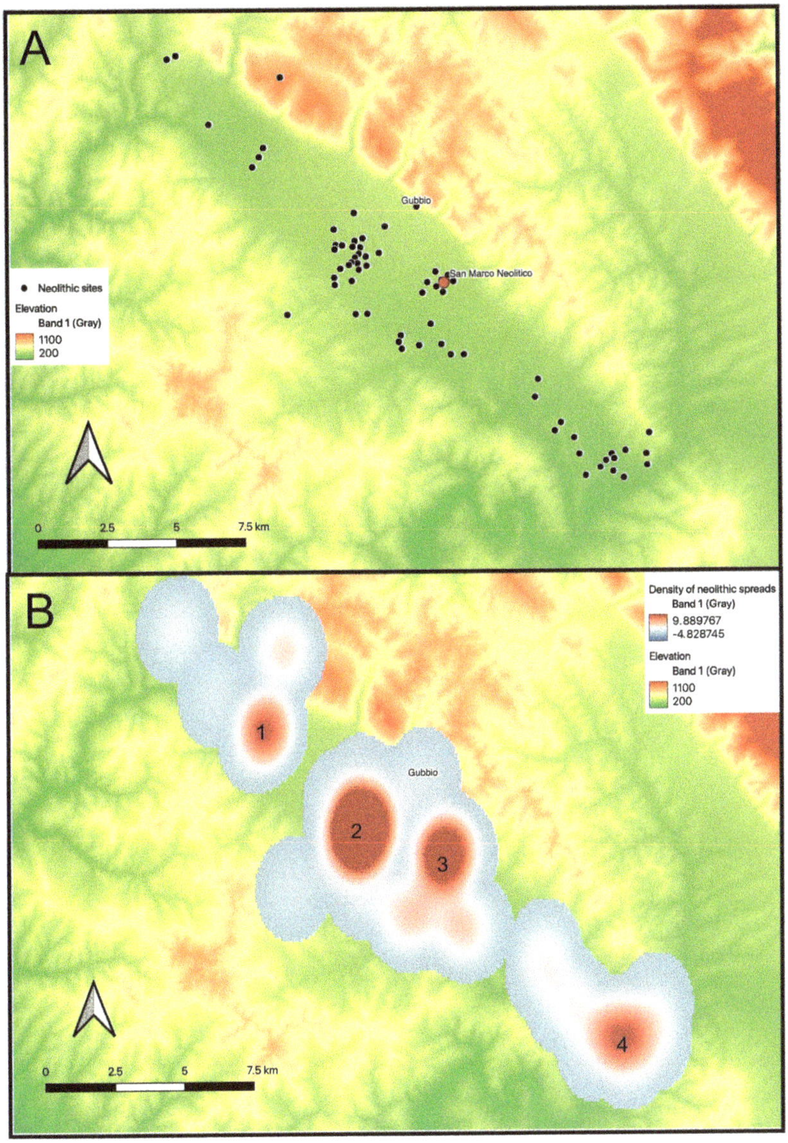


**Figure 5.** (**A**) Distribution of Neolithic sites in the valley; (**B**) Heatmap of Neolithic activity based on the spreads of lithic and ceramic material recorded during the Gubbio Survey, using kernel densities estimation. A. Red dot is San Marco. 'Numbers 1–4 in B refer to clusters of sites cited in the text.

The analysis of the distribution of radiocarbon dates (Figure 6) shows that the Neolithic of the Gubbio valley coincided with the latter part of the peak of Neolithic activity in Central Italy. This is a pattern that makes considerable sense for the arrival of the Neolithic in this inland and relatively upland part of the region. Relatively little evidence has, however, been uncovered for the Chalcolithic (3500–2300 BC) (Figure 7), a pattern that also seems to be true of the other neighboring valleys, whereas it is a period that is much better represented in Tuscany and the Tyrrhenian coastal areas of Central Italy. Since the valley was systematically surveyed, this appears to be a radically different trajectory for the valley compared with regions further west.

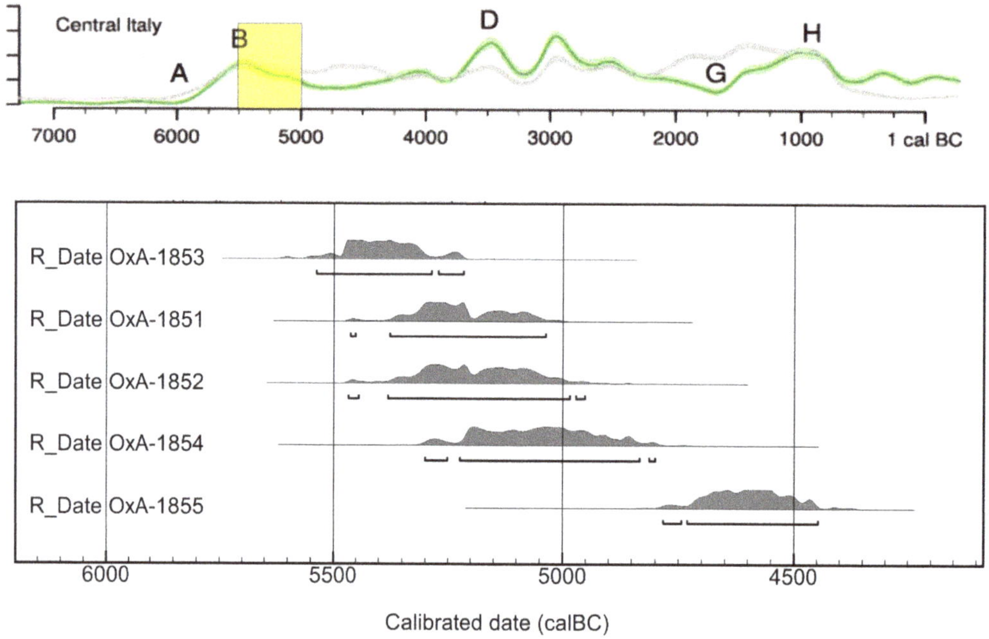


**Figure 6.** Distribution of San Marco Neolithic dates relative to the regional picture presented here [9] Calibration data based on [10,11]. A–H refer to significant changes in the "demographic" curve.

### 1.3. Bronze Age

In the Middle to Final Bronze Age (1400–1200 BC) (Figure 8), the valley was once more integrated within the patterns of Central Italy, reflecting the inferred demographic increase throughout the region as interpreted from radiocarbon frequencies (Figure 9). The character of the distribution of settlement shares much with Monte Cetona to the west, where there was clear evidence of clustering of settlement. At Gubbio, the most important site of reference was Monte Ingino, or more specifically the Rocca Posteriore on the peak looking back towards the Marche in the direction of the Sentino valley. This site bridged the Middle, Recent and Final Bronze Age (Figure 9), and provided a focal, elevated point of reference for the community. We plan to submit further radiocarbon samples to establish a tighter chronology of this transition than those measured in the 1980s. Finds excavated from this prolific site included around 30,000 pottery sherds, small metalwork fragments (e.g., over thirty dress pins), very few lithics and 3000 carbonized seeds [12]. In addition, more than 25,000 fragments of animal bone were recovered (principally sheep, cow and pig), representing an estimated 75% of all the published animal bone of this period so far recovered from Central Italy. If we can interpret wealth as vested in flocks of animals conspicuously consumed in a prominent position intervisible with other contemporary sites, this would have been a substantial, communal, display: a prominent example of ritual embedded within social practice. These flocks may have been drawn from a wide territorial area, perhaps coincident with what is visible from the summit. This idea is being explored scientifically through isotopic and genetic studies of the animals to examine the range over which they were pastured, with results currently pending. Within the deposits, we can identify signs not only of consumption and burning of meat and grain, but also micro-rituals of deliberate fragmentation and bending of metal items. The structured deposition also included decorated, drinking and storage, ceramics as well as a smaller number of rarer artifacts including a decorated bone comb, an amber bead, animal figurines and glass beads. These last artifacts were seemingly distinct from the communal rituals, belonging much more to the personal sphere. This combination of finds once again underlines the embedded nature of rituals within social life during this period of the Bronze Age.

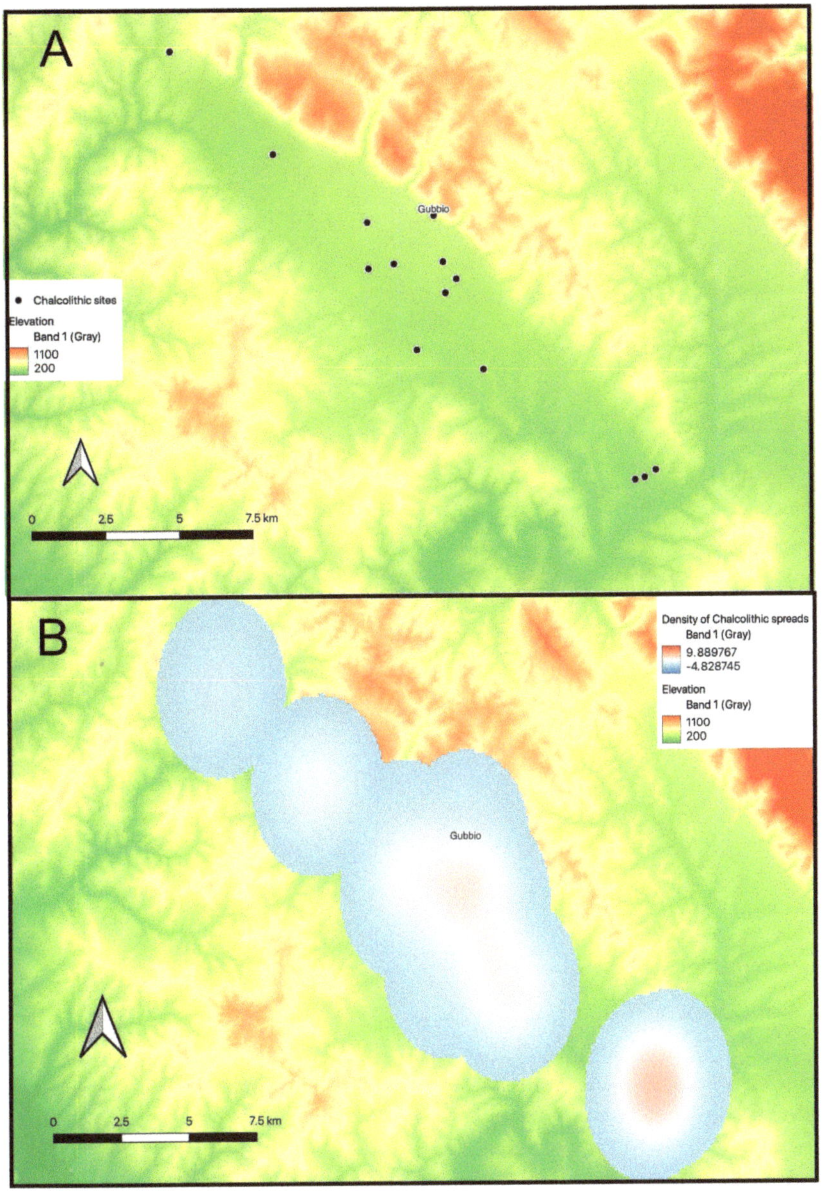


**Figure 7.** (**A**) Distribution of Chalcolithic sites in the valley; (**B**) Heatmap of Chalcolithic activity based on the spreads of lithic and ceramic material recorded during the Gubbio Survey, using kernel densities estimation.

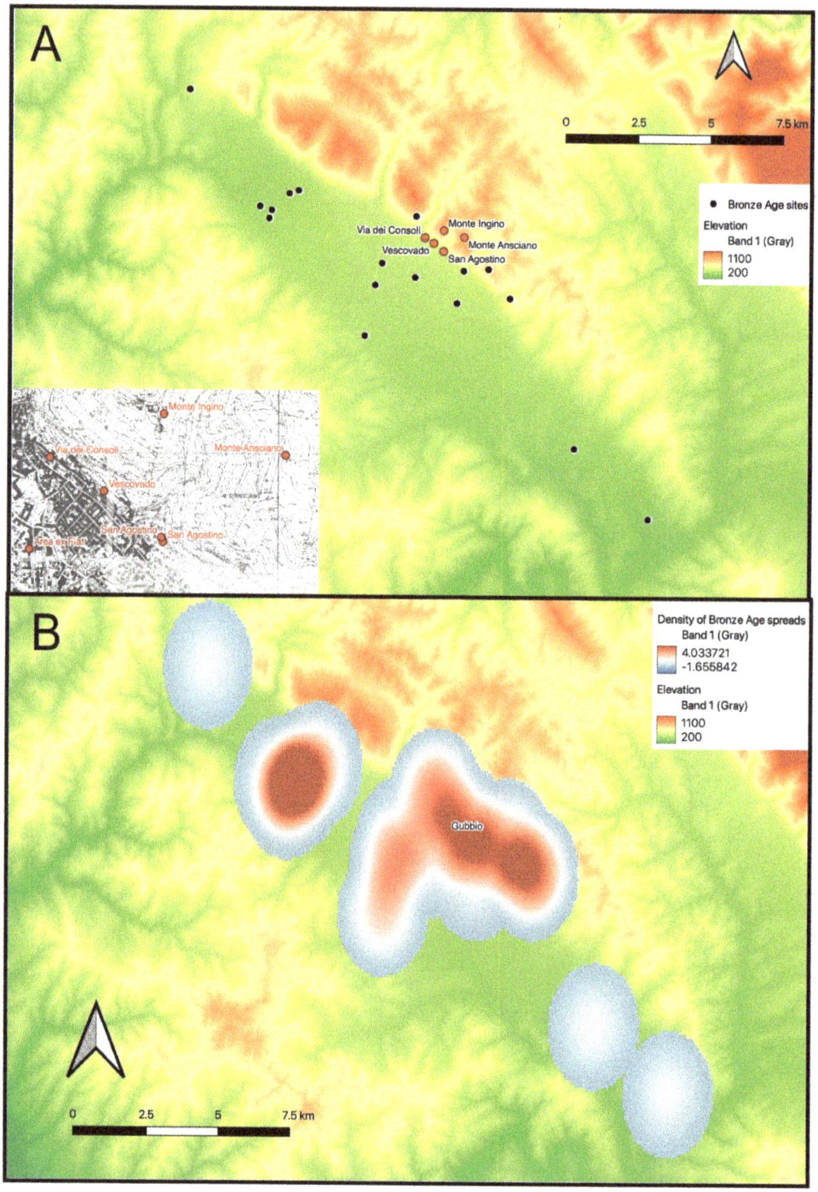


**Figure 8.** (**A**) Distribution of Bronze Age sites in the valley and (inset) excavated sites around Gubbio; (**B**) Heatmap of Bronze Age activity based on the spreads of ceramic material recorded during the Gubbio Survey, using kernel densities estimation.

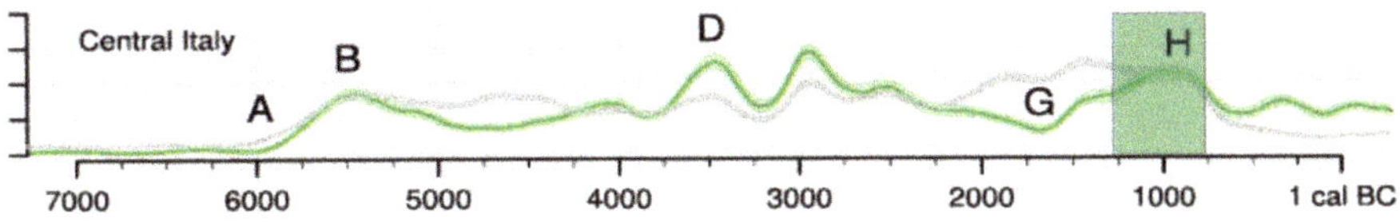


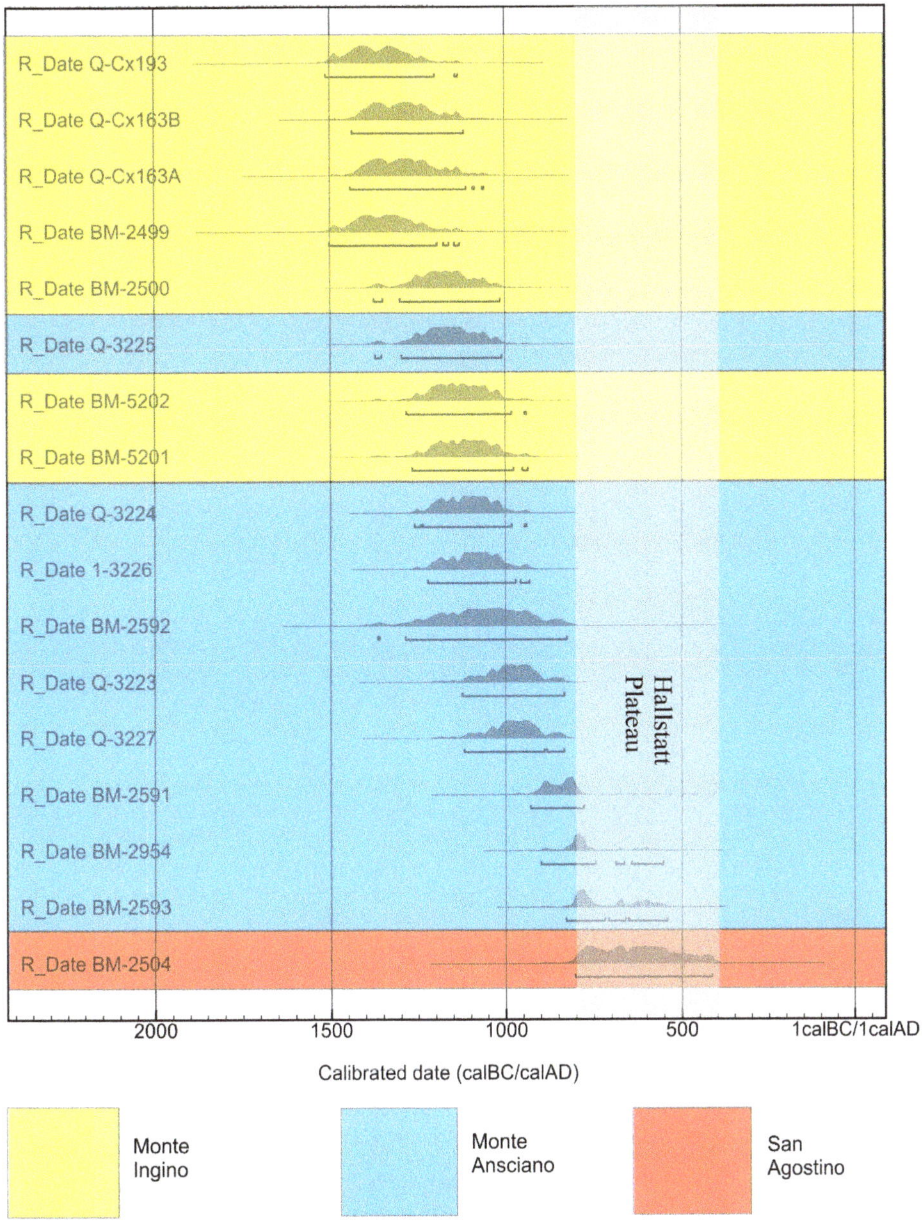


**Figure 9.** The Bronze and Early Iron Age radiocarbon dates of Gubbio in the regional context of Central Italy presented here [9], showing the clear transition between the sites of Monte Ingino, Monte Ansciano

and San Agostino, allowing for the complexities of the Hallstatt plateau in the calibration curve. Calibration data based on [10,11]. A–H refer to significant changes in the "demographic" curve.

These finds of the Middle, Recent and Final Bronze Age (1400–1200 BC) from Monte Ingino combine with other sites to form the impression of an important, expanding Bronze Age polyfocal settlement in the central part of the valley. In 2007, the rescue excavations in Via dei Consoli [13] brought to light evidence for another Middle Bronze Age settlement on the mid-slopes of Monte Ingino that continued into the Recent Bronze Age. During the Final Bronze Age, settlement, broadly in the same area, was accompanied by a cremation cemetery of at least 40 individuals. The settlement continued at Via Dei Consoli 103 and extended to Via Dei Consoli 105 and towards the southeast, where related material has been recovered in the excavation of the Vescovado. Final Bronze Age material has also been uncovered further down on the alluvial fan below Monte Ingino at the ex-Fiat site [14]. The mountain summit continued to present signs of embedded ritual activity during the early phases of the Final Bronze Age, but it is likely that the necropolis was built next to the principal settlement in a more sheltered location. This evidence on the summit and on the slope indicates an integrated complementary system of occupation of the mountain. The exposed location and harsh climate at 900 m, makes it plausible to suggest that the mountain top was used for rituals embedded in seasonal occupation of the upper reaches of the landscape. This seasonality of occupation is supported by the available evidence from the faunal remains of sheep and pigs mentioned above which on the basis of teeth wear [12] suggest a spring occupation.

### 1.4. *Iron Age*

With the Final Bronze Age and beginning of the Iron Age (1200–1000 BC), it is also important to consider Monte Ansciano, also overlooking Gubbio to the East of Monte Ingino, on the summit of which a second site was excavated during the Gubbio Project. This excavation also recovered many faunal remains (c. 12,000 fragments) from the late Bronze Age and Early Iron Age, including sheep, cow and pig, repeating the pattern of ritual consumption embedded in social life, although, based on the finds alone, without the same level of personal items. The only exception was the deliberate placement of an elaborate transitional Bronze Age/Early Iron Age fibula as one of the final acts of structured deposition above the midden deposit. The area in this period had a more developed built environment including an enclosure wall and an oval hut located in the saddle of the hill just below the summit.

The most distinctive feature of Monte Ansciano was the presence of structures and finds associated with its later use as a sanctuary in the Archaic period. These included a drystone platform placed above the Bronze Age enclosure, 65 schematic bronze figurines, of typically Umbrian style and datable to the 5th to 3rd Centuries BC, and an *aes rude* probably of the 3rd century BC. A representative sample of all these objects is now displayed in the Comune Museum of the Palazzo dei Consoli in Gubbio. The drystone platform separates the ritual embedded in social practice of the later Bronze Age and Early Iron Age from the more formalized collective rituals of the Archaic period practiced some vertical distance from the enlarged settlement which was by this stage located solely on the mid and lower slopes of the limestone escarpment of Monte Ansciano (including San Agostino) and at the ex-Fiat property [14] on the lower part of the alluvial fan.

The ritual characterization of the tops of the mountains is also emphasized by the fact that these peaks and others used in Umbria for deposition and other rituals, were all intervisible between each other [15]. For example, looking from Monte Tezio, to the north of Perugia, towards the northeast it is possible to see both Monte Ingino and Monte Ansciano as well as Monte Foce to the West of Monte Ingino. This visual interlinking may indicate an enduring territorial strategy of communities between the late second millennium BC and the early first millennium AD (the period hanging between prehistory and history which will be discussed below) actively displaying their presence to other groups, through the use of fire and feasting which result in the deposits on the peaks. This indicates that even though the rituals were transformed from those embedded in social practice of the late second millennium to more formalized rituals by the 5th century BC, clearly distinct from their settlement areas, marked by standardized figurines and more elaborate structures, they shared a sense of a regional community through shared ritual practices and intervisible ritual locations. Once again, the examination of the isotopic signatures (initially C/N and subsequently Sr) on the animal bones may give a sense of the breadth of the territory where the animals grazed, providing an indication of the territory managed by the human community.

After Roman integration, more intriguing material of a probably ritual nature was recovered from Monte Ansciano in the form of fragments of imperial period Roman lamps and simple drinking cups. Despite an apparent hiatus of 300+ years, these are of great interest, presenting the possibility of continuity of ritual focus without the construction of any additional associated structure. Thus, while the deposited lamps were Roman, their deposition in that location drew on Umbrian and Iguvine ritual practice, which was transferred down the centuries embedded in the collective ritual memory of the people and their relationship with the landscape and especially the mountain tops, which dominated the settlement.

## 2. The Distribution of Settlements between Protohistory and History

The entire *Gubbio Survey* data have been digitized and the later pottery has been restudied to enhance the analysis and interpretation of data in the valley from pre-Roman (Umbrian) to Roman times. To understand the phasing and distribution of the sites, a system of six phases was created (Table 1). A phase was defined according to the material culture, and in particular the fineware, present. For example, "imitation bucchero" and "late bucchero" were defined as Hellenistic and therefore mainly characterize Umbrian sites of Phase 2. In the interests of the longue duree focus of this article, these phases have been simplified into three density distributions to illustrate the broad trends within the valley. Employing this system and the quality and quantity of pottery recovered in each field, maps of the principal Iron Age, Umbrian and Roman sites in the valley can be created (Figures 10–12). This is where advances have been made in comparison with the 1994 publication which concentrated on the details of the prehistoric period. Further differentiation will be made in future publications, but here it is the longue duree at a scale comparable to the prehistoric period that has been favoured, without relating the details of new ceramic chronologies.

**Table 1.** Phasing of pottery and correspondence to density maps in the later periods.

<table>
<tr><th>Phase</th><th>Dates</th><th>Description</th><th>Map</th></tr>
<tr><td>1</td><td>5th C. BC</td><td>Iron Age, Archaic, Umbrian</td><td>10</td></tr>
<tr><td>2</td><td>4th and 3rd C. BC</td><td>Early Hellenistic, Umbrian</td><td rowspan="2">11</td></tr>
<tr><td>3</td><td>2nd C. BC</td><td>Hellenistic, phase of transition</td></tr>
<tr><td>4</td><td>1st C BC to first third 1 C AD</td><td>Republican/Early Imperial, phase of transition</td><td rowspan="3">12</td></tr>
<tr><td>5</td><td>2nd third 1 C AD to 3rd C AD</td><td>Full Imperial Roman</td></tr>
<tr><td>6</td><td>4th to 5th C AD</td><td>Late Roman</td></tr>
</table>

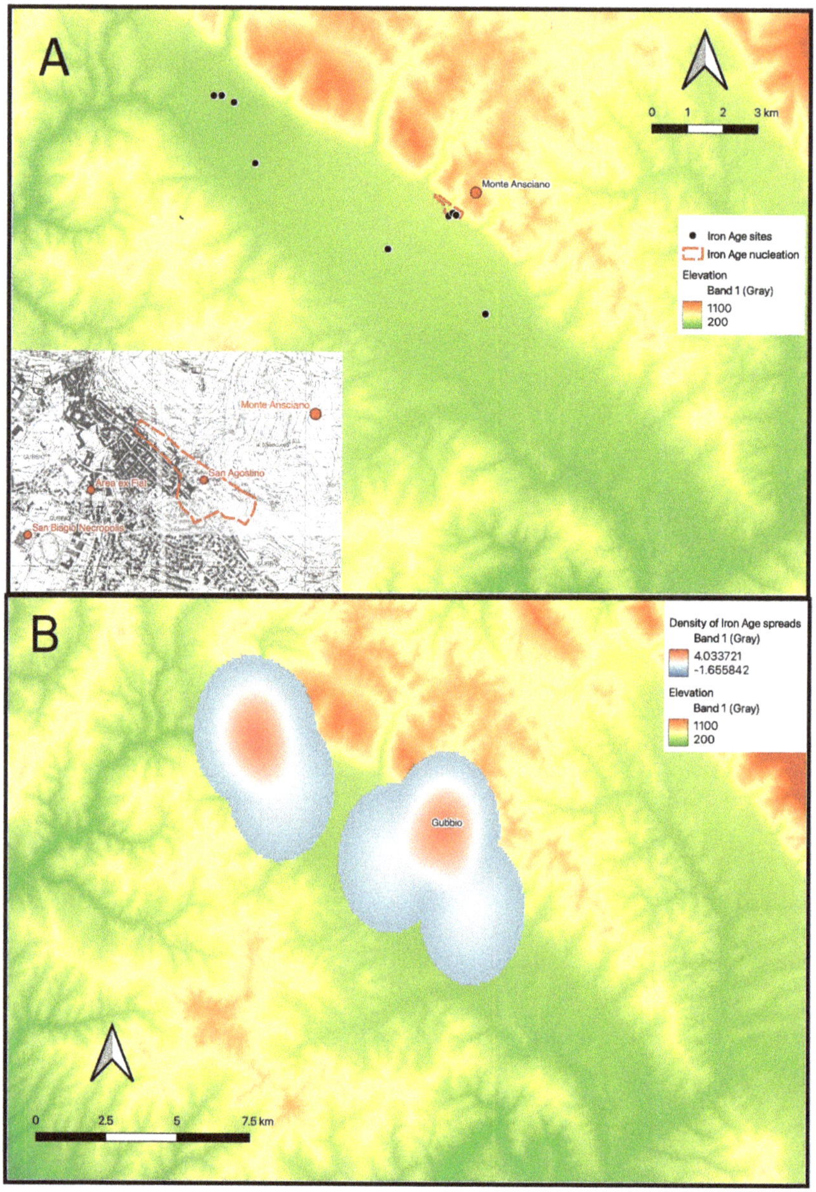


**Figure 10.** (**A**) Distribution of Iron Age sites in the valley; (**B**) Heatmap of Iron Age activity based on the spreads of ceramic material recorded during the Gubbio Survey, using kernel densities estimation. The same scale of density of pottery is employed for Figures 11 and 12.

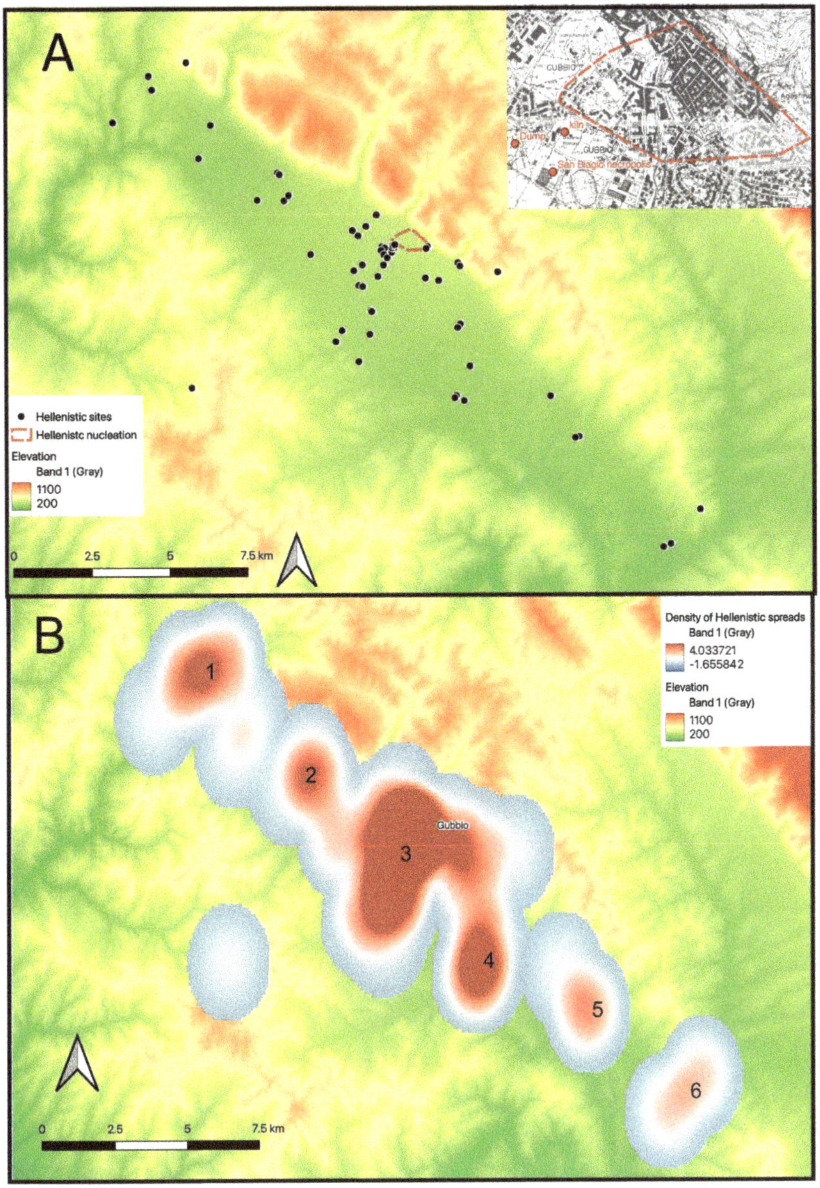


**Figure 11.** (**A**) Distribution of Hellenistic sites in the valley with inset showing the Gubbio area; (**B**) Heatmap of Hellenistic activity based on the distribution of ceramic material recorded during the Gubbio Survey, using kernel densities estimation. The same scale of density of pottery is employed for Figures 10 and 12. Numbers 1–6 refer to clusters of sites cited in the text.

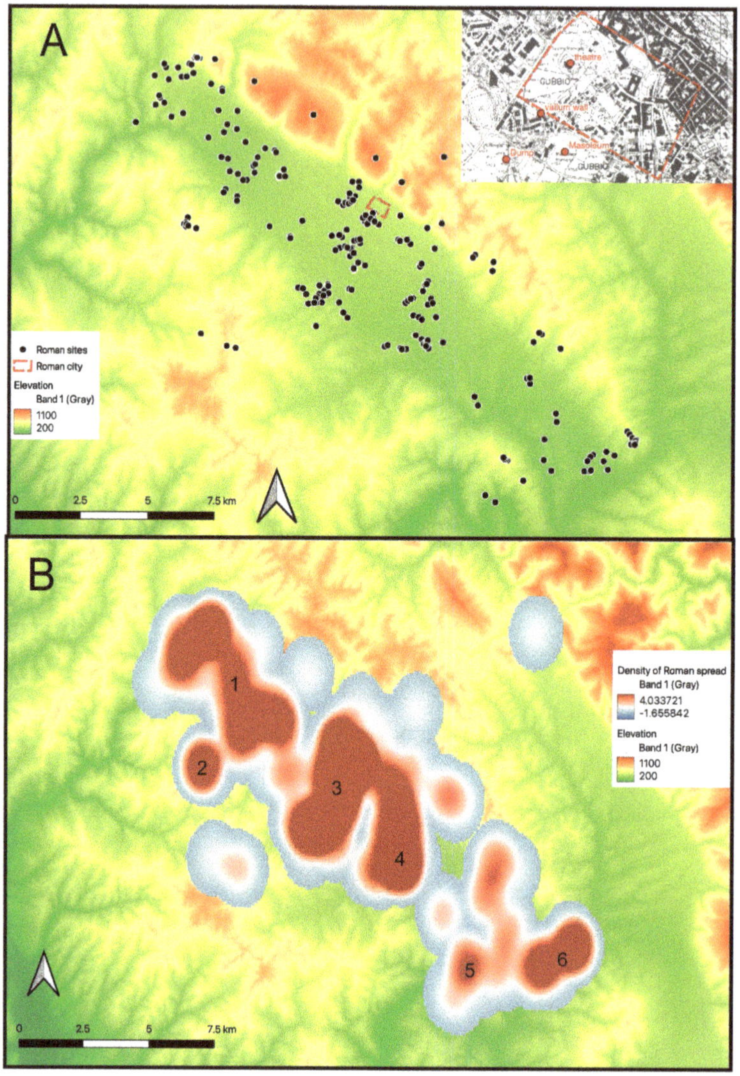


**Figure 12.** (**A**) Distribution of Roman sites in the valley (with inset showing sites in the Gubbio area); (**B**) Heatmap of Roman activity based on the spreads of ceramic material recorded during the Gubbio Survey, using kernel densities estimation. The same scale of density of pottery is employed for Figures 10 and 11. Numbers 1–6 refer to clusters of sites cited in the text.

Pottery density maps of the valley have thus been created for the three merged ceramic phases. These three maps (Figures 10–12) are collectively useful in understanding the distribution of sites and density of activity in the valley and show an appreciable increase in rural activity from the Iron Age to the Hellenistic, reaching an almost complete saturation in the northern part of the valley during the full Roman period, although more detailed analysis, not represented here, shows a decline in evidence for the later Roman phases. Nevertheless, it is clear that, whereas the whole valley, including the most rural areas, was exploited, the density in some areas was much greater, such as around Mocaiana (Figures 11 and 12, concentration 1), which corresponds to the fields in the northernmost part of the valley and Raggio just below. Interestingly the southeast of the valley, Branca (Figures 11 and 12, concentration 6), although closer to the important communication route of the Via Flaminia seems to have had a lower density of pottery, plausibly in response to the much heavier Plio-Pleistocene clays of the area.

As would be expected, the densest material concentrations are around the town of Gubbio and likely represent several discrete areas of activity. The Gubbio Survey data suggest a high concentration of Umbrian activity around the urban area of Gubbio and the suburban area of San Biagio, where there was a *vernice nera* kiln datable between the end of the third century and the beginning of the second century BC, and a dump, which produced a meaningful number of many local vases [16]. The quantity of material in this area suggests the existence of a settlement, coinciding with Manconi's identification of the location of Umbrian Gubbio in this area [17]. The material of Hellenistic date recovered on the surface appears to be too significant simply to result from the presence of the dump. According to Manconi, the dump was directly connected with the existence of the kiln, since the great majority of the material can in fact be identified as locally produced [18], although Sisani disagrees, considering the dump to be a consequence of urbanization during Roman times [18]. Despite the debate about the origin of the dump, the presence of the kiln and abundant ceramic material is evidence of the importance of the area before the Roman conquest.

The main Umbrian rural sites, based on the pottery that was recovered during the survey, are at Mocaiana and Casa Regni (Figure 11, concentration 1). The proximity of these two sites, indicates that rural settlement was well distributed and definitely developed in the *territorium* before the arrival of the Romans. The distribution of sites could, moreover, be an indication of a landscape pattern that is defined as typically Umbrian, drawn from analysis of the Iguvine tables by philologists Alex Mullen and James Clackson [19]. This proposes that there was more focus on the community and less on boundaries, where according to the Iguvine tables the only relevant boundary was the upper sky limits of the sanctuaries on the upper reaches of the landscape. From the data presented in this research, and especially the occupation of the rural areas, it is clear that the use of the landscape was not so tightly restricted to the main community of Gubbio itself, but extended into other smaller nucleations in the wider landscape.

Mocaiana in the north provides an interesting case study, because there are many Roman sites, but also one main Umbrian site. The Umbrian sites show continuity of occupation into the Roman period and must have had an influence on the wider Roman period settlement of the area. Also, since Mocaiana is at the foot of Monte Loreto, another Iron Age hilltop sanctuary, memory may have had a role in shaping the landscape. In addition to Mocaiana under Monte Loreto, there were other sites in the Gubbio area, which appear to be close to or associated with ritual spaces, in particular San Agostino under Monte Ansciano and San Biagio, close to a series of necropoleis.

On this basis, we can also briefly set the valley of Gubbio within the regional context of Central Italy. The new information presented here modestly revises our previous view of the Umbrian pre-Roman countryside [20–22], that there was relatively little pre-Roman rural settlement in the Gubbio valley. However, it remains true that the concentration of fully archaic and preceding orientalizing material was progressively greater in the direction of the territories of Perugia, Chiusi and, even more emphatically, the great cities of the Tyrrhenian coast [23,24]. Even in the Roman period, when the rural occupation of the Gubbio valley reached a substantially new level of density, it was not as great as that reported in the more central provinces of Central Italy [25].

## 3. Roman? Religious Landscape

The Iguvine tables were discovered in 1444 and, although complex and difficult to interpret, are fundamental to understanding the particular embedded religiosity and complex rituals of the valley. The seven tables are written partly in the Umbrian and partly in the Latin alphabets, probably reflecting the date of transcription onto bronze and underlining how the rituals described are pre-Roman, deriving from a religion which is intrinsic to Gubbio and Umbria and that was modified with the arrival of the Roman culture. During the Roman period in Gubbio, religion can be understood on three levels; there is the official political level, the local social and more Umbrian level, and finally the personal and private level. This makes an interesting parallel with the different scales (personal, social and inter-regional) already identified for earlier periods.

The first official level is the easiest to consider because its symbols and meanings were spread in the same way across Italy and the provinces. We know of typical Roman temples and sanctuaries in Gubbio through archaeological remains and inscriptions. For example, the foundations of the Temple of the Guastaglia, in a Roman quarter of Gubbio, have been excavated and an artistic reconstruction made showing a traditional Roman temple form [26]. An inscription in the Palazzo dei Consoli names the *quattuovir Gnaeus Satrius Rufus* as the person who paid for the restoration of the Temple of Diana in the 1st century AD (*CIL* XI 5820) [27] (*p.* 98). The donation of *Gnaeus Satrius Rufus* is firmly part of the official level of religion having as much a political as a religious purpose; the inscription, which had been installed as a pair on the theatre complex and was, therefore, highly visible also records his restoration of the theatre and his other generous donations. This is a typical example of the process of *romanization* which involved mostly elites and politicians who, through visible local patronage could enter the Roman political class.

Another apparently similar inscription records that *Iavolenus Apulus* paid for the restoration of the Temple of Mars Cyprius, a rural temple, and also dedicated a statue of Mars, currently in Florence with a copy displayed in the Palazzo dei Consoli. Here the position seems very different, however, as it does not appear to have been highly visible or politically motivated. The inscription, in a rural temple on a statue base would have been rarely seen, but it has a religious element. The text, includes the formula *votum solvit*, indicating a direct relationship with the deity and, despite still being part of the level of official of religion, there is also a personal element on display. It is also interesting that the deity, Marte Cyprio combines Roman Mars and probably Cupra, a local/picene deity.

In two other fascinating inscriptions, *Vittorius Rufus* is named in a funerary inscription as *avispex extispicus sacerdos publicus et privatus* (*CIL* XI 5824) [27] (*p.* 97), with *Sestus Vetiarius Surus* possessing a similar title. That we have a record of two people in Gubbio who interpret the flight of birds and entrails and manage public and private rituals appears significant. While not an especially common profession in the Roman world, it was important in the Umbrian/Etruscan environment and played a very important role in the Iguvine tables. These individuals appear, therefore to provide a direct connection to an Umbrian past with its landscape, birds and celestial boundaries.

The Iguvines, taking advantage of their position in the Roman Empire, imported some cults from Egypt, in particular the cult of Isis and her son Harpocrates, of whom a statuette is now displayed in Palazzo dei Consoli. Moreover, in the Roman cemetery of the Vittorina, where 237 tombs were excavated by the Soprintendenza, there are images of Harpocrates in two tombs, 80 and 232, the latter a child's tomb in which was an apotropaic pendant of the God. In the same cemetery, tomb 117, a supine inhumation, is that of the 'priestess of Isis', who was found holding a *sistrum*, typical of the cult, in her left hand and with an incense burner in each of the four corners of the tomb, now displayed in the Antiquarium of Gubbio next to the Roman theatre. These incense burners take the shape of a stemmed carinated kylix with finger impressed decoration on the rim and were a common find during the Gubbio Survey, in the excavation of San Marco Romano and in other tombs of the Vittorina. While present in other areas of Italy, they are a much more common find in the Gubbio valley and may be seen as a marker for some rural Iguvine ritual practice. Similarly, 14 small fragments of terracotta figurines were found in the rural areas of the valley during the survey. While 14 is not a large number in absolute terms, because of their size and characteristics, the probability of finding them during survey was very low and therefore the original number in the landscape must have been correspondingly very high. We are not aware of similar figurines being found on survey elsewhere in Italy, indicating a specific local, rural ritual practice. These figurines are likely, therefore to represent a much larger number that was once deposited in the rural landscape of Gubbio and were probably markers for a personal level of ritual or religious festival.

## 4. Discussion

The *Gubbio Revisited* project is significantly enhancing our understanding of the Gubbio valley's historical and cultural landscape. We have presented here the broad trends of occupation of the Gubbio valley, showing not only a general trend of increasing intensity culminating in the Roman period, but also phases of relative inactivity. We have also given a general sense of comparison with the wider trends of development in the broader region of Central Italy, showing where the valley is in and out of alignment, drawing on comparisons of settlement and radiocarbon density. At the same time, through a reanalysis of the archaeological evidence collected during the original Gubbio Project, and through other textual, epigraphic and archaeological evidence, this research demonstrates the levels of continuity and evolution of ritual practices from the Neolithic to the Roman periods. In the earlier stages, there was a pattern of ritual practice embedded in social life. In the later stages, the ritual practice embedded in social life was complemented by more formal rituals within the urban form. In this later period, the persistence and development of indigenous religious practices, in parallel to growing Roman influence, presents a unique blend of local and imperial religiosity. The full re-examination not only sheds light on the sacral significance of the Gubbio valley, but also emphasizes the integral role of ritual in its societal evolution, offering a richer, more nuanced perspective on the valley's past.

The extensive data, particularly regarding ceramic evidence from later periods, provides a solid basis for understanding density patterns. However, since these data rely on lithic and ceramic distributions documented in paper records from the 1980s, it cannot be entirely accurate. Data from earlier periods are less precise because of the absence of re-analysis of pottery, which would have provided a more detailed understanding of site dating. Methodologically, the reanalysis has leveraged modern techniques such as advanced radiocarbon dating and GIS mapping, offering a more precise understanding of settlement patterns and chronological frameworks. These methodologies have addressed some limitations of the original data, though inherent potential inaccuracies in the earlier records remain a challenge. Consequently, the Gubbio Revisited Project generalizes earlier periods more than later ones. Nonetheless, survey data from earlier periods are complemented by systematic excavations, particularly for the Neolithic, Bronze Age, and Iron Age, which provide detailed insights into settlement patterns and both domestic and ritual activities.

Ritual activity is a fundamental feature of the valley, intrinsic to and embedded in the landscape, influencing settlement patterns. The valley's particular geography, delimited by its upper reaches, means that principal settlements, including Gubbio itself, were founded on the slopes of these sacral places. Rural sites were distributed throughout the valley across all periods, although with varying densities and phases of inactivity, highlighting a specific pattern that persists through different ages in the valley. The importance of sacrality in the valley also suggests that religious and social activities were deeply intertwined, influencing not only daily life, but also the broader societal structures. This blend of local and Roman religious practices likely played a crucial role in the community's adaptation to changing political and cultural landscapes. In conclusion, the Gubbio Revisited project provides a detailed and nuanced understanding of the valley's past, emphasizing the integral role of ritual in societal evolution.

**Author Contributions:** Conceptualization, M.N., N.W. and S.S.; methodology, M.N., N.W., S.S. and C.M.; software, M.N.; validation, S.S. and N.W.; formal analysis, M.N.; investigation, N.W., C.M. and S.S.; resources, S.S.; data curation, M.N.; writing—original draft preparation, M.N.; writing—review and editing, N.W. and S.S.; visualization, M.N.; supervision, S.S.; project administration, S.S.; funding acquisition, S.S. All authors have read and agreed to the published version of the manuscript.

**Funding:** This research was funded by the British Academy, grant number SRG22\220926 and The APC was funded by the University of Cambridge, marginally discounted by a review undertaken by SS. Further funding was provided by Magdalene College and the McDonald Institute for Archaeological Research of the University of Cambridge. The town council of Gubbio welcomed the project with outreach opportunities.

**Data Availability Statement:** The raw data supporting the conclusions of this article will be made available by the authors on request.

**Acknowledgments:** The Gubbio Revisited project would like to thank the British Academy, the McDonald Institute and Magdalene College, Cambridge for recent financial support. The project is centered around Marianna Negro's PhD project, supported by Letizia Ceccarelli and Nicholas Whitehead for the identification of material culture and Simon Stoddart and Caroline Malone for background information. Caroline Malone, Nicholas Whitehead and Simon Stoddart all participated in the original 1980s project.

**Conflicts of Interest:** The authors declare no conflicts of interest. The funders had no role in the design of the study; in the collection, analyses, or interpretation of data; in the writing of the manuscript; or in the decision to publish the results.

## References

1. Bintliff, J.L.; Snodgrass, A.M. Off-site pottery distributions: A regional and inter-regional perspective. *Curr. Anthropol.* **1988**, *29*, 506–512. [CrossRef]
2. Stoddart, S.; Whitehead, N. Cleaning the Iguvine stables: Site and off-site analysis from a Central Mediterranean perspective. In *Interpreting Artefact Scatters: Contributions to Ploughzone Archaeology*; Schofield, A.J., Ed.; Oxbow: Oxford, UK, 1991; pp. 141–148.
3. Terrenato, N. Visibility and Site Recovery in the Cecina Valley Survey, Italy. *J. Field Archaeol.* **1996**, *23*, 91–109. [CrossRef]
4. Moroni, A.; Abati, F.; Baldanza, A.; Coltorti, M.; De Angelis, M.C.; Mancini, S.; Moroni, B.; Pieruccini, P. L' alto e Medio Bacino del Tevere durante il Paleolitico Medio. Ipotesi sul Popolamento e la Mobilità dei Gruppi di Cacciatori—Raccoglitori Neanderthaliani. In *Bollettino di Archeologia on Line. Direzione Generale per le Antichità 2 (2011/2-3)*; 2011; pp. 171–201, Figure 17. Available online: https://bollettinodiarcheologiaonline.beniculturali.it/numero-2-3-2011-anno-ii/ (accessed on 12 August 2024).
5. Hunt, C.; Malone, C.; Sevink, J.; Stoddart, S. Environment, soils and early agriculture in Apennine central Italy. *World Archaeol.* **1990**, *22*, 34–44. [CrossRef]
6. Malone, C.; Stoddart, S.; Barker, G.; Coltorti, M.; Costantini, L.; Giorgi, J.; Clark, G.; Harding, J.; Hunt, C.; Reynolds, T.; et al. The Neolithic site of San Marco, Gubbio (Perugia), Umbria: Survey and excavation, 1985–1987. *Pap. Br. Sch. Rome* **1992**, *60*, 1–69. [CrossRef]
7. Matarese, I.; Spagnolo, V.; Moroni, A.; Stoddart, S.; Malone, C. The Neolithic Settlement of San Marco di Gubbio (Perugia). A Reassessment in the Light of Recent research. In *Atti dell'IIPP, Perugia, Gubbio*; Studi Etruschi: Florence, Italy, 2024; *in press*.
8. Bradley, R. *An Archaeology of Natural Places*; Routledge: London, UK, 2000.
9. Parkinson, E.W.; McLaughlin, T.R.; Esposito, C.; Stoddart, S.; Malone, C. Radiocarbon Dated Trends and Central Mediterranean Prehistory. *J. World Prehistory* **2021**, *34*, 317–379. [CrossRef]
10. Reimer, P.J.; Austin, W.E.N.; Bard, E.; Bayliss, A.; Blackwell, P.G.; Bronk Ramsey, C.; Butzin, M.; Cheng, H.; Edwards, R.L.; Friedrich, M.; et al. The IntCal20 Northern Hemisphere Radiocarbon Age Calibration Curve (0–55 cal kBP). *Radiocarbon* **2020**, *62*, 725–757. [CrossRef]
11. Bronk Ramsey, C. OxCal 4.4. 2021. Available online: http://c14.arch.ox.ac.uk/oxcal (accessed on 1 July 2024).
12. Malone, C.; Stoddart, S. Territory, Time and State. In *The Archaeological Development of the Gubbio Basin*; Cambridge University Press: Cambridge, UK, 1994.
13. Cenciaioli, L.; Germini, F. Gubbio. Il sepolcreto di via dei Consoli. In *Gli Umbri in età Preromana. Atti del XXVII Convegno di Studi Etruschi e Italici*; Bettini, M.C., Ed.; Istituto di Studi Etruschi ed Italici: Firenze, Italy, 2013; pp. 485–520.
14. Germini, F. L'area ex-Fiat a Gubbio. In *Bollettino di Archeologia on Line. Direzione Generale per le Antichità 2 (2011/2-3)*; Gangemi: Roma, Italy, 2011; pp. 105–120.
15. Stoddart, S.; Redhouse, D. The Umbrians: An archaeological perspective. In *Entre Archéologie et Histoire: Dialogues sur Divers Peuples de L'italie Préromaine*; Aberson, M., Biella, M.C., Wullschleger, M., Di Fazio, M., Eds.; Université de Genève-Faculté des Lettres-Département des Sciences de l'Antiquité: Genève, Switzerland, 2014; pp. 107–124.
16. Pisano, F. Gli scavi di via Bruno Buozzi. In *Gubbio. Scavi e Nuove Ricerche. Vol. 1:Gli Ultimi Rivenimenti*; Manconi, D., Ed.; Edimond: Città di Castello, Italy, 2008; pp. 65–78.
17. Manconi, D. Gli ultimi rinvenimenti. In *Gubbio. Scavi e Nuove Ricerche. Vol. 1: Gli Ultimi Rivenimenti*; Manconi, D., Ed.; Edimond: Città di Castello, Italy, 2008; pp. 1–23.
18. Sisani, S. *Tuta Ikuvina. Sviluppo e Ideologia Della Forma Urbana di Gubbio*; Edizioni Quasar: Roma, Italy, 2001.

19. Stoddart, S.; Baroni, M.; Ceccarelli, L.; Cifani, G.; Clackson, J.; Ferrara, F.; della Giovampaola, I.; Fulminante, F.; Licence, T.; Malone, C.; et al. Opening the Frontier: The Gubbio—Perugia frontier in the course of history. *Pap. Br. Sch. Rome* **2012**, *80*, 257–294. [CrossRef]
20. Stoddart, S. Changing views of the Gubbio landscape. In *L'Étrurie et l'Ombrie avant Rome. Cité et Territoire. Actes du colloque International Louvain-la-Neuve Halles Universitaires, Sénat Académique 13–14 février 2004*; Fontaine, P., Ed.; Belgisch Historisch Instituut te Rome: Rome, Italy, 2010; pp. 211–218.
21. Stoddart, S. Between text, body and context: Expressing 'Umbrian' identity in the landscape. In *Landscape, Ethnicity and Identity in the Archaic Mediterranean Area*; Cifani, G., Stoddart, S., Eds.; Oxbow Books: Oxford, UK, 2012; pp. 173–186.
22. Cifani, G. Approaching ethnicity and landscapes in pre-Roman Italy: The middle Tiber Valley. In *Landscape, Ethnicity and Identity in the Archaic Mediterranean Area*; Cifani, G., Stoddart, S., Eds.; Oxbow Books: Oxford, UK, 2012; pp. 144–162.
23. Stoddart, S. *Power and Place in Etruria: The Spatial Dynamics of a Mediterranean Civilization: 1200–500 B.C.*; Cambridge University Press: Cambridge, UK, 2020.
24. Zeviani, C. Invisible Etruscans. A study on Rural Landscape and Settlement Organisation during the Urbanisation of Etruria (7th–5th Centuries BC). Ph.D. Thesis, University of Cambridge, Cambridge, UK, 2023.
25. Launaro, A. *Peasants and Slaves: The Rural Population of Roman Italy (200 BC to AD 100)*; Cambridge University Press: Cambridge, UK, 2011.
26. Grassigli, G.L.; Cecconi, N.; Nati, D. *Gubbio Guastuglia*; Bretschneider: Rome, Italy, 2024.
27. Matteini Chiari, M. *Museo Comunale di Gubbio: Materiali Archeologici*; Electa Editori Umbri Associati: Perugia, Italy, 1995.

*land*

MDPI

*Article*

# Exploring the Gobi Wall: Archaeology of a Large-Scale Medieval Frontier System in the Mongolian Desert

**Dan Golan [1], Gideon Shelach-Lavi [1,*], Chunag Amartuvshin [2], Zhidong Zhang [1], Ido Wachtel [1], Jingchao Chen [1], Gantumur Angaragdulguun [1], Itay Lubel [1], Dor Heimberg [1], Mark Cavanagh [3], Micka Ullman [4] and William Honeychurch [5]**

1 Department of Asian Studies, Hebrew University, Jerusalem 9190501, Israel; daniel.golan@mail.huji.ac.il (D.G.); zhidong.zhang@mail.huji.ac.il (Z.Z.); ido.wachtel@mail.huji.ac.il (I.W.); jingchao.chen@mail.huji.ac.il (J.C.); angaragd.gantumur@mail.huji.ac.il (G.A.); itay.lubel@mail.huji.ac.il (I.L.); dor.heimberg@mail.huji.ac.il (D.H.)

2 Department of Anthropology and Archaeology, National University of Mongolia, Ulaanbaatar 14200, Mongolia; amartuvshin.ch@num.edu.mn

3 Institute of Archaeology, Tel-Aviv University, Tel-Aviv 6139001, Israel; markgordonc@mail.tau.ac.il

4 Institute of Archaeology, Haifa University, Haifa 3103301, Israel; micka.ullman@mail.huji.ac.il

5 Department of Anthropology, Yale University, New Haven, CT 06520, USA; william.honeychurch@yale.edu

* Correspondence: gideon.shelach@mail.huji.ac.il

**Abstract:** The Gobi Wall is a 321 km-long structure made of earth, stone, and wood, located in the Gobi highland desert of Mongolia. It is the least understood section of the medieval wall system that extends from China into Mongolia. This study aims to determine its builders, purpose, and chronology. Additionally, we seek to better understand the ecological implications of constructing such an extensive system of walls, trenches, garrisons, and fortresses in the remote and harsh environment of the Gobi Desert. Our field expedition combined remote sensing, pedestrian surveys, and targeted excavations at key sites. The results indicate that the garrison walls and main long wall were primarily constructed using rammed earth, with wood and stone reinforcements. Excavations of garrisons uncovered evidence of long-term occupation, including artifacts spanning from 2nd c. BCE to 19th c. CE. According to our findings, the main construction and usage phase of the wall and its associated structures occurred throughout the Xi Xia dynasty (1038–1227 CE), a period characterized by advanced frontier defense systems and significant geopolitical shifts. This study challenges the perception of such structures as being purely defensive, revealing the Gobi Wall's multifunctional role as an imperial tool for demarcating boundaries, managing populations and resources, and consolidating territorial control. Furthermore, our spatial and ecological analysis demonstrates that the distribution of local resources, such as water and wood, was critical in determining the route of the wall and the placement of associated garrisons and forts. Other geographic factors, including the location of mountain passes and the spread of sand dunes, were strategically utilized to enhance the effectiveness of the wall system. The results of this study reshape our understanding of medieval Inner Asian imperial infrastructure and its lasting impact on geopolitical landscapes. By integrating historical and archeological evidence with geographical analysis of the locations of garrisons and fortifications, we underscore the Xi Xia kingdom's strategic emphasis on regulating trade, securing transportation routes, and monitoring frontier movement.

**Keywords:** Gobi Desert; Mongolia; Gobi Wall; Tangut; Xi Xia; Western Xia; medieval wall system

## 1. Introduction

The Medieval (10th to 13th century CE) Wall System (MWS) stretches approximately 4000 km across extensive regions in northern China and Mongolia, as well as shorter sections in Russia (Figure 1). It represents one of the most extensive yet enigmatic architectural features in East Asia. In recent years The Wall Project, funded by the European Research Council, as well as other projects, has extensively studied and published on different sections of this wall line. Such research demonstrated that this extensive system of earthen walls was built by different empires from c. the 10th to the 13th centuries CE [1–6]. Among the different sections of the MWS, the wall section located in the southern Mongolia's Gobi Desert is the least explored and still poorly understood. This study focuses on a 321 km-long segment of this wall line, located in Ömnögovi province of Mongolia, that we refer to as the Gobi Wall (Figure 2).

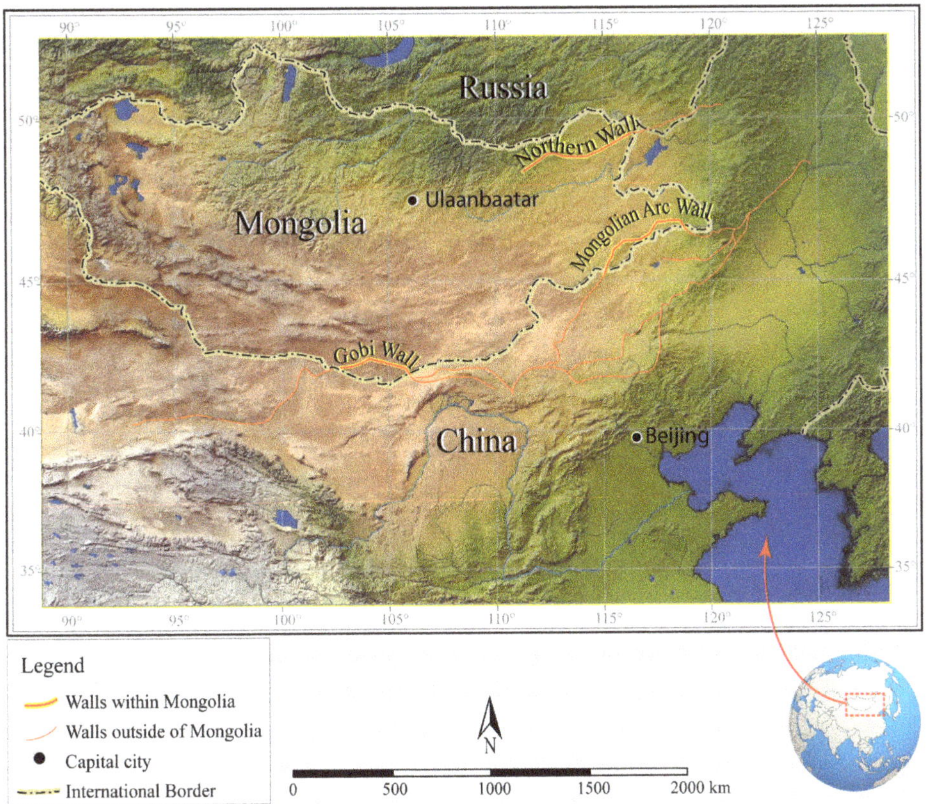


**Figure 1.** Map of the spatial distribution of different sections of the Medieval Wall System.

The medieval period in East Asia represents the zenith of dynastic rule by nomadic and semi-nomadic peoples. Following the collapse of the Tang dynasty (618–907) and the subsequent rise of the Song dynasty (960–1279) in the Yellow and Yangzi River basins, the Khitan established the Liao dynasty (907–1125), which controlled extensive territories across North China, Inner Mongolia, and Mongolia. In regions that now comprise western China, Inner Mongolia, and southern Mongolia, the Tangut founded the Western Xia (Xi Xia) dynasty (1038–1227). In the early 12th century, the Jurchen people of Manchuria overthrew their Khitan overlords and established the Jin dynasty (1115–1234), ultimately conquering the Yellow River basin from the Chinese Song. The Mongols emerged as a dominant force in the early 13th century, destroying the Xi Xia in 1227 and the Jin in 1234, before extending their conquests across China and establishing a vast transcontinental empire. Extensive wall systems and demarcations were constructed by the Liao [4], Jin [1], and Xi Xia. This paper focuses on the Xi Xia wall system.

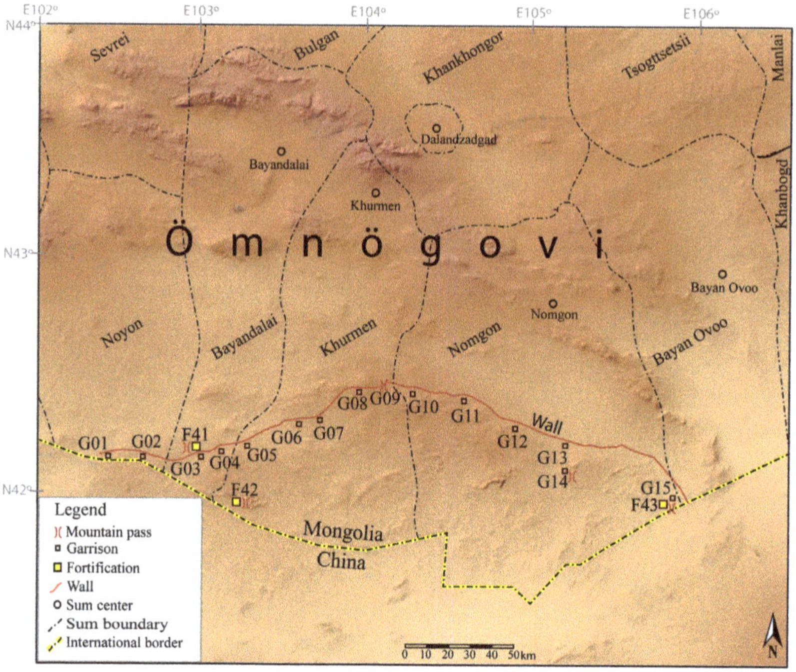


**Figure 2.** Map of the Gobi Wall system in Ömnögovi province, Mongolia, illustrating the wall's alignment and the distribution of garrison positions and mountain fortifications. Contemporary sum (district) administrative boundaries are marked.

The main goals of this research are to better understand the construction techniques used in building the walls, garrisons, and fortifications; to date their construction and periods of use; and to determine the purposes for which they were built. We also aim to situate this particular section within its broader environmental context and in relation to medieval Inner Asian political and economic dynamics. The study addresses several key research questions: When were the wall-line and associated structures constructed and used? Were there different phases of renovation or expansion of the wall system? What was the historical context for its construction and renovation? What construction techniques were employed, and how does the construction and design of the Gobi Wall compare to other sections of the MWS? What was the nature of activity along and across the wall, and how did the associated structures function within this system? How did different geographic and ecological parameters affect the design and functioning of the wall system? By addressing these questions, this study aims to contribute to our understanding of medieval Inner Asian political frontier management, long-distance trade networks, and the complex interplay between nomadic and sedentary societies in this critical borderland region.

## 2. Previous Research

Previous investigations of the MWS include large-scale surveys of wall sections located in China [7,8], but few excavations have accompanied these surveys. In Mongolia most intensive survey and excavations focused on the North-Wall as marked on Figure 1

[2–4,9,10], with some attention paid to the section we call the Mongolian-Arc [1,5]. These investigations have challenged traditional interpretations of such structures as purely defensive installations, suggesting more complex roles in monitoring the movement of people and herds and regulating trade. The Gobi Wall, however, presents unique challenges and opportunities for research due to its remote location and distinctive environmental context.

Early observations of the Gobi Wall date back to European travelers in the 18th century, with more systematic studies initiated in the mid-20th century [11,12]. Mongolian archeologist Kh. Perlee [13] (pp. 270–280) conducted pioneering fieldwork in the 1950s examining several structures along the wall in Nomgon sum. These foundational investigations, undertaken in logistically challenging conditions with limited technological resources, established the initial methodological framework and chronological hypotheses that continue to inform contemporary scholarship. Subsequent research by Baasan [14] provided extensive research of all wall sections in Mongolia, with a description of the Gobi Wall based on sections documented in Soviet topographic maps. More recently, the Gobi Wall in Mongolia was surveyed by a Japanese team [15], and by a Russian–Mongolian team headed by Kovalev and Erdenebaatar [11,12]. Both teams attributed the wall to the Xi Xia dynasty 1038–1227 CE (also known as Western Xia) which was founded by the Tangut people (Figure 3). The evidence supporting this periodization is the best available to date; however, we aim to enhance chronological resolution and further refine these previous conclusions.

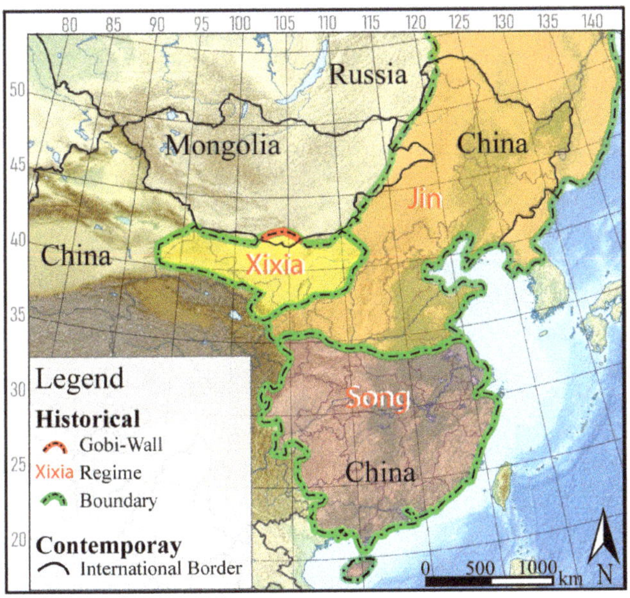


**Figure 3.** Map depicting the Gobi Wall, marked in red, as part of the Xi Xia boundary, in year 1142 CE. Map after Tan [16] (p. 72).

The section we call the Gobi Wall extends eastwards into China, Inner Mongolia province. Some Chinese researchers working on this part, including the editors of the atlas of the cultural heritage (Inner Mongolia volume), attributed this line and associated structures to the Han period (206 BCE–220 CE) [17] (pp. 268–269). Others, such as Li [18] and Xing [19] (pp. 78, 81), reconstructed a more complex scenario whereby the wall was initially constructed during the Han period but may have been reused by later dynasties such as the Jin, and Xi Xia as well. These ongoing debates, as well as the poor understanding of pertinent questions about the function for which the wall and associated structures were built, the identity and lifeways of the people who were stationed along this wall line, and the techniques used in its construction, underscore the need for a comprehensive, multidisciplinary approach to study the Gobi Wall.

## 3. Methodology

The research methodology employed for this investigation of the Gobi Wall integrated multiple analytical approaches to document and interpret this archeological landscape. The methodological framework comprised four principal components: (1) geospatial analysis and remote sensing, (2) systematic archeological survey, (3) targeted excavations, and (4) laboratory analyses of recovered materials.

### 3.1. *Geospatial Analysis and Remote Sensing*

Analytical protocols delineated by Storozum [6] were implemented as the foundational framework for our geospatial investigation. Historical cartographic sources—including the Historical Atlas of China edited by Tan [16] and Soviet-era topographic maps—were systematically examined to establish baseline geographical data. In parallel, high-resolution satellite imagery from multiple platforms (Google, Bing, ACME, ESRI, Planet, Umbra, Airbus) was analyzed to delineate the entire 321 km Gobi Wall segment within contemporary Mongolia, enabling the identification of associated wall structures. The complete geospatial dataset was subsequently processed using ArcGIS Pro by ESRI, version 3.4 (released November 2024). For elevation analysis, we employed the "標高 (Japanese terrain)" digital elevation model by ESRI JP Content (updated 28 March 2023), further refining our spatial understanding of the study area.

We georeferenced and digitized twenty Soviet topographic maps at a 1:100,000 scale (Figure 4) to extract features relevant to the frontier system. These 20 maps include detailed symbology [20]. They were surveyed between 1943 and 1945 and printed in the 1970s and offer a valuable yet underutilized geospatial resource for archeological research in East Asia [21–23]. Their production predates major anthropogenic alterations such as modern industrial development and climate change, thereby offering a mid-20th-century environmental baseline that is particularly useful for contextualizing long-term landscape dynamics.

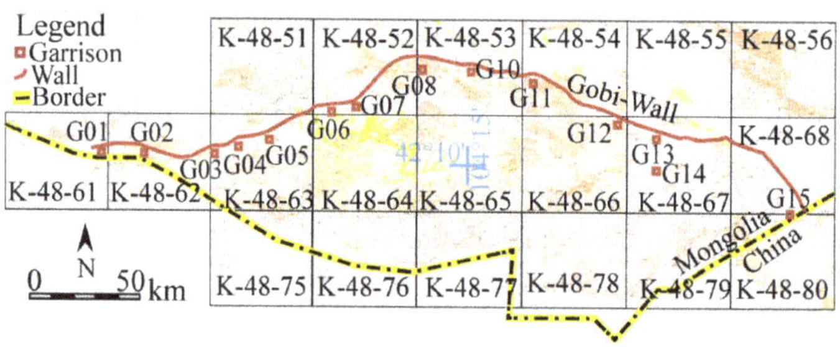


**Figure 4.** Distribution of 20 Soviet topographic map sheets covering the study area. Sheet numbers follow the standard Soviet nomenclature system (K-48-XX). The overlay depicts the wall line and the Mongolian Chinese border for general orientation.

While we recognize that the Soviet topographic maps do not capture environmental conditions from the medieval period with precision, their applicability is supported by paleoenvironmental studies demonstrating relative ecological continuity in this region. Pollen records from northern Asia, including the Gobi Desert, indicate the dominance of arid-adapted shrub and grassland vegetation across the late Holocene, with no significant increase in arboreal species over the past millennia [24]. Similarly, aeolian landforms in the Gobi Desert exhibit comparatively low mobility and dust flux [25] (p. 10). Dune systems in the region show long-term spatial persistence despite episodic reactivation. Optically Stimulated Luminescence (OSL) dating from the Gobi Khongoryn Els dune field confirms Late Holocene remobilization events while retaining broad spatial consistency [26] (p. 112). These findings indicate that dune outlines recorded in the Soviet-era maps can serve as meaningful proxies for understanding past environmental constraints.

Additional support for the precision of the Soviet-era maps comes from our field observations during the 2024 pedestrian survey, where the only wood species depicted on maps—poplar trees and saxaul shrubs—were also dominant in the landscape and recovered from wood samples dated to the medieval period.

All these indicators demonstrate the value of using Soviet-era maps as interpretive tools for analyzing medieval landscape configurations. However, we emphasize the importance of treating these maps as proxies, to be used with appropriate methodological caution and contextualized through paleoenvironmental and field-based data. The resulting dataset, derived from digitizing the Soviet maps, includes wall lines, well locations (n = 439), well depth and volumetric measurements (n = 61), individual poplar trees (n = 279), saxaul shrub polygons (n = 158, covering 711 km$^2$), and sand dune polygons (n = 45, covering 1500 km$^2$).

### 3.2. Systematic Archeological Field Survey

Based on remote sensing results, a comprehensive survey strategy was formulated for ground truthing and documenting all types of features potentially associated with the wall system. Aerial documentation employed drone photography using DJI Mavic 3T (includes an optional thermal camera) to capture wall sections and associated structures from multiple perspectives and elevations. This approach generated high-resolution orthophotography mosaics that facilitated detailed morphological analysis of architectural features.

A systematic pedestrian survey was conducted following standard archeological protocols [27], with the objective of identifying visible archeological material such as potsherds, lithics, and metal objects. Survey areas were defined in advance by marking polygons on satellite imagery of structures, encompassing each targeted structure and extending 30 m beyond its outer perimeter. Surveyors were spaced at 3 m intervals and walked in straight transects at a controlled pace set by the team leader. All surface finds were flagged and recorded, ensuring comprehensive retrieval rather than selective sampling. This visual inspection was immediately followed by a metal detector survey over the same area to enhance artifact recovery, map metal distributions, and identify potential activity zones based on artifact clustering. Soil augering was conducted in triplicate at locations of high artifact density to evaluate subsurface stratigraphy and determine promising areas for targeted excavation.

### 3.3. Targeted Excavations

Excavations focused on two selected garrisons (G05 and G10) and specific wall sections. Rather than exposing large areas or entire structures, excavation strategy prioritized the recovery of datable materials, investigation of construction methods, and collection of environmental and activity-related samples. This targeted approach maximized data acquisition while minimizing site disturbance. Small-scale excavations were conducted with fine stratigraphic control and with all contexts and features documented three-dimensionally. All excavated soil was processed through 5 mm mesh sieves to ensure recovery of small artifacts. The spatial position of stratigraphic layers, features, and individual artifacts was documented using Real-Time Kinematic (RTK) equipment, generating precise spatial data that was integrated into the site and excavation area maps.

### 3.4. Laboratory Analyses

Radiocarbon ($^{14}C$) analysis was conducted on twelve wood and charcoal samples collected from five distinct locations representing six feature typologies. These analyses were conducted at the W.M. Keck Carbon Cycle Accelerator Mass Spectrometry (KCCAMS) Facility, University of California, Irvine, following established protocols [28].

Wood samples were submitted to the Laboratory of Archaeobotanical and Ancient Environments at Tel Aviv University for species identification. Microscopic examination along three observational planes (transverse, tangential, and radial) was conducted using a stereoscopic Carl Zeiss SteREO Discovery.V20 microscope (Oberkochen, Germany), at magnifications up to 360×. Macerations of wood samples were examined using a Nikon Eclipse e100 light microscope, (Tokyo, Japan), at magnifications up to 40×, enabling detailed analysis of cellular structures for taxonomic identification.

Artifacts recovered through survey and excavation, particularly ceramics and metal objects, were subjected to typological analysis and comparative study. Numismatic specimens were identified and dated based on established reference collections [29,30] with special attention given to Xi Xia coinage [31,32].

## 4. Archeological Research of the Gobi Wall System

During the preparatory phase for fieldwork and analysis of extant resources, including cartographic materials, atlases, and remote sensing data, four distinct types of anthropogenic features were identified (Table 1) within the research area: 1. Wall-sections; 2. Large rectangular enclosures (garrisons). 3. Small hill-top fortifications overlooking mountain passes and mountain passes (Figure 2).

**Table 1.** Spatial distribution of features along the Gobi Wall, noting the research type.

| Site Label | District (Sum) | Coordinates Latitude, Longitude | Research Type |
|---|---|---|---|
| G01 | Noyon | 42.17226, 102.41749 | Satellite observation |
| G02 | Noyon | 42.19049, 102.65412 | Satellite observation |
| G03 | Bayandalai | 42.15132, 103.01183 | Field survey |
| G04 | Bayandalai | 42.19178, 103.14206 | Field survey |
| G05 | Khurmen | 42.23697, 103.28883 | Survey and Targeted excavations |
| G06 | Khurmen | 42.33963, 103.61582 | Field survey |
| G07 | Khurmen | 42.34569, 103.73799 | Field survey |
| G08 | Khurmen | 42.47144, 104.03172 | Field survey |
| G09 | Nomgon | 42.48703, 104.17385 | Field survey |
| G10 | Nomgon | 42.47315, 104.32389 | Survey and Targeted excavations |
| G11 | Nomgon | 42.43807, 104.66517 | Field survey |
| G12 | Nomgon | 42.31544, 104.99789 | Field survey |
| G13 | Nomgon | 42.22180, 105.33192 | Satellite observation |
| G14 | Nomgon | 42.10577, 105.29337 | Satellite observation recorded for the first time |
| G15 | Bayan-Ovoo | 41.97812, 105.74893 | Satellite observation |
| F41 | Bayandalai | 42.19413, 102.97452 | Field survey |
| F42 | Bayandalai | 42.01897, 103.26631 | Field survey |
| F43 | Bayan-Ovoo | 41.97702, 105.74585 | Satellite observation |

### 4.1. Wall Construction and Typology

During fieldwork operations, the wall line was traversed from G03 to G12, including F41, F42, (Figure 2) with examinations conducted at various points to document wall preservation and construction techniques as they were naturally exposed. In some locations, we conducted a more thorough observation and at one section, north of garrison G05, we excavated two trenches through the wall. The construction of the Gobi Wall exhibits sophisticated engineering adapted to the local geography and available resources. The wall is currently preserved at varying heights ranging from almost ground level to 2.6 m (Figure 5A), and its width (including deterioration debris) measures between 2.5 and 3 m. Three main building materials were used, depending on their geographical availability along the wall line: earth, stone and wood. In certain instances, unworked stone material was utilized for the exterior (northern) façade of the wall, while wooden elements constituted the interior (southern) façade, with compacted earth forming the structural core. In numerous locations, the sole discernible feature comprises slightly elevated earthen material and a linear arrangement of stones (Figure 5B). The original incorporation of wood elements in these wall sections remains indeterminate. The wall is much better preserved in the western part of the section we surveyed. The highest preserved section we identified, at coordinates 42.19775 N, 103.13390 E, is 2.60 m high and its construction made extensive use of wooden branches (Figure 5A). Small stones scattered north of it could indicate the use of stone to face the outer side, but without excavations, this was not clear.

**Figure 5.** Wall sections. (**A**) Wall section between G03 and G04 the highest wall section. (**B**) Deteriorated wall section near G07 (Height up to 0.5 m) revealing the stones used.

A wall section composed entirely of dark unworked stone was identified in proximity to garrison G03, at a location where the defensive structure traverses a prominent peak (Figure 6). This section, constructed primarily from local black basalt stones, represents a deliberate choice in routing, as topographic analysis indicates alternative pathways were available through adjacent plains (Figure 6B). The decision to construct over challenging terrain suggests this section served broader strategic purposes, likely as a visual deterrent and display of defensive engineering capability.

Excavations conducted along the wall alignment adjacent to G05 yielded substantive insights regarding the structural composition and constructional methodologies implemented at this particular locality. Figure 7 illustrates two test pits excavated in this area. Test pit a-a′ (Figure 7C) represents a cleaned section where a small stream had cut through the wall, while test pit b-b′ (Figure 7B) was positioned on top of the wall. At this location, the wall was preserved to approximately 2 m in height, with stone elements and wooden branches visible on the surface.

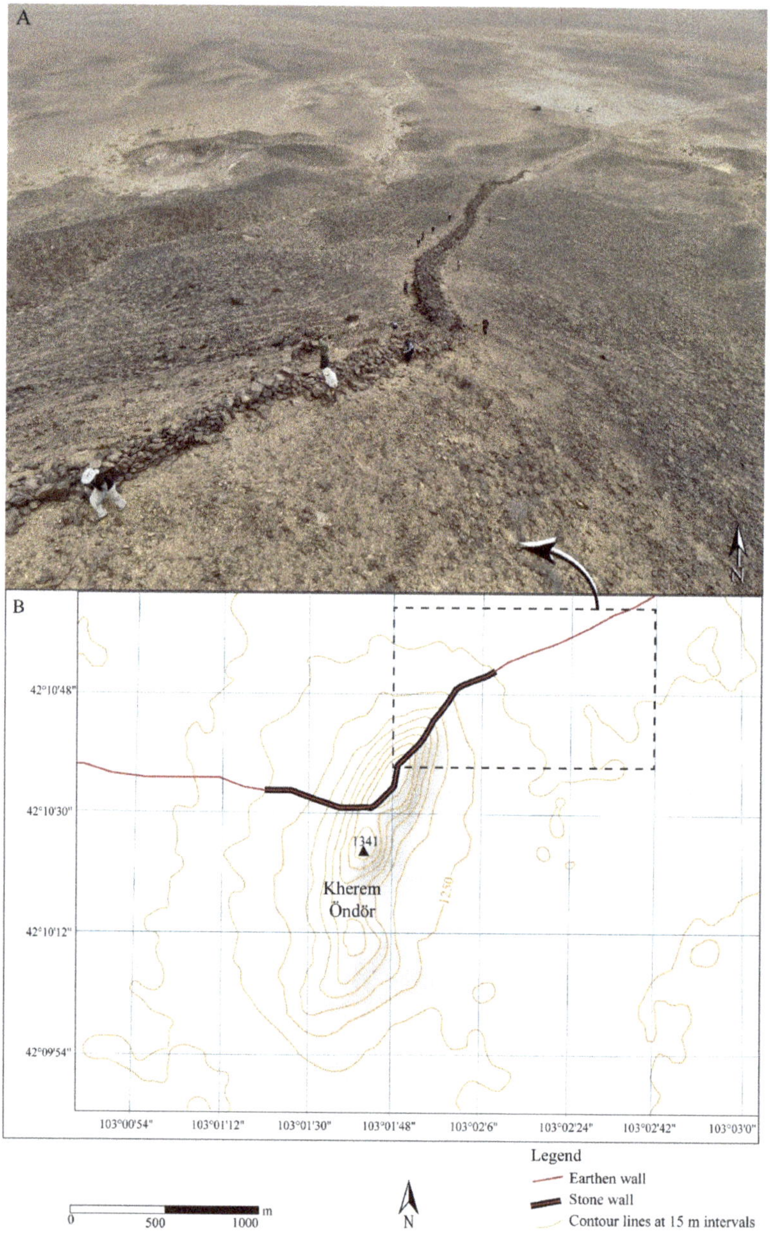


**Figure 6.** Black stone wall section. (**A**) Aerial drone photograph showing the preserved stone wall section traversing the hillside. (**B**) Topographic map illustrating the wall's strategic placement across the Kherem Öndör peak, with differentiated earthen (red line) and stone wall (black) segments.

The complete cross-sectional exposure revealed a structural core comprising compacted brown/reddish sediment interspersed with gray strata containing fragmented stone inclusions. The northern (exterior) face of the wall was reinforced with unworked stone. Between this stone facing and the compacted core, sandy material was utilized as fill. On the interior side, the earthen core was supported by a layer of wooden branches (Figure 7B). This multifaceted construction methodology, incorporating diverse materials and distinct construction phases, indicates deliberate engineering considerations to ensure structural integrity.

**Figure 7.** Wall excavation adjacent to G05. (**A**) Satellite imagery of the wall with superimposed contour lines, indicating the locations of cross-sectional excavations a-a′ and b-b′. (**B**) Cross-section b-b′, top view of excavation demonstrating the integration of stone and wood elements in the wall construction. (**C**) Cross-section a-a′ showing the stone exterior face, wood interior face, and stratigraphic layers of the wall.

In the area of b-b′ (Figure 7B), a 2 × 2.5 m test pit was established and excavated to a depth of 60 cm to expose the structure of the wall's upper portions. The exterior stone facing was clearly visible on the northern side of this square. The remaining width of the wall to the south consisted entirely of wooden branches with earth filling the interstices. This extensive utilization of wood appears to provide support for the entire upper section of the wall, a feature also observed in excavations of the enclosing walls at garrison G05.

The integrated use of multiple construction materials within individual wall segments represents a sophisticated engineering approach that maximized structural integrity while optimizing resource utilization. The typical configuration documented near garrison G05 featured an earthen core, stone facing on the northern (external) wall face, and wooden reinforcement on the southern (internal) face. This tripartite construction method enhanced both structural stability and defensive capabilities, with the stone exterior providing protection against erosion and assault, while the wooden reinforcement prevented slumping and collapse of the earthen core.

### 4.2. Garrison Structure

Prior to field investigations, rectangular architectural complexes were identified through remote sensing methodologies and preliminarily classified as garrison structures. During the field research, we visited 10 of those locations (G03–G12). Among those, at

G09 we did not identify enclosure walls. At the other nine locations, we surveyed and took drone photos. The perimeter walls enclosing all identified garrison structures exhibit rectangular configurations, accompanied by external ditches. The four corners are wider and higher than the rest of the wall, suggesting the existence of corner towers. Morphological depressions observed in the eastern or southeastern perimeter walls of several garrison structures indicate the probable location of gates. The garrisons varied in size between 4745 $m^2$ to a maximum of 14,560 $m^2$. Previous scholars have argued that garrisons' corners are oriented to the cardinal directions [11], but our analysis shows that their orientation varied between north −45 to North +45 (Figure 8). It is more likely that the exact orientation of each garrison was determined by their association with the nearby wall section as well as with the specific topographic conditions at the site of their construction.

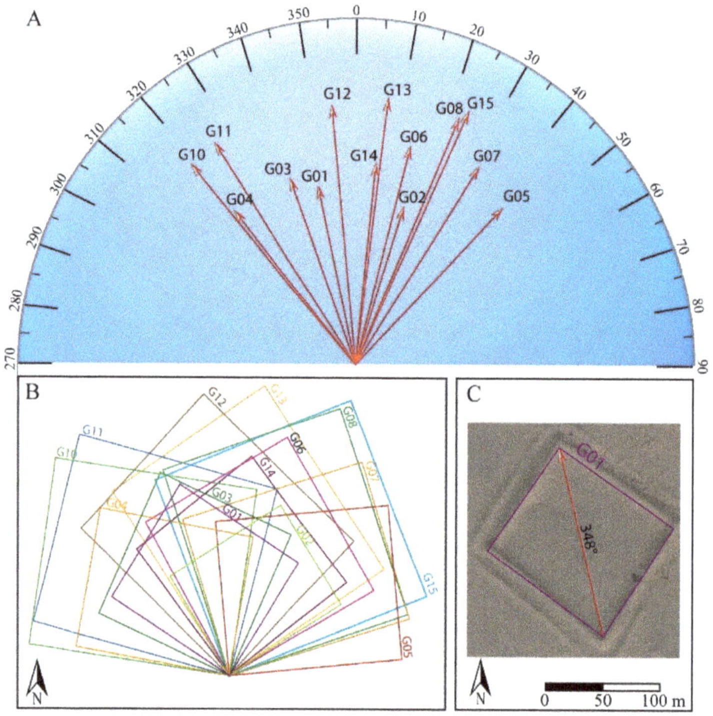


**Figure 8.** The orientations and sizes of garrisons. (**A**) Radial plot depicting garrisons' orientations. Red line lengths represent sizes of garrisons, while azimuths indicate orientation from the southernmost corner to its opposite. (**B**) Outline comparisons of garrison structures, size and orientation, colors are random for clarity. (**C**) Example of garrison G01, illustrating the method for determining orientation.

Among the garrisons which we did not survey, G14 stands out because of its location far from the wall line (Figure 2). It is possible that this garrison had a different function or perhaps it is of an earlier date.

The preservation conditions of garrison walls exhibited substantial variability across the survey area. Some, especially those located in the eastern part, are highly eroded, while others are preserved up to 2.5m high. Analogous to the wall lines discussed above, the construction of garrison enclosures demonstrates heterogeneity in both material and architecture. Both earthen walls and the combination of earth stone and wood construction were observed.

The predominant construction methodology employed was rammed earth (Chinese hangtu), a technique documented in China since the Neolithic period [33]. The importance of this method in the construction of city walls and fortification during the Medieval time period is corroborated by the Yingzao Fashi 'State Building Standards' compiled in 1100 CE by the Directorate of Buildings and Construction of the Song dynasty. This source describes the use of an implement with a wooden handle and a stone head (shi hang tou 石夯头) (Figure 9) to pound the thin earth layers [34]. The importance of the rammed earth technique was made clear by the stone head uncovered in our excavations at area C in garrison G10 (Figure 9C). This limestone artifact measures 9 cm in height with a diameter of 13 cm, featuring a rounded side and a central cavity several centimeters deep on the opposite broken face. An elongated piece of wood found close to it was, perhaps, part of the tool's wooden pole. Similar artifacts, though shaped in a slightly different fashion, were discovered in our excavations along the Northern Line (Figure 9B) and the Mongolian Arc [5].

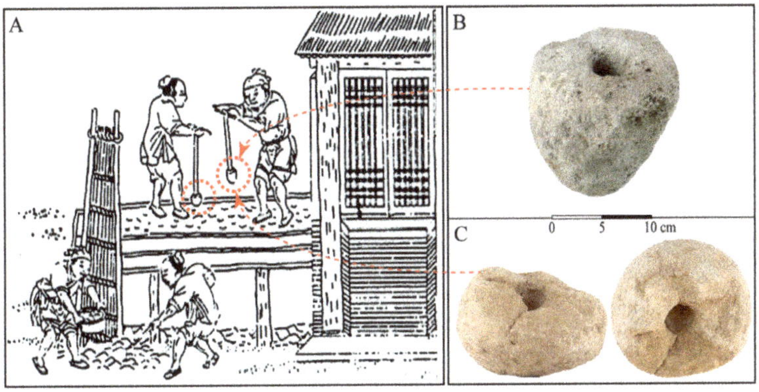


**Figure 9.** Stone-headed tool used for rammed earth wall construction. (**A**) Historical representation of construction of rammed earth wall using stone-headed tamping implement, adapted from Needham [35] (p. 39). (**B**) Stone head recovered from cluster 23 at the Northern Wall line. (**C**) Stone head recovered from garrison G10, area C, top and side views. Red circles highlight the stone head component of each tool.

Observations and excavations at multiple garrison sites revealed substantial evidence for the utilization of wood materials in perimeter wall construction. This was clear from our survey at G06 (Figure 10) where surface evidence suggests that the garrison's walls were made of an outer and inner wood structure (identified as Saxaul) and an internal earthen core, probably of rammed earth.

Wooden stake elements (identified as Poplar), measuring approximately 20–30 cm in length (Figure 11), were identified along the perimeter of the G03 garrison enclosure at approximately mid-elevation within the wall structure, positioned at regular 3 m intervals. Each stake exhibits a sharpened terminus that was inserted into the wall and a mechanically fragmented superior end, consistent with damage resulting from percussive force. The function of those stakes is unclear, but it is suggested that they marked plans to expand or strengthen the garrison's walls.

Small-scale excavations were conducted at two garrisons: G05 and G10 (Figure 12). The goal of our excavations was mainly to recover samples for dating and analysis as well as to gain some insights on the structure of the garrisons' walls and on the occupation inside them. Among other things, these excavations confirmed our survey observations regarding the different construction techniques used at different garrisons. At G10, our excavation of the northeastern corner tower at area (B) revealed the use of the rammed earth construction method. A few wooden beams found in the excavations may have been part of the upper structure built above.

**Figure 10.** The wall surrounding garrison G06. (**A**) enclosure wall depicting wood use on both sides. (**B**) exposed section of the interior part of the enclosure wall, depicting extensive use of wood.

**Figure 11.** Archeological specimens of wooden stakes recovered from garrison G03, exhibiting consistent morphological characteristics.

**Figure 12.** Excavation units. (**A**) Excavation units at garrison G05: (A) corner tower base (B) elevated area (C) two units excavated on and through the long wall. (**B**) Excavation units at garrison G10: (A, C, D) areas with concentrations of artifacts (B) Corner tower base.

At G05, excavations of the southeastern tower at area (A), revealed extensive use of wood in the construction (Figure 13). The branches used here were thicker (4 to 15 cm in diameter) than those excavated in the long wall near this garrison at excavation area (C). They were tightly aligned one next to the other, generally in an east–west orientation, and were embedded in red/brown and green sediments. These sediments are different from the sediments currently seen on the surface of the site, suggesting that they were specifically selected for the construction of the garrison's walls. A small test pit excavated inside the wall of G05 at area (B), did not reveal evidence for structures or activity. At G10, excavations of area (C) uncovered a garbage pit (or midden) with a relatively dense concentration of animal bones and charcoal. In area (D), excavated on a slightly elevated ground east of area (C), occupational remains were uncovered, including at least two hearths, one of them equipped with three standing stones possibly to support the placement of cooking pots (Figure 14).

**Figure 13.** Excavation at G05 at area (A). Large branches exposed at the elevated garrison corner. (Photo: M. Ullman, team archaeologist).

**Figure 14.** A hearth with three standing stones, and charcoal in between them at G10, area (D) (Photo: M. Ullman, team archaeologist).

Structural evidence of permanent habitation architecture was not identified, potentially indicating that garrison occupants utilized impermanent dwellings such as Gers (nomadic tents also known as Yurts). However, the limited spatial extent of the excavation areas precludes definitive conclusions regarding the type of dwelling used at this and other garrisons. Our excavations of structures along the Northern wall line and the Mongolian Arc wall line did find evidence for permanent habitation structures made of wood and stones [9,36] thus suggesting the possibility that similar structures may have existed at garrisons of the Gobi Wall as well.

Excavations and surface collections at various garrisons yielded ceramic assemblages of differing densities (Figure 15). The recovered materials are predominantly high-fired ceramic fragments, typically in light-colored or greyware forms. Several examples feature black-glazed surfaces, a hallmark of Tangut (Xi Xia) material culture [36]. Notably, Artifacts 4, 5, 10, and 11 in Figure 15 exhibit stylistic traits characteristic of Xi Xia-period ceramics, most likely produced at the Lingwu Kiln located within the Xi Xia territory about 500 km south of the Gobi Wall or the Hui Lane Kiln which is also located within the Xi Xia territory [37–39].

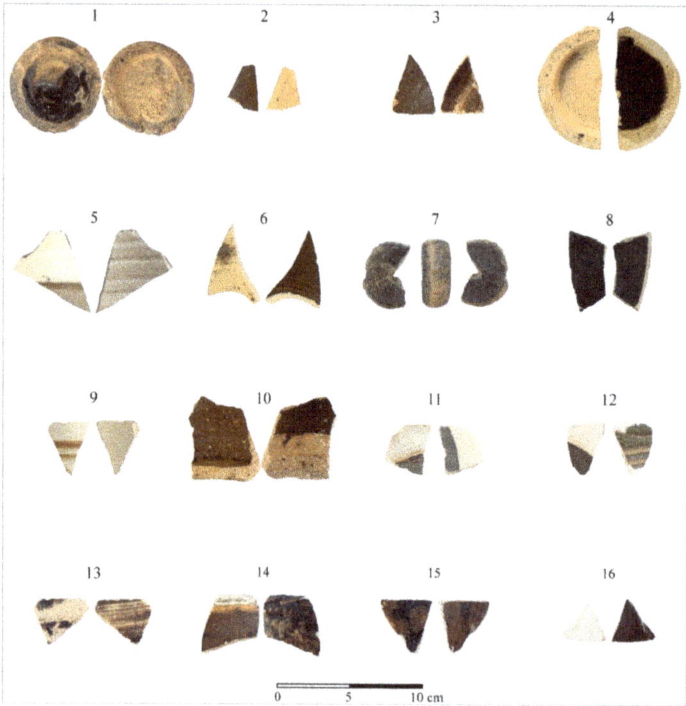


**Figure 15.** Selected shards found, face and back image: 1 at G03, 2 and 3 at G04, 4 and 5 at G05, 6 and 7 at G06, 8 and 9 at G07, 10 at G08, 11–16 at G10.

Metallurgical artifacts were recovered through systematic pedestrian surveys and notably, through metal detection survey protocols implemented at garrisons G05 and G10 (Figure 16). The assemblage includes different types of artifacts, including coins, projectile points, architectural fasteners, fragmentary wheel hubs for carts, and utilitarian implements. Particularly dense concentrations of metal artifacts were uncovered by our metal detector survey of G10.

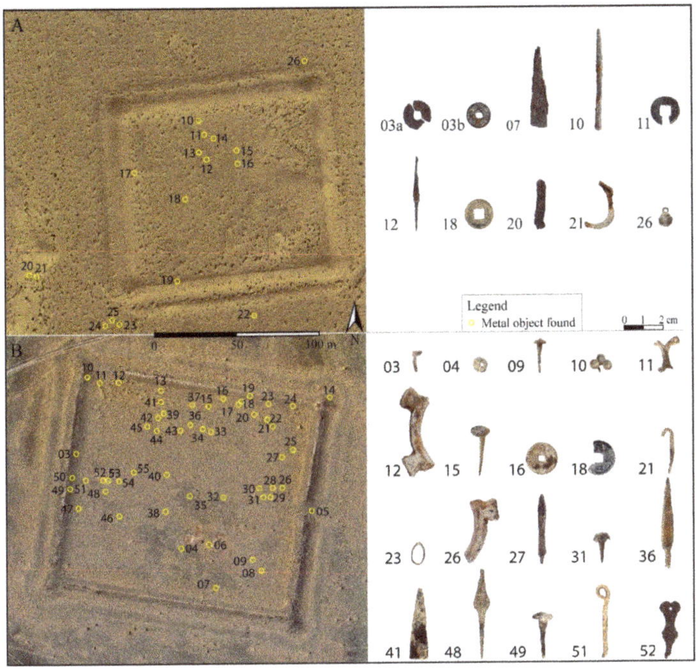


**Figure 16.** Spatial distribution of metal objects found, with photos of selected items. (**A**) At garrison G05. (**B**) At garrison G10.

### *4.3. Mountain Pass Fortifications and Strategic Points*

Three elevated fortifications, initially documented by Kovalev and Erdenebaatar [12], were incorporated into the remote sensing survey protocol and labeled F41-F43. Among them, we surveyed two fortifications—F41 and F42. The two are small oval-shaped structures located on high ground near a mountain pass. F41, situated north of the wall near Kherem Öndör peak 1310 m ASL (Figure 17). This site has internal dimensions of 13 × 4.5 m, and is entirely built of stones. On the inside the walls are preserved to c. 2 m high but they are much higher on the outside. The view from the site is more than 180° to the Northwest and South and they seemed to function as observational posts over the entrance to the mountain pass. No artifacts were found inside or around F41 and no materials for dating were recovered.

F42, positioned south of the wall at 1311 m ASL is situated on the western side of a dry river that cut through the mountain ridge at a position that controls the surrounding landscape (Figure 18C). Internally it measures 14 × 7 m, unlike F41 the internal and external faces of the walls are visible on the drone image. The wall is some 2 m wide and stands about 3m high at the highest places. Unlike the walls of F41, relatively large wooden beams were integrated inside the unworked stone construction of the walls of F42 (Figure 18B). Here, as well, no artifacts were found but samples of wood were obtained for dating.

An additional mountain pass traversing the frontier perimeter was identified through survey methodologies and Geographic Information System (GIS) analyses. This topographical feature is situated in proximity to the location initially designated as G09; however, structural remnants of a garrison installation were not detected at this site (Figure 19). The geographical positioning of this location demonstrates significant strategic value, being situated at a mountain pass where multiple traversable pathways converge before subsequently diverging outward. Soviet maps indicate a demarcated trail extending southwestward from the G09 position toward the subsequent mountain pass at fortification F42. Based on topographical and strategic analyses, the G09 location appears to represent a point of defensive vulnerability along the fortification perimeter.

**Figure 17.** Drone photo of the mountain pass and fortification F41.

**Figure 18.** Mountain pass fortifications. (**A**) Wood integrated into the stone construction of fortification F42. (**B**) Drone image of fortification F42 overlooking the entrance to the mountain pass. (**C**) Drone image of fortification F42 from above.

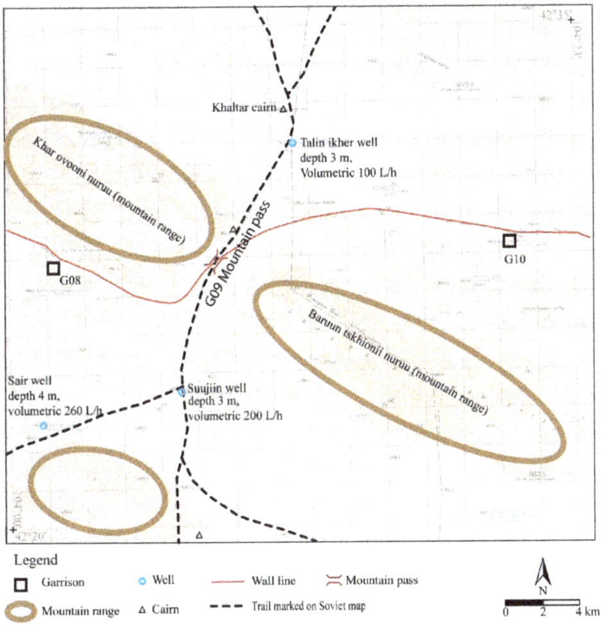


**Figure 19.** Annotated Soviet topographic map (section from sheet K-48-53) depicting topographical and anthropogenic features at the G09 mountain pass.

## 5. Chronology

While ceramics found at some of the garrisons suggested that they were used during the Xi Xia period (Figure 15), coins provide additional concrete information.

### *5.1. Coins*

Coins are an important source for dating sites because they often have a narrow chronological range of minting dates. However, the range of their circulation and use time is sometimes very long, so we should not treat them as absolute chronological markers. Our expedition recovered nine coins at three garrisons: four at G05, three at G06, and two at G10 (Figure 20).

Interestingly, the chronological range of the assemblage spans approximately two millennia, from the Han dynasty to the Qing dynasty. This is a good indication of the complex history of this region, but it cannot, by itself, help us identify the time of construction and use of the Gobi Wall system. Our identification and dating of the coins are based on collections of Chinese coins [29,30]. Additional sources were used for Xi Xia coins, which are of special interest [31,32].

The earliest numismatic evidence is two Han dynasty wuzhu (五銖) coins, both found at G05 (Figure 20—6 and 7). These are common Han period coins, which were first minted in 118 BCE but continued to be minted and used during the Eastern Han period (25–220 CE) and even as late as 621 CE. However, based on the style of the two coins, particularly because of the four extended lines from the square on the reverse side, we estimate that they were minted during the time of Emperor Lingdi (R. 157–189 CE) of the Eastern Han dynasty. However, the discovery of Han coins in other, clearly dated Xi Xia sites, suggests that such coins could have been in circulation for a much longer time and may have been used by the Tangut [31] (pp. 88–109).

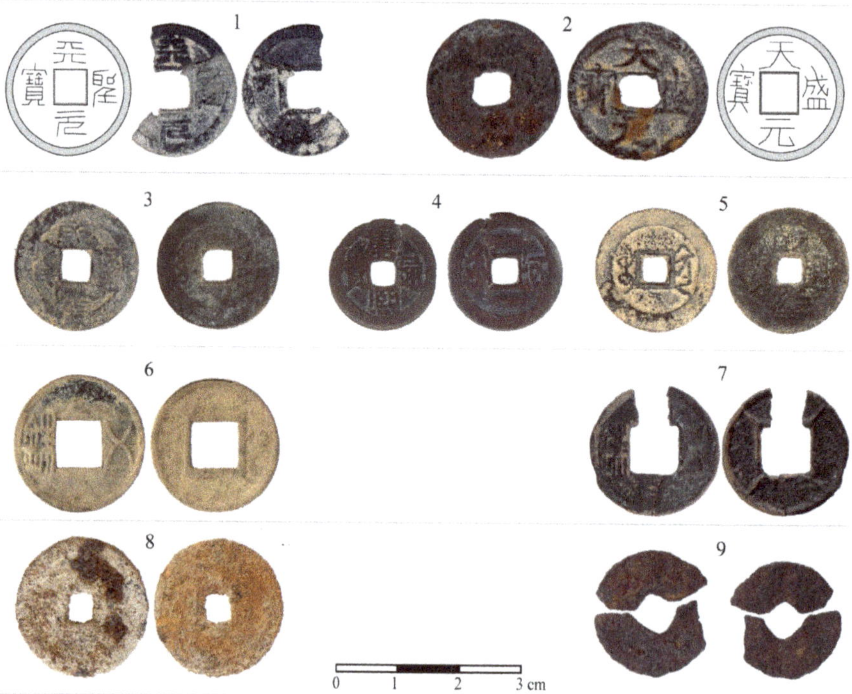

**Figure 20.** Archeological assemblage of nine recovered coins, obverse and reverse: (1) Song dynasty Tiansheng (1023–1032) coin with a reconstruction image; (2) Xi Xia Tiansheng (1149–1169) coin with a reconstruction image; (3, 4 and 5) Qing dynasty coins from the Kangxi (4) and Qianlong eras (3 and 5); (6 and 7) Han dynasty Wuzhu coins; (8 and 9) unidentified iron coins, probably Manchu coins.

The three Qing period coins were all found at G06. Those include one Kangxi Tongbao (康熙通寶) (Figure 20—4), dated to the time of the Kangxi Emperor (1654–1722). The Chinese character he (河) and Manchu word ho (ᡥᠣ), inscribed on the revers, show that it was minted in Kaifeng, Henan. Another two coins discovered in the same garrison are Qianlong Tongbao coins (乾隆通宝) (Figure 20—3 and 5) dated to the time of the Qianlong Emperor (1711–1799). On the reverse of the first coin (no. 3), the mint mark is crusted and hard to read; our tentative reading of the Manchu words (in Mongol script) is boo jin, (ᠪᠣᠣ ᠵᡳᠨ). If this mint mark is correct, then the coin was minted in Taiyuan, Shanxi. The mint mark on the second coin (no. 5) reads boo giyan (ᠪᠣᠣ ᡤᡳᠶᠠᠨ), suggesting that it was minted in Guizhou, southern China. The presence of Qing period coins in this region is by itself not surprising because it was part of the Qing territory. This by itself is not evidence that the Qing constructed or used any of the garrisons.

A Northern Song dynasty Tiansheng (天聖) coin was found in G10 (Figure 20—1). It is dated to the Tiansheng period (1023–1032) but such coins were used long after the fall of the Northern Song Dynasty including by the Xi Xia dynasty. In fact, Song coins were the main currency circulating in Xi Xia, even more numerous than coins minted by the Xi Xia state [31] (pp. 88–109), [32] (p. 58). Direct evidence for Xi Xia activity is a Xi Xia minted Tiansheng (天盛) coin dated to 1149–1169 CE (Figure 20—2) which we uncovered in G05. During the first century of the Xi Xia dynasty, the polity barely minted coins, hence the extensive use of Northern Song coins. Coins such as the ones we found, which were minted during the second half of the 12th century, are the most common Xi Xia coins (minted both in bronze and iron).

### 5.2. Radiocarbon Dating Results

The attribution of the Gobi Wall to the Xi Xia state, initially posited by Kovalev and Erdenebaatar [11,12], was predicated upon radiocarbon analysis conducted at the Institute for the History of Material Culture (IHMC RAS) in St. Petersburg during 2007–2009. Wood samples collected during a 2005 survey established a preliminary chronological framework spanning the 11th to 14th centuries CE. Although Kovalev and Erdenebaatar interpreted selected results as evidence that the wall line and associated features were constructed by the Xi Xia dynasty, the variability of the radiocarbon dates necessitates further verification. To refine the chronological assessment of the Gobi wall line, twelve samples were collected from diverse archeological contexts—the wall line itself, garrison walls, anthropogenic strata within garrisons, and a hilltop fortification—comprising five discrete locations (G03, G05, G06, G10, F42). This methodological approach was implemented to establish a more thorough and systematic sampling across multiple structural features and depositional contexts. A botanical assessment of the wood samples identified almost all woods as black saxaul with the one exception of the wooden stakes collected from the G03 fortification wall which are identified as poplar (see below for details). As of yet, we have no definitive analysis for the charcoal samples although carbonized saxaul is likely.

AMS analyses were conducted on wood (n = 7) and charcoal (n = 5) and full results are provided in Table 2. Comparing across sigma-2 probability ranges, the majority of dates fall between the mid-11th and early 13th century CE, although there are two dates clearly earlier than the rest (UCIAMS-1354 and 1357). Both early dates are on saxaul wood and may exhibit an old wood effect; however, upon reflection, it is probable that all wood and charcoal samples collected do not correspond exactly to the period of human use and deposition that we intend to date. Black saxaul has an average lifespan of 45–50 years [40,41], poplar can live up to 100+ years [42], and even older relict pieces of wood could have been obtained from the local ground surface. As such, these samples are likely characterized by 'inbuilt age' as discussed by Dee and Bronk Ramsey [43] and can be evaluated using an exponential probability distribution as implemented by the Exponential Outlier method in OxCal Bayesian modeling software [44]. Underlying this approach is the assumption that a majority of radiocarbon results on old wood and charcoal samples will date near to the period of human use, while a smaller number will be moderately older than the target period, and fewer still may be much older than the time of human use and deposition [43] (pp. 83–85).

To evaluate the start and end dates, as well as the span of use for the Gobi wall line, we modeled each of our twelve dates using the Exponential Outlier method in OxCal 4.4 (see Supplementary Materials for code and details). We obtained posterior results as shown in Figure 21 with modeled 68% and 95% Credible Intervals (CI) for each individual date as well as for start and end sequence boundaries. Given the historical period in question, the narrower 68% CI ranges will be discussed, although complete modeling results can be found in the Supplementary Materials. Based on available dates and our modeling approach, the start date for the construction of the wall line and associated garrison settlements falls between 971 and 1043 CE (68% CI) and the end date falls between 1206 and 1252 CE (68% CI). The span of occupation and use along the South Gobi wall line is 181 to 270 years long (68% CI, Figure 22). Given that the historical date for the rise in the Xi Xia dynasty is 1038 CE and the Mongol destruction of the Xia state occurred in 1227 CE, these radiocarbon dates and our Bayesian modeling lend support to the hypothesis that the Gobi wall was a Xi Xia frontier construction. It is important to note that the Tangut state existed prior to its constitution as a Chinese style dynasty, although the height of its territorial expansion primarily occurred during the early to mid-11th century CE. Therefore, an intensive Xi Xia occupation of the Gobi region at this time would be in keeping with our understanding of the historical record.

**Table 2.** AMS $^{14}C$ analysis of samples excavated from the South Gobi wall line and garrisons (OxCal 4.4, IntCal20 software [44,45].

| Lab No. | Sample Type | Site and Context | $^{14}C$ Age (BP) | Error | 2 Sigma (CE) |
|---|---|---|---|---|---|
| UCIAMS 1354 | Wood, saxaul | F42, Hilltop fortification | 1040 | 15 | 992–1025 (95.4%) |
| UCIAMS 1358 | Wood, saxaul | Garrison G05, Area C wall line | 920 | 15 | 1041–1109 (59.6%)<br>1114–1174 (35.8%) |
| UCIAMS 1353 | Wood, saxaul | Garrison G05, Area C wall line | 850 | 15 | 1164–1228 (95.4%) |
| UCIAMS 1357 | Wood, saxaul | Garrison G05, Area A garrison corner | 1050 | 15 | 988–1026 (95.4%) |
| UCIAMS 1359 | Charcoal | Garrison G05, Area A garrison corner | 955 | 15 | 1031–1054 (18.2%)<br>1075–1158 (77.3%) |
| UCIAMS 1364 | Charcoal | Garrison G10, Area D | 950 | 15 | 1033–1054 (16.1%)<br>1064–1068 (01.1%)<br>1073–1158 (78.2%) |
| UCIAMS 1363 | Charcoal | Garrison G10, Area D | 885 | 15 | 1053–1073 (05.4%)<br>1156–1219 (90.1%) |
| UCIAMS 1360 | Wood, saxaul | Garrison G10, Area D | 925 | 15 | 1040–1165 (95.4%) |
| UCIAMS 1361 | Charcoal | Garrison G10, Area C | 890 | 15 | 1053–1076 (12.3%)<br>1156–1216 (83.2%) |
| UCIAMS 1362 | Charcoal | Garrison G10, Area C | 870 | 15 | 1162–1219 (95.4%) |
| UCIAMS 1355 | Wood, saxaul | Garrison G06 | 925 | 15 | 1040–1165 (95.4%) |
| UCIAMS 1356 | Wood, poplar | Garrison G03 | 855 | 15 | 1165–1224 (95.4%) |

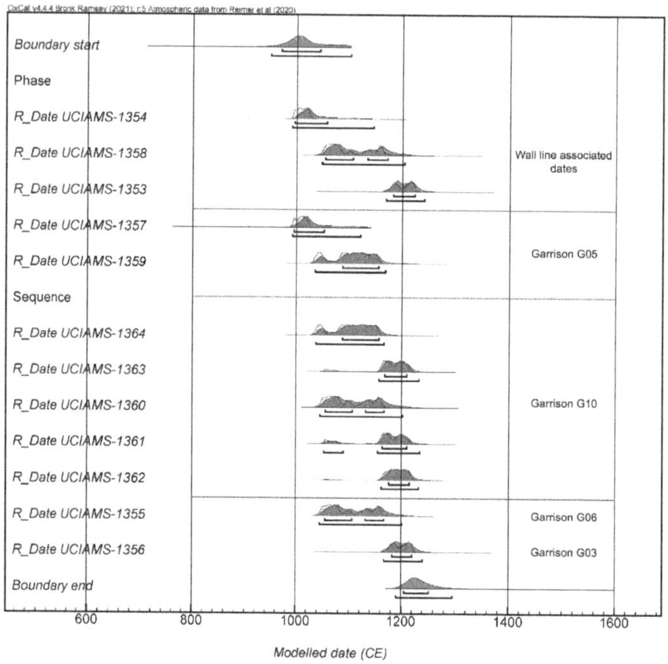


**Figure 21.** Bayesian modeled radiocarbon dates and boundaries using the Exponential Outlier method [44–46].

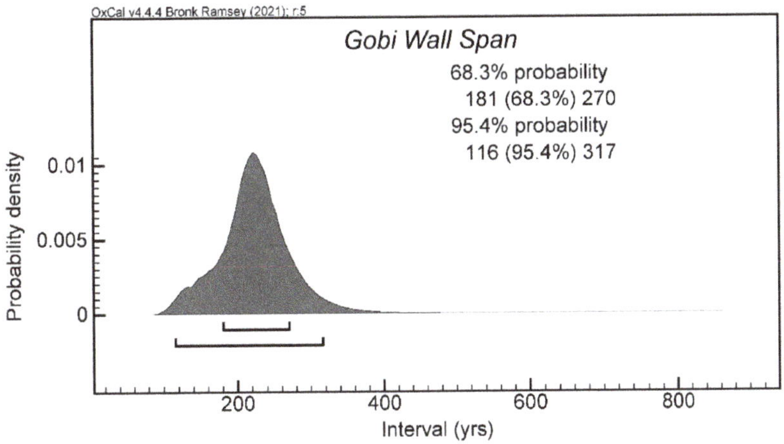


**Figure 22.** Timespan estimates for the Gobi wall line and associated structures [44].

## 6. Analysis and Discussion

### 6.1. Topographical and Ecological Analysis

Our geographic and environmental analysis suggest that the route of the Gobi Wall was determined by a combination of political, topographic, and ecological constraints, in addition to the functional requirements for which it was constructed. One important geographic element that the wall builders had to take into consideration was the morphology of the Altai Mountain terminal ridge. The shape and location of this ridge directly influences the characteristic shape of the wall line. Elevation analysis reveals that the highest point occurs between garrisons G08 and G10, with gradual slopes descending toward the piedmont on either side (Figure 23). This topographical configuration suggests strategic utilization of natural elevation advantages, maximizing defensive capabilities while minimizing construction requirements.

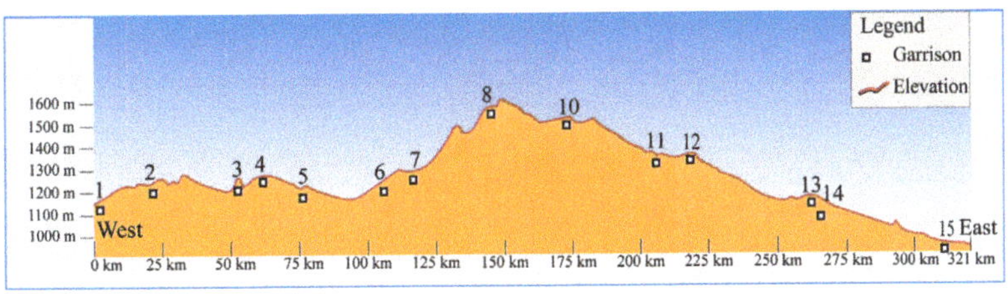


**Figure 23.** Topographic elevation profile along the Gobi Wall system, indicating the location of garrisons.

Analysis indicates that wall architects deliberately avoided areas with substantial dune coverage, while strategically aligning numerous segments—particularly in the eastern sector—just south of these sandy formations (Figure 24). Such positioning not only facilitated construction on firmer terrain but may also have enhanced defensive efficacy by integrating dune fields as natural barriers. Establishing the wall system primarily south of the mapped dune belts likely increased the difficulty of approach from the north. This configuration suggests that the design optimized both environmental adaptation and defensive performance. However, the analysis is constrained by the dynamic nature of dunes; Late Holocene remobilization episodes in southern Mongolia are well documented [26], and localized dune movement could have occurred after the initial construction of the wall.

While major dune systems often exhibit spatial persistence over centuries [25]. Our use of Soviet-era topographic maps as proxies assumes relative stability in dune extent over the past millennium. Future studies incorporating satellite image time series (e.g., Corona, Landsat) or wind regime modeling may offer further insight into the temporal variability of dune positions in relation to the ancient frontier system. Still, the apparent alignment of the wall with the southern fringe of major dune systems reflects a resource-efficient strategy that capitalized on existing geomorphic barriers [47].

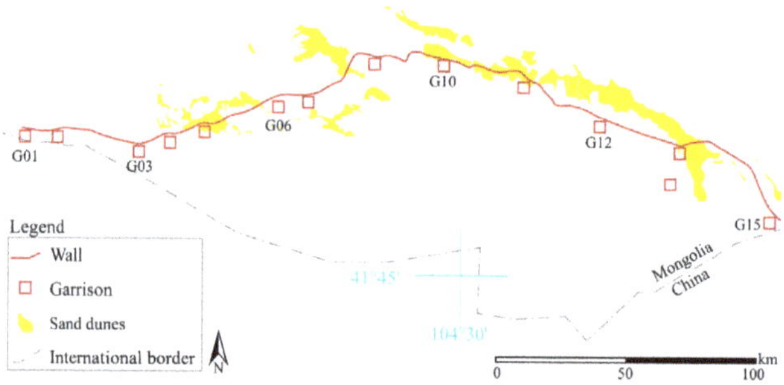


**Figure 24.** Position of sand dunes marked on Soviet maps, vis-à-vis the wall line.

Other ecological factors that potentially influenced the wall route and garrison placement include availability of construction materials, predominantly wood and stone, and access to critical resources, particularly water, which represents the most crucial element in arid desert environments. To evaluate water availability along the Gobi Wall and near its associated garrisons, Soviet-era topographic maps were analyzed, focusing on well distribution, depth, and discharge rates. A total of 439 wells were digitized across the region, with volumetric and depth data available for 60 wells documented on 20 map sheets containing visible wall remains. These wells typically range in depth from 1 to 4 m (Mean = 2.0 m $\pm$ 0.6 m) and their discharge rates span from 50 to 500 L/h (Mean = 169 $\pm$ 100 L/h).

For spatial analysis, a defined "wall buffer zone" was established, spanning 8 km both north and south of the wall's route (8 km is about 2 h walk from the wall to a well). This buffer zone creates a 640 km$^2$ buffer zone within each map sheet containing 1600 km$^2$. A clear spatial pattern emerges from this analysis: wells are densely concentrated within and south of the corridor, while their presence north of the wall is markedly sparse. This pattern is exemplified on map sheet K48-053 (Figure 25), which shows a pronounced disparity in well distribution.

Although the documented wells do not necessarily date back to the medieval period, their locations highlight zones where shallow aquifers are accessible, thus revealing hydrological features likely present during the wall's construction. Both topographical prominence and the existence of seasonal waterways mark the presence of reliable water availability. The alignment of the wall with this aquifer system suggests a deliberate strategy: selecting a route that ensures reliable water access for defensive forces while simultaneously limiting access for potential adversaries approaching from the drier northern side. Additional analysis of 11 wells positioned near garrisons along the full length of the wall—each with complete depth and discharge data—reinforces this pattern (Figure 26). Most of these wells are about 2 m deep and yield between 200 and 400 L/h, supporting the conclusion that the wall was intentionally routed through areas with high groundwater availability.

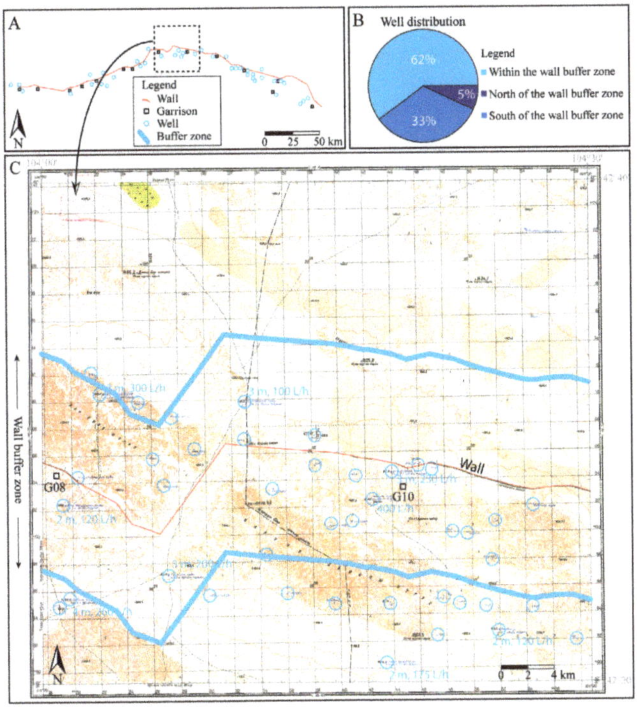


**Figure 25.** Spatial and quantitative analysis of well distribution along the Gobi Wall buffer zone using Soviet map sheet K48-053. (**A**) Location of K48-053 within the Gobi Wall area. (**B**) Statistical distribution of well locations. (**C**) Annotated map with well positions and associated data.

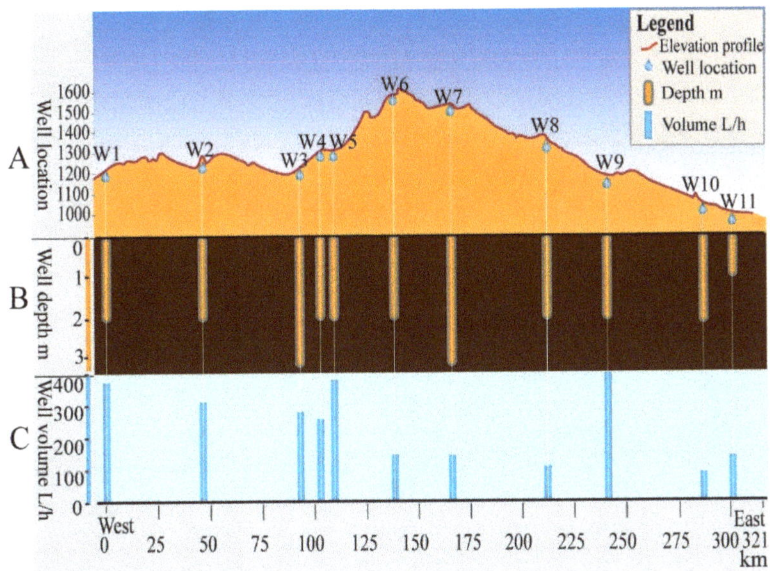


**Figure 26.** Analysis of 11 selected wells near Gobi Wall garrisons. (**A**) Spatial distribution of wells (W1–W11). (**B**) Well depths (meters). (**C**) Water discharge rates (L/h).

These findings point to consistent patterns of groundwater accessibility along the wall. The underlying hydrological system consists of a shallow streambed aquifer formed in permeable Quaternary alluvium, with water tables ranging from 0.5 to 7 m below ground. Groundwater is primarily of the $Na(HCO_3)$ type, indicating relatively young, recently recharged water. This recharge occurs mainly through the topographical prominence of the region, where geomorphological features effectively capture and channel the limited precipitation into the groundwater system [48–50]. Despite minimal rainfall in the Gobi region, these distinctive landforms facilitate reliable infiltration through the sandy substrate into the underlying water table. Field investigations have demonstrated that sufficient groundwater reserves are maintained throughout the year, thereby enabling garrison personnel to sustain a continuous presence in the wall region.

Two distinct woody species were identified in archeological samples, *Haloxylon ammodendron* (Saxaul) and *Populus euphratica* (Poplar) [51] (p. 126). Analysis of the wall-line route in relation to Poplar (*Populus euphratica*) tree distribution (Figure 27) reveals that the wall is positioned mainly at the northern periphery of local poplar tree populations. The deliberate positioning of the wall-line in relation to the poplar suggests a calculated assessment of subsurface hydrological conditions, as regions supporting *Populus euphratica* growth inherently indicate the presence of accessible groundwater resources. Poplar exhibits specific hydrological requirements that provide valuable archeological insights. As an obligate phreatophyte, this species relies on permanent groundwater access, typically requiring water table depths between 2 and 3 m for optimal growth [52]. The presence of poplar wood in the defensive structures provides important environmental context for understanding site selection and resource utilization.

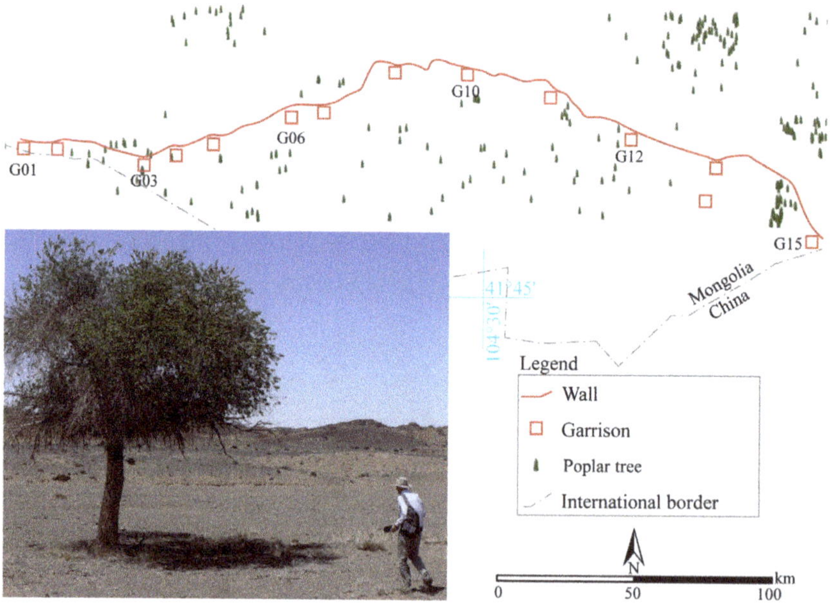


**Figure 27.** Poplar individual trees marked on Soviet maps, with a field photo of a Poplar tree near garrison G12.

The distribution of Saxaul shrubs demonstrates a strong correlation with construction material utilization. Garrisons G05, G06, and G07 are situated in areas where Saxaul shrubs currently exhibit the highest density (Figure 28). These locations precisely correspond to where the most substantial evidence has been collected regarding the use of wooden branches in the construction of both the wall line and the enclosing walls of the garrisons. This correlation suggests deliberate resource proximity planning, minimizing transportation logistics while ensuring material availability for construction and maintenance operations.

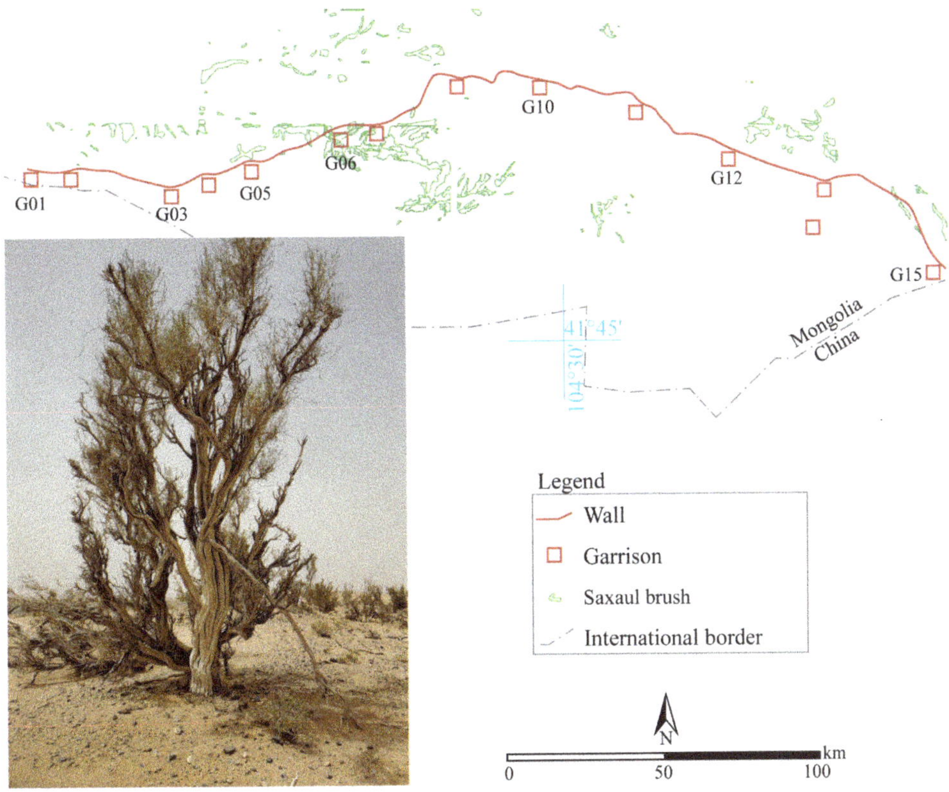


**Figure 28.** Saxaul shrubs polygons marked on Soviet maps, with field photo of Saxaul from garrison G06 (Photo D. Wolin, team archaeologist).

### *6.2. Functional Analysis of the Defensive System*

The Gobi Wall system is a combination of several different kinds of structures all meticulously integrated with features of the local landscape. To better understand the way the wall system functioned we focused our analysis on the western section of the wall line where many of those structural elements come together in a relatively small area (Figures 29 and 30). The interconnected defensive infrastructure in this area includes fortification F41, which guards a mountain pass north of the wall-line; the stone wall traversing Kherem Öndör peak, garrison G03; and fortifications F42, positioned to defend a southern mountain pass. The collective presence of these defensive structures provides evidence that this location constituted a strategically significant point along the defensive perimeter. The route delineated on the map, connecting F41 and F42, served as a critical corridor for human mobility, diplomatic exchanges and commercial activities during periods of peace. However, this same area simultaneously presented a potential infiltration route for hostile military forces seeking to penetrate Xi Xia territorial boundaries.

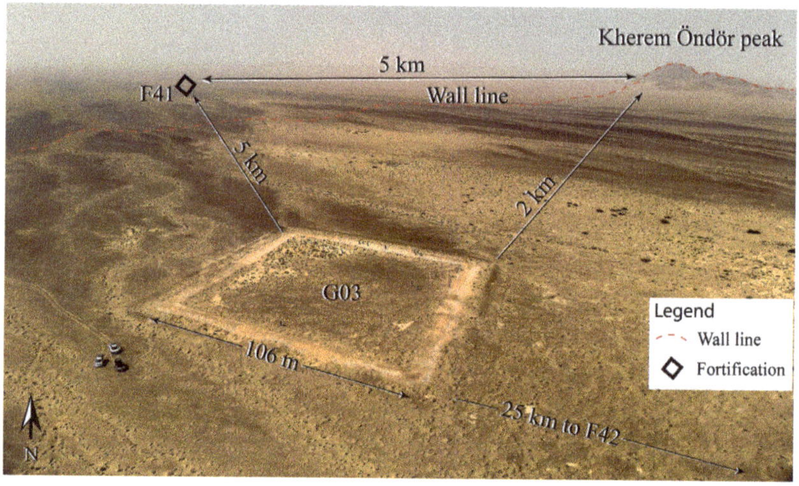


**Figure 29.** Drone documentation of garrison G03, illustrating spatial relationships with Kherem Öndör Mountain and fortification F41 within the integrated defensive network.

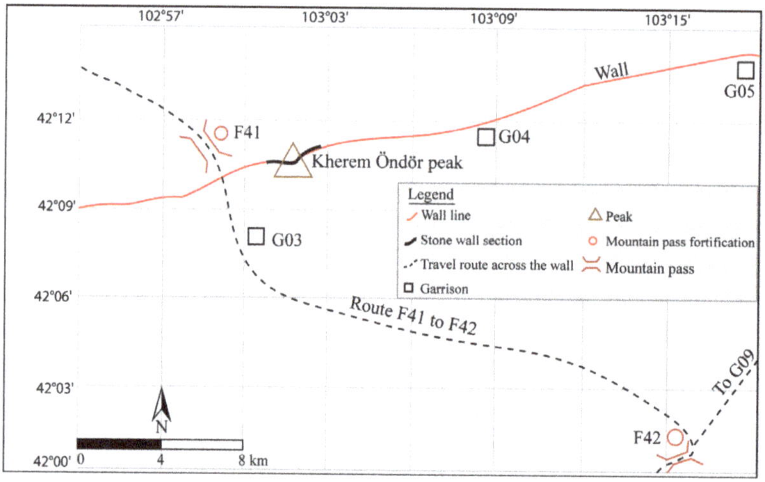


**Figure 30.** Schematic map depicting the strategic positioning of fortified elements along a critical wall crossing.

Garrison G03 (Figure 29) was a pivotal element within a systematically integrated defensive network, functioning in strategic conjunction with fortifications F41 and F42, both of which guard two critical mountain passes along a route traversing the wall. This configuration establishes a defensive perimeter that maximizes territorial control while optimizing resource allocation. The complementary positioning of these elements demonstrates sophisticated military engineering principles, wherein each installation's defensive capabilities are enhanced through spatial relationships with the others, evidencing complex strategic planning designed to control movement through topographically defined transit corridors.

The Gobi Wall system represents a multifunctional frontier infrastructure that transcends purely military applications. Analysis of its spatial organization, architectural features, and associated artifacts reveals a complex system designed for territorial demarcation, resource management, movement control, and administrative projection of imperial authority. The wall's alignment along the northern margin of shallow aquifers, as demonstrated through well distribution analysis, indicates a strategic design that prioritized sustainable occupation rather than absolute defensive positioning. The wells documented on Soviet maps suggest a deliberate selection of terrain that could support long-term garrison occupation as documented in similar frontiers [53].

The arrangement of garrisons along the wall line reveals a systematic approach to spatial organization. This systematic distributional pattern optimized territorial control while acknowledging logistical constraints, creating an interconnected network of administrative nodes capable of monitoring movement across the frontier zone. The internal spatial organization of garrison sites, particularly the evidence of habitation and military activities at G10, suggests these installations functioned as permanent settlements rather than temporary military outposts.

The material assemblage recovered from garrison sites provides insight into their functional role. The diversity of artifacts, including utilitarian ceramics as well as more expensive wares, and imperial coinage, indicates multifaceted activities encompassing military, administrative, and economic functions. The recovery of Northern Song and Xi Xia coins at garrisons G05 and G10 suggests integration with broader economic networks and administrative systems. This material evidence aligns with historical documentation in the Tiansheng legal code, which describes frontier installations as nodes in an imperial administrative network responsible for regulating trade, collecting customs duties, and monitoring population movements [54] (pp. 211–216).

The strategic positioning of mountain fortifications F41 and F42 at critical passage points provides evidence for a sophisticated border management system. These small fortifications, situated to monitor key mountain passes, facilitated controlled movement across the frontier rather than hermetic sealing of the border. This archeological evidence corresponds with historical documentation indicating that Xi Xia frontier policy regulated rather than prohibited cross-border movement, maintaining designated crossing points for diplomatic and commercial traffic while restricting unauthorized passage [55] (pp. 117–123).

### 6.3. Historical Context and Implications

The spatiotemporal distribution of archeological features across the Gobi Wall system reveals significant patterns in construction, occupation, and defensive strategies throughout the Xi Xia period. Bayesian analysis of radiocarbon dating from garrison sites suggests the frontier system originated in the early to mid-11th century CE during the expansion under emperor Yuanhao (r. 1038–1048) and his immediate successors, who sought to secure newly claimed territories against competing polities. While our sample cannot determine if the entire wall system was constructed during this early phase, evidence indicates most elements existed by the mid-12th century—a period of intensified conflicts between Xi Xia and neighboring Jin and Song dynasties [56] (p. 50).

Particularly notable are the continuous charcoal deposits at garrison G10 (Figure 21 and Table 2), which provide compelling evidence for uninterrupted occupation throughout the Xi Xia period. This settlement stability, despite harsh desert conditions, challenges prevailing assumptions that nomadic populations in this region were highly mobile, lacked permanent structures, and did not invest in fixed territorial markers or infrastructure [57]. Instead, the evidence suggests substantial investment in territorial control. For example, stakes recovered from garrison G03 may indicate repair activities, pointing to ongoing maintenance of fortifications. These finds may represent anticipatory measures preceding documented invasion threats, including the confrontations with Mongols in the early 13th century [58].

The discovery of a Xi Xia-minted Tiansheng coin dated to 1149–1169 CE at garrison G05 provides direct numismatic evidence of occupation during the dynasty's final years. This finding aligns chronologically with the Revised and Newly Endorsed Law Code of Tiansheng Era, (hereafter LC), compiled in 1169 CE, which detailed frontier administrative protocols during a period of administrative consolidation coinciding with physical fortification efforts [54] (p. 197).

The substantial construction investment indicates significant logistical capabilities and resource mobilization—consistent with the documented administrative capacity of the Xi Xia state [56] (pp. 19–106). Garrison enclosures exhibit standardized design with consistent rectangular configurations featuring corner bastions and single entrance gates. This standardization suggests adherence to imperial military architectural protocols, while the integration of local materials demonstrates adaptation to regional resource constraints. The architectural evidence aligns with written sources describing standardization of frontier fortifications under Xi Xia administrative regulations [54] (pp. 194–197).

Who could cross the wall line, either from or to the territory under the Xi Xia, represents a crucial aspect in our understanding of this wall system. The frontier walls served as territorial demarcation beyond their military functions. The LC prohibited Xi Xia subjects from living or working beyond designated limits, while forbidding subjects of neighboring states from crossing into Xi Xia territory for settlement, herding, or hunting [54] (pp. 210–211). The Tiansheng legal code placed frontier elements under the jurisdiction of supervisory military administrations (監軍司), directly accountable to the Xi Xia court [54] (p. 197), [11] (p. 70). The code distinguished between authorized and unauthorized border crossings, channeling legitimate traffic—diplomatic missions, commercial caravans, and authorized migrants—through designated checkpoints staffed by inspection officials [56] (pp. 117–123). The archeological identification of mountain pass fortifications provides the physical infrastructure for implementing these policies.

The border-crossing regulations differentiated between two categories: potential threats (spies, raiders, enemy forces) and those attempting to smuggle prohibited goods or escape Xi Xia territory [54] (pp. 199–206, 209, 212–222, 274–281, 283–287). Notably, the system maintained selective permeability—defectors from enemy states and political refugees were granted asylum [54] (pp. 268–270), [55] (p. 122), [59] (pp. 73, 76, 98), [60] (pp. 131–132). This indicates the frontier management system functioned as an active tool of statecraft and diplomacy calibrated to advance specific geopolitical objectives rather than merely as a passive barrier.

The LC further prohibited soldiers from relocating their posts, even to avoid enemies or seek better resources [54] (p. 196), highlighting the strategic importance of maintaining specific defensive positions like fortification F42. Historical documentation characterizes these garrisons as nodes in an imperial network regulating trade, collecting customs duties, and monitoring population movements [54] (pp. 211–216).

The frontier garrisons were staffed by male, female, and auxiliary soldiers, supplemented by corvée laborers from diverse origins, including convicted criminals [54] (pp. 114–116, 194–197, 393–397, 611). The LC delineated commanders' responsibilities and prescribed penalties for violations such as post abandonment [54] (pp. 194–199).

The fortifications operated as an integrated defense system, with guards detecting threats and dispatching messengers to alert neighboring watches and higher-level installations [54] (pp. 212–216), [61] (p. 23).

Despite its sophistication and resource investment, the Gobi Wall appears not to have presented a significant barrier against Genghis Khan's Mongol invasion in 1226. The *Secret History of the Mongols* [59] does not mention these fortifications in its description of the campaign against Xi Xia, suggesting they were not considered a major obstacle. This aligns with interpretations of other medieval wall systems in Mongolia and China, whose primary purpose was not to halt large invading armies but rather to control border areas, pacify small-scale conflicts, and regulate the movement of people and commodities between states [3,5].

## 7. Conclusions

This investigation of the Gobi Wall system offers substantial new insight into medieval frontier management strategies and architectural practices in Inner Asia. By integrating remote sensing, field surveys, targeted excavations, and historical documentation, we have developed a nuanced interpretation of this complex and multifunctional infrastructure.

Chronological evidence—especially the radiocarbon dating of wooden structural elements and numismatic finds—supports the attribution of the wall's primary construction and utilization to the Xi Xia period (1038–1227 CE). This dating is further corroborated by the architectural configurations observed at the garrison sites, which align with formal construction patterns documented in Xi Xia administrative texts. Excavation data from key garrisons indicate long-term occupation, reinforced by the recovery of iron tools and state-grade construction implements, including a stone tamping tool from G10. These finds illuminate the practical application of rammed-earth building techniques and suggest continued state support for frontier maintenance and control.

The wall's alignment reveals deliberate environmental adaptation. Its positioning near shallow aquifers—evidenced through well distributions and poplar tree growth—reflects strategic planning to ensure sustained year-round occupation. This careful integration of environmental knowledge facilitated not only garrison sustainability but also effective monitoring and control of trans-frontier movement. Furthermore, the extensive use of local trees and shrubs in construction appears unique to this wall line and is not documented in other wall systems in Mongolia. Given the ecological sensitivity of the region, this feature raises important questions about medieval land-use practices and environmental impact, which remain key subjects for future research.

This research supports a broader reconceptualization of medieval frontiers—not merely as static defensive barriers, but as dynamic administrative infrastructures. The Gobi Wall exemplifies a mode of Xi Xia statecraft that used architectural investments to manage re-sources, population movement, and territorial boundaries. This understanding aligns with theoretical models framing frontiers as zones of control and interaction, rather than rigid dividing lines [57], and invites broader comparative analysis across Eurasian contexts.

## 8. Future Research Directions

Further research should expand the methodological toolkit used in this study. While the current investigation relied on satellite imagery, drone photography, and pedestrian surveys, incorporating high-resolution (sub meter) multispectral remote sensing combined with SAR and LiDAR would allow detection of subtle topographic features, such as eroded wall traces, depressions, and platforms that are often undetectable through conventional means [62].

In addition, the architectural and spatial organization of the Gobi Wall appears to follow a shared structural logic observed in other medieval frontier systems in Mongolia. These include the Mongolian Arc constructed by the Jin dynasty [1] and the Northern Wall attributed to the Liao dynasty [4]. All three exhibit a common pattern of long wall segments interspersed with regularly spaced garrisons and supported by natural obstacles. Future comparative analysis should explore whether these configurations reflect a wider East Asian tradition of imperial frontier infrastructure, possibly grounded in a shared conceptual model. Extending this lens to include examples such as the Roman Limes may offer broader insights into how linear architectures served multifunctional, ideological, and territorial roles in diverse imperial contexts. In future research, it would be beneficial to extend the comparison to other frontier zones worldwide. For instance, the volume edited by Escalona and Reynolds [63] explores how scale and scale change shaped landscapes and social organization in medieval Europe, offering a valuable framework for understanding how empires restructured frontier regions across different historical contexts.

Finally, the analytical framework developed in this study—including its emphasis on interdisciplinary integration and environmental analysis—offers a model for investigating other historical frontier systems within and beyond Inner Asia. Comparative studies that incorporate both technical methods and theoretical perspectives will be essential to understanding how empires across time and space transformed marginal zones into managed landscapes.

**Supplementary Materials:** The following supporting information can be downloaded at: https://www.mdpi.com/article/10.3390/land14051087/s1, Figure S1. Distribution of modeled dates upon the Intcal20 calibration curve (Note that UCIAMS-1360 and 1355 are entirely overlapping); Table S1. Output from Oxcal model.

**Author Contributions:** Research Conceptualization, G.S.-L., D.G., C.A. and W.H.; Writing—original draft, D.G. and G.S.-L.; Review and amendment of original draft, W.H., I.W., M.U., J.C., M.C. and Z.Z.; Maps and figures, D.G.; GIS analysis, I.W., D.G. and G.A.; Analysis of the radiocarbon dates, W.H.; Data acquisition, M.C., M.U., D.H., I.L. and G.A.; Historical Resources, Numismatics, J.C., G.A., C.A. and Z.Z.; Supervision, C.A., G.S.-L. and W.H.; Project administration, W.H., G.S.-L., C.A. and G.A.; Funding acquisition, G.S.-L. All authors have read and agreed to the published version of the manuscript.

**Funding:** This research was funded by the European Research Council (ERC) under the European Union's Horizon 2020 research and innovation program ("The Wall" project, grant agreement No. 882894).

**Data Availability Statement:** The data presented in this study will be made available upon completion of the ERC funded project and can be obtained from the corresponding author upon request.

**Acknowledgments:** The authors sincerely acknowledge Tal Rogovski for his expertise in acquiring and post-processing all (unless otherwise stated) drone and camera photographs presented in this manuscript. His technical proficiency and meticulous execution were essential to the photographic documentation. All photos (unless otherwise stated) were taken by Tal Rogovski. All figures (unless otherwise stated) were created by Dan Golan. Participants in the 2024 Gobi Wall expedition: Gideon Shelach-Lavi, Chunag Amartuvshin, William Honeychurch, Dan Golan, Tal Rogovski, Jingchao Chen, Angaragdulguun Gantumur, Gideon Avni, Ido Wachtel, Mika Ullman, Nachem Doron, Yotam Toib, Itay Lubel, Dor Heimberg, Daniela Wolin, Byambatseren Batdalai, Munkhbaatar Munkh-Urchral.

**Conflicts of Interest:** The authors declare no conflicts of interest.

## References

1. Fung, Y.T.; Gantumur, A.; Wachtel, I.; Chunag, A.; Zhang, Z.; Fenigstein, O.; Golan, D.; Shelach-Lavi, G. Unraveling the Mongolian Arc: A field survey and spatial investigation of a previously unexplored wall system in eastern Mongolia. *J. Field Archaeol.* **2024**, *49*, 225–242. [CrossRef]
2. Kradin, N.N.; Prokopets, S. The Great Wall of Khitan: North Eastern Wall of Chinggis Khan. Nauka—Vostochnaya Literatura. 2019. Available online: https://www.academia.edu/43375401/The_Great_Wall_of_Khitan_North_Eastern_Wall_of_Chinggis_Khan_%D0%92%D0%B5%D0%BB%D0%B8%D0%BA%D0%B0%D1%8F_%D0%BA%D0%B8%D0%B4%D0%B0%D0%BD%D1%8C%D1%81%D0%BA%D0%B0%D1%8F_%D1%81%D1%82%D0%B5%D0%BD%D0%B0_%D0%A1%D0%B5%D0%B2%D0%B5%D1%80o_%D0%B2o%D1%81%D1%82o%D1%87%D0%BD%D1%8B%D0%B9_%D0%B2%D0%B0%D0%BB_%D0%A7%D0%B8%D0%BD%D0%B3%D0%B8%D1%81_%D1%85%D0%B0%D0%BD%D0%B0 (accessed on 14 May 2025).

3. Shelach-Lavi, G.; Honeychurch, W.; Chunag, A. Does extra-large equal extra-ordinary? The 'Wall of Chinggis Khan' from a multidimensional perspective. *Humanit. Soc. Sci. Commun.* **2020**, *7*, 22. [CrossRef]
4. Shelach-Lavi, G.; Wachtel, I.; Golan, D.; Batzorig, O.; Amartuvshin, C.; Ellenblum, R.; Honeychurch, W. Medieval long-wall construction on the Mongolian Steppe during the eleventh to thirteenth centuries AD. *Antiquity* **2020**, *94*, 724–741. [CrossRef]
5. Shelach-Lavi, G.; Amartuvshin, C.; Heimberg, D.; Wolin, D.; Angaragdulguun, G.; Rogovski, T.; Chen, J.; Fenigstein, O.; Steiner, T.; Honeychurch, W. Life along the medieval frontier: Archaeological investigations of the southeastern long wall of Mongolia. *Antiquity*, 2025, *in press*.
6. Storozum, M.; Golan, D.; Wachtel, I.; Zhang, Z.; Lotze, J.S.; Shelach-Lavi, G. Mapping the medieval wall system of China and Mongolia: A multi-method approach. *Land* **2021**, *10*, 997. [CrossRef]
7. Danda'er. The Typological Research of the Fortified Towns on the Sides of Jin Jiehao. Master's Thesis, Inner Mongolia Normal University, Hohhot, China, 2013.
8. Xie, D.; Zhang, Y.; Li, Y. *Defense System and Military Settlements of the Jin Great Wall*; China Architecture & Building Press: Beijing, China, 2020.
9. Hanks, B.K.; Shelach-Lavi, G.; Bermann, M.; Eklund, E.; Greaves, A.; Amartuvshin, C.; Tserendash, N.; Goldsmith, Y.; Burentogtokh, J.; Erdenebat, U. Examining medieval long-wall frontier systems (11th–12th centuries AD) through archaeological geophysics in the eastern Mongolian steppe region. *J. Archaeol. Sci. Rep.* **2024**, *60*, 104860. [CrossRef]
10. Amartuvshin, C.; Shelach-Lavi, G.; Honeychurch, W.; Byambatseren, B.; Shamir, O.; Munkhtur, U.; Wolin, D.; Wang, S.; Shamir, N. An elite grave of the pre-Mongol period, from Dornod Province, Mongolia. *Archaeol. Res. Asia* **2024**, *39*, 100537. [CrossRef]
11. Kovalev, A.A.; Erdenebaatar, D. The northern border of the Tangut state, XiXia: According to the archaeological evidence and written sources. *Archaeol. J. Kanazawa Univ.* **2021**, *80*, 49–77.
12. Kovalev, A.A.; Erdenebaatar, D. Reconsideration about Great wall of Xi Xia and Shouxiangcheng fortress of the Western Han in Ömnögov' aimag, Mongolia. *Inn. Mong. Cult. Relics Archaeol.* **2008**, *2008*, 101–110.
13. Perlee, K. Archaeological Exploration Done on the Gobi Steppe in Ömnögovi and Övör Khangai Aimag (1962). In *Erdem Shinzhilgeenii Ögüüllüüd*; Perlee, K., Ed.; Mongolian Academy of Sciences: Ulaanbaatar, Mongolia, 2001; Volume 1, pp. 270–280.
14. Baasan, T. *What Are the Chinggis Walls?* Admon: Ulaanbaatar, Mongolia, 2006.
15. Moriya, K.; Shiraishi, N.; Odsuren, D.; Buyanhishig, T.; Batbold, G.; Souma, H. Survey Report of the Great Wall and Castle Ruins in South Gobi Province, Mongolia. In *Proceedings of the International Conference on the Archaeology of Ancient Settlements and Cities in Northern Asia*; Zhang, Y., Chen, L., Eds.; Science Press: Beijing, China, 2014; pp. 75–80.
16. Tan, Q. *The Historical Atlas of China*; Cartographic Publishing House: Beijing, China, 1982; Volume 6.
17. National Cultural Heritage Administration. *The Atlas of the Cultural Heritage of Inner Mongolia*; Xi'an Ditu Chubanshe: Xi'an, China, 2003.
18. Li, Y. The long walls of the Northern frontier of China. *Neimenggu Wenwu Kaogu* **2001**, *2001*, 1–51.
19. Xing, Y. "Remote observation" of the fortress ruins along the southern line of the Han Great Wall in Inner Mongolia. *Perspect. Anc. Mod. Hist.* **2020**, *34*, 78–118.
20. Department of the Army. Soviet topographic map symbols (Technical Manual 30-548). U.S. Government Printing Office. 1958. Available online: https://s3.eu-west-2.amazonaws.com/redatlas/TM30-548_1958.pdf (accessed on 14 May 2025).
21. Aylmer, C. Soviet military maps of China. *Cartogr. J.* **2022**, *59*, 376–388. [CrossRef]
22. Cruickshank, J.L. The evolution of Soviet topographic maps as revealed by their published supporting documentation. *Cartogr. J.* **2021**, *59*, 280–299. [CrossRef]
23. Rondelli, B.; Stride, S.; García-Granero, J.J. Soviet military maps and archaeological survey in the Samarkand region. *J. Cult. Herit.* **2013**, *14*, 270–276. [CrossRef]
24. Cao, X.; Tian, F.; Li, F.; Gaillard, M.J.; Rudaya, N.; Xu, Q.; Herzschuh, U. Pollen-based quantitative land-cover reconstruction for northern Asia covering the last 40 ka cal BP. *Clim. Past* **2019**, *15*, 1503–1536. [CrossRef]
25. Li, H.; Yang, X.; Scuderi, L.A.; Hu, F.; Liang, P.; Jiang, Q.; Buylaert, J.P.; Wang, X.; Du, J.; Kang, S.; et al. East Gobi megalake systems reveal East Asian Monsoon dynamics over the last interglacial-glacial cycle. *Nat. Commun.* **2023**, *14*, 2103. [CrossRef]
26. Hülle, D.; Hilgers, A.; Radtke, U.; Stolz, C.; Hempelmann, N.; Grunert, J.; Felauer, T.; Lehmkuhl, F. OSL dating of sediments from the Gobi Desert, Southern Mongolia. *Quat. Geochronol.* **2010**, *5*, 107–113. [CrossRef]
27. Banning, E.B. *Archaeological Survey (Manuals in Archaeological Method, Theory and Technique)*; Springer: Berlin/Heidelberg, Germany, 2002. [CrossRef]
28. Southon, J.; Santos, G.M.; Druffel-Rodriguez, K.; Druffel, E.R.M.; Trumbore, S.; Xu, X.; Griffin, S.; Ali, S.; Mazon, M. The Keck Carbon Cycle AMS Laboratory, University of California, Irvine: Initial operation and a background surprise. *Radiocarbon* **2004**, *46*, 41–49. [CrossRef]
29. Thierry, F. *Ancient Chinese Coins: From Origins to the End of the Empire*; Les Belles Lettres: Paris, France, 2017; ISBN 978-2-251-44686-8.

30. Yu, W. *The Complete Collection of Chinese Ancient Coins*; Southwestern University of Finance and Economics Press: Chengdu, China, 1997.
31. Niu, D.; Niu, Z. Discoveries and research of Xi Xia coins. *Tangut Res.* **2013**, *2013*, 88–109.
32. Wang, X. Archaeological Research on the Coins of the Western Xia Dynasty. Master's Thesis, Inner Mongolia Normal University, HohHot, China, 2023.
33. Shelach-Lavi, G. *The Archaeology of Early China: From Prehistory to the Han Dynasty*; Cambridge University Press: Cambridge, UK, 2015. [CrossRef]
34. Guo, Q. Yingzao Fashi: Twelfth-century Chinese building manual. *Archit. Hist.* **1998**, *41*, 1–13. [CrossRef]
35. Needham, J. *Science and Civilisation in China: Volume 4, Physics and Physical Technology, Part 3, Civil Engineering and Nautics*; Cambridge University Press: Cambridge, UK, 1971.
36. Zou, X.; Wang, R. Design of Xi Xia flat pot and change of nomadic culture. *J. Ceram.* **2024**, *45*, 1276–1284. [CrossRef]
37. Institute of Archaeology, Chinese Academy of Social Sciences. *Excavation Report of the Lingwu Kiln in Ningxia*; Zhongguo Da Baike Quanshu Chubanshe: Beijing, China, 1995.
38. Ma, W. *Lingwu Kiln in Ningxia*; Palace Museum Press: Beijing, China, 1988.
39. Ningxia Archaeological Institute. Excavation of the Hui Lane Kiln in Lingwu City, Ningxia. *Archaeology* **2002**, *8*, 59–68.
40. Shamsutdinova, E.; Shamsutdinov, Z. Biological features and forage performance of black saxaul (Haloxylon Aphyllum (Minkw) Iljin) in the Central Asian Desert. *BIO Web Conf.* **2022**, *43*, 01023. [CrossRef]
41. Fan, B.; Zhang, A.; Yang, Y.; Ma, Q.; Li, X.; Zhao, C. Long-term effects of xerophytic shrub Haloxylon ammodendron plantations on soil properties and vegetation dynamics in Northwest China. *PLOS ONE* **2016**, *11*, e0168000. [CrossRef]
42. Thomas, F.; Jeschke, M.; Zhang, X.; Lang, P. Stand structure and productivity of Populus euphratica along a gradient of groundwater distances at the Tarim River (NW China). *J. Plant Ecol.* **2017**, *10*, 753–764. [CrossRef]
43. Dee, M.; Bronk Ramsey, C. High-precision Bayesian modeling of samples susceptible to inbuilt age. *Radiocarbon* **2014**, *56*, 83–94. [CrossRef] [PubMed]
44. Bronk Ramsey, C. OxCal (Version 4.4.4) (Computer Software). 2021. Available online: https://c14.arch.ox.ac.uk/oxcal.html (accessed on 14 May 2025).
45. Reimer, P.J.; Austin, W.E.N.; Bard, E.; Bayliss, A.; Blackwell, P.G.; Bronk Ramsey, C.; Butzin, M.; Cheng, H.; Edwards, R.L.; Friedrich, M.; et al. The IntCal20 Northern Hemisphere radiocarbon age calibration curve (0–55 cal kBP). *Radiocarbon* **2020**, *62*, 725–757. [CrossRef]
46. Bronk Ramsey, C. Dealing with outliers and offsets in radiocarbon dating. *Radiocarbon* **2009**, *51*, 1023–1045. [CrossRef]
47. Shulyk, V.; Shevchenko, L.; Cherniyavskyi, V.; Davydov, A.; Boborykin, O. About the concept of improving border defense lines by means of landscape architecture. *Landsc. Archit. Art* **2024**, *24*, 89–96. [CrossRef]
48. Altantungalag, D.; Buyankhishig, N.; Altangadas, E.; Purevsuren, U. Groundwater recharge estimation in Undai watershed area, southern Mongolia [Poster presentation]. In Proceedings of the EGU General Assembly 2020: Sharing Geoscience Online, Online, 4–8 May 2020. [CrossRef]
49. Bayanzul, B.; Nakamura, K.; Machida, I.; Watanabe, N.; Komai, T. Construction of a conceptual model for confined groundwater flow in the Gunii Khooloi Basin, Southern Gobi Region, Mongolia. *Hydrogeol. J.* **2019**, *27*, 1581–1596. [CrossRef]
50. Batsukh, K.; Zlotnik, V.A.; Nasta, P. Analysis of groundwater recharge in Mongolian drylands using composite vadose zone modeling. *Front. Water* **2022**, *4*, 802208. [CrossRef]
51. Amartuvshin, N. *The Family Chenopodiaceae vent. in Mongolia: Phylogeny, Historical Development, Range, Distribution, Ecology, Taxonomy*; Toonot Print: Ulaanbaatar, Mongolia, 2021.
52. Zhu, Y.; Ren, L.; Skaggs, T.H.; Lü, H.; Yu, Z.; Wu, Y.; Fang, X. Simulation of Populus euphratica root uptake of groundwater in an arid woodland of the Ejina Basin, China. *Hydrol. Process.* **2009**, *23*, 2460–2469. [CrossRef]
53. Lattimore, O. *Studies in Frontier History: Collected Papers, 1928–1958*; Oxford University Press: Oxford, UK, 1962.
54. Shi, J.; Nie, H.; Bai, B. *Revised and Newly Endorsed Law Code of the Tiansheng Era*; Falü Chubanshe Law Press: Beijing, China, 2000.
55. Yang, R. Some discussions on the Silk Road of Xi xia. *J. Chin. Hist. Geogr.* **2003**, *18*, 117–123.
56. Dunnell, R.W. The rise of Chinggis Khan and the United Empire, 1206–1260. In *The Cambridge history of the Mongol Empire (Volume 1: Political History, pp. 19–106)*; Biran, M., Kim, H., Eds.; Cambridge University Press: Cambridge, UK, 2023.
57. Standen, N. *Unbounded Loyalty: Frontier Crossings in Liao China*; University of Hawai'i Press: Honolulu, HI, USA, 2007.
58. Dunnell, R. Xi Xia: The First Mongol Conquest. In *Genghis Khan and the Mongol Empire*; Fitzhugh, W., Rossabi, M., Honeychurch, W., Eds.; Arctic Studies Center, Smithsonian Institution: Washington, DC, USA, 2013; pp. 152–159.
59. de Rachewiltz, I. *The Secret History of the Mongols: A Mongolian Epic Chronicle of the Thirteenth Century*; Brill: Boston, MA, USA, 2004; Volume 1–2.
60. Du, J. *Relationship between Xixia and Neighbors*; Gansu Cultural Publishing House: Lanzhou, China, 2017.
61. Shi, J. *Research on Xixia Military Documents*; Gansu Cultural Publishing House: Lanzhou, China, 2021.

62. Frachetti, M.D.; Berner, J.; Liu, X.; Bevan, A.; Hritz, C.; Williams, T. Large-scale medieval urbanism traced by UAV–lidar in highland Central Asia. *Nature* **2024**, *634*, 1118–1124. [CrossRef] [PubMed]
63. Escalona, J.; Reynolds, A.J. (Eds.) *Scale and Scale Change in the Early Middle Ages: Exploring Landscape, Local Society, and the World Beyond*; in The Medieval Countryside; Brepols: Turnhout, Belgium, 2011; Volume 6.

land

MDPI

Article

# New Realizations at the Archaeological and Funereal Park of Takino Cemetery in Hokkaido (Japan)

**Joseph Cabeza-Lainez [1,*], Victor Marquet-Saget [1], Inmaculada Rodriguez-Cunill [2], Cesar Puchol-Barcina [3] and Miguel Gutiérrez-Villarubia [1]**

1 Department of Composition, School of Architecture, University of Sevilla, 41012 Sevilla, Spain; vmarquetsaget@us.es (V.M.-S.); miggutvil@alum.us.es (M.G.-V.)
2 Department of Painting, Faculty of Fine Arts, University of Sevilla, 41003 Sevilla, Spain; cunill@us.es
3 Department of Archaeology and Cultural Anthropology, Jeonbuk National University, Jeonju 54896, Republic of Korea
* Correspondence: crowley@us.es

**Abstract:** For decades, in historical research, archeological vestiges have been linked with geomancy and cults of the mythical ancestors of a group of the population. This is particularly true in Eastern Asia and especially in China, Korea, and Japan. A fundamental problem of Japanese archeology is that few of the remnants were realized in stone. One of the most important parts of archeological sciences is the study of Necropolises or ancient interments. From the 1970s onwards, in the relatively "new" and promising land of Hokkaido, cemeteries were built with the concept of landscape in mind; this is also due to the lavish vegetation features of this northernmost island of Japan. In the case of the Takino cemetery on the plains of Sapporo, Hokkaido, whose construction began in 1982, solemnity and religiousness were incorporated by producing exact stone replicas of famous funerary landmarks from antiquity as such materials were inexistent in the Nipponese Isles. This trend to grant eternity included traditional Buddhist funereal monuments like the Stupa, Seokguram grotto, and Kamakura sites, but at a certain and exuberant point, under the influence of Isamu Noguchi, it reached Stonehenge in England and the Moai from Easter Island in Polynesia (being after all located in a remote isle of the Pacific Ocean). In this article we will outline such process of generation and overall conception, analyzing the inclusion and architectural assembly of the different compounds and the recent and extraordinary additions projected and built by the celebrity architect Tadao Ando. We expect, in this manner, to facilitate the comprehension of the significance of venerable landscape sublimated through archeology for the Nipponese modern civilization.

**Keywords:** landscape archeology; heritage of the land; ancient stone monuments; Asian architecture; Easter Island; hypogeal chambers; architectural retrofits; Tadao Ando

## 1. Introduction: The Fascination of an Unknown Island

For a variety of reasons, the Japanese archipelago cannot be considered an utterly ancient hinterland [1]. Even from a geological point of view, its variegated igneous nature, still active, seems to eschew extreme antiquity [2,3]. The northernmost of the four large islands, Hokkaido 北海道, passes as even newer from an anthropological point of view [4], as it was only inhabited by aboriginals called Ainu (i.e., people) [5], who remained anchored on the verge of the Bronze Age, until well beyond the 19th century (Figure 1). The main foundational texts of the Japanese civilization, like the Kojiki [6], 古事記, strive to convey the grandiosity attributed to immemorial times. Nevertheless, the anvil of history, by virtue of Chinese sources, endorsed with recent genetic analyses, has revealed that the now prevailing Yamato people only harks back from the 3rd century C.E. at the most [7,8]. The Ainu aboriginals or their ancestors termed Emishi 蝦夷, commonly a synonym of barbarian, suggest a much older pre-existence on these isles, a fact often disregarded or diminished by partisan historians [9].

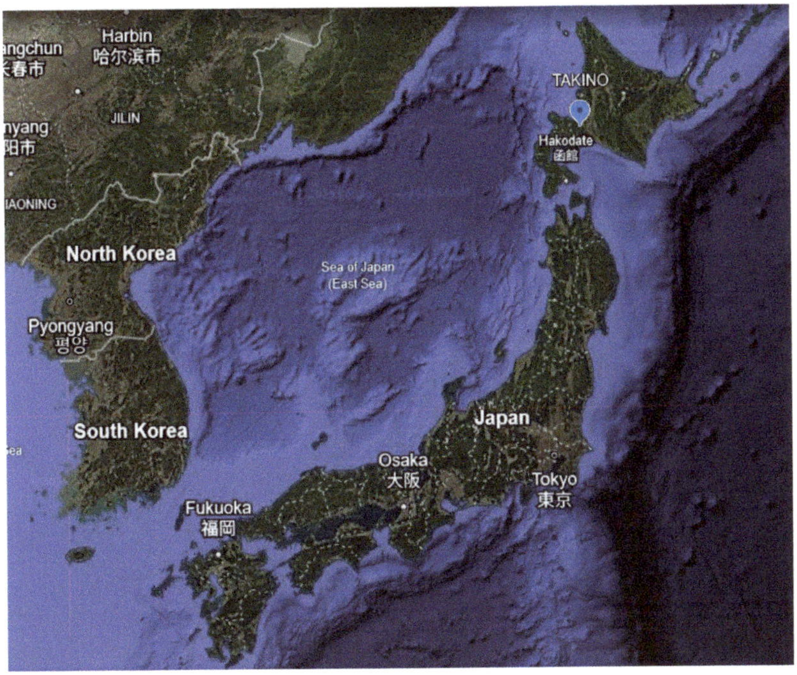


**Figure 1.** Map of Japan with Takino near Sapporo in Hokkaido.

It is generally accepted that the tumulus or ceremonial mound is the preferred construction on the land or soil erected by many civilizations [10], and particularly by the peoples of East Asia, with the paramount examples of Korean geomantic tumuli and Japanese Kofun 古墳. These generally consist of a set of tombs of ancient rulers in which the terracotta statuettes, named haniwa 埴輪, were deposited [11,12]. The kofun echoed Indian stupa, although it is not a tomb and later gave way to the stupa's rebirth in variegated forms such as the Tahoto 多宝塔 or the hall of many treasures [13]. Such artificial hills were completely absent from Hokkaido's landscape as the Ainu had other forms of religious worship more centred on living nature and not related to man-made constructions. Moreover, Japanese archeology faces the important issue that, as construction in stone was rare, very few remnants can be spotted by traditional excavation techniques. Therefore, for designers of ceremonial spaces at all epochs, it was imperative to imbue a sense of antiquity and, moreover, of eternity to the places where the ancestors and the deceased were revered, by performing volumetric alterations to the land [13,14], as we can read in Horace's Ode 3.30 [15].

Consider in a parallel with Christendom, the verses of 1808 by William Blake: "And did de Countenance Divine/Shine forth upon our clouded hills?/And was Jerusalem builded here/Among these dark Satanic Mills" [16]. These outlandish lines certainly evoke the concept by virtue of which, even on the heathen lands of England, the erection of a sacred monument, particularly a funereal one, seems to have demanded a hallowed although natural precinct much in the image of the seven hills of Rome or other orographic formations [17].

Only ten years later, in 1818, Percy Shelley wrote about the huge statues of Ramses II at Abu Simbel: "My name is Ozymandias, King of Kings:/Look on my works, ye Mighty, and despair!"/No thing beside remains. Round the decay/Of that colossal wreck, boundless and bare/The lone and level sands stretch far away [18,19].

Thus, clearly confirming the inspirational power of archeological vestiges on human memory and behaviour as an invariant for the erection of sacred sites and hierothesions [20].

In the case of the Sapporo area, the first precedent, which inaugurated this trend for Japanese architectural design, seems to appear in the Moerenuma Park (Figure 2) in the northern zone of the city and more specifically the scenic work known as Play Mountain [21].

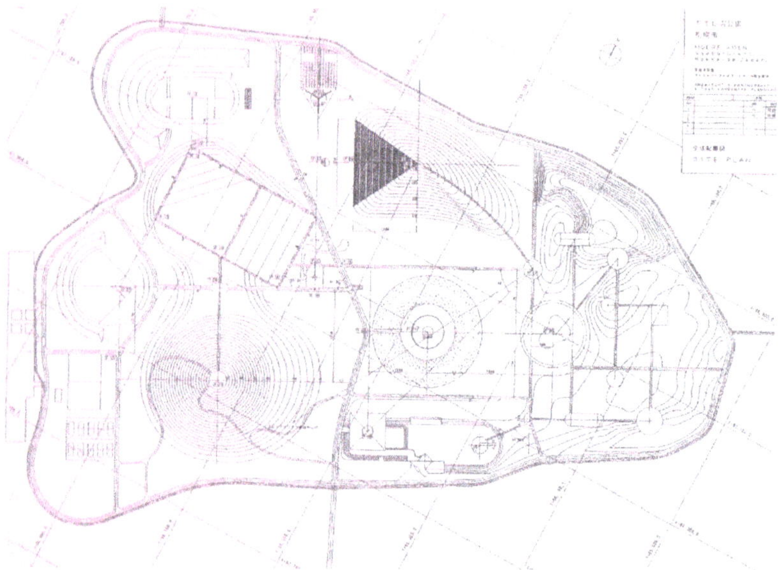

**Figure 2.** General plan of the Moerenuma Park designed by Isamu Noguchi.

In it, the Japanese-American sculptor and landscape artist Isamu Noguchi takes advantage of the verdant features of Hokkaido to recreate the striking imagery of Maya-like pyramids, as Kukulkan at Chichen Itza or Palenque [21,22] (Figure 3). We need to bear in mind that Noguchi lived for some time in Mexico where the precedent of pre-Hispanic cultures, brought to him by artist acquaintances among which ranks Frida Kahlo, exerted an indubitable attraction if not a pervasive influence [23]. At the so-called Play Mountain, by means of bare stripes of stone superposed on the sloped turf, he creates the vision that the full potency of an ancient pyramid lies covered under thick vegetation (Figure 4), to the point that the visitors, which can occasionally be seen on its summit, play the role of acting as king-priests of a forsaken temple [24] (Figure 5).

From an ecological point of view, such pervasive creation is only natural if we realize that, as previously mentioned, grass is more abundant in Japan [25] than stone ashlars are.

To prove the previous statement, let us quote a famous haiku by Basho on the lost capital of Hiraizumi: The Summer grass leaves/Ruins of the Dream/of the Warriors of Old (Trans. by author). 夏草や兵どもが夢の跡. It denotes that the only remnants of a once proud capital are the ephemeral leaves of the summer.

**Figure 3.** The pyramid of Palenque. Mexico. Source: author.

**Figure 4.** Play Mountain at Moerenuma by Isamu Noguchi. Source: Author.

**Figure 5.** Detail of the play Mountain with stripes resembling ancient stones. Source: Author.

In compounds like the Karesansui 枯山水 (stone miniature landscape) or at Katsura detached palace, ashlars are considered almost as venerable relics [20,21]. The dictum of the artist Joseph Beuys comes instantly to our minds at this point: "It is to us only, to imbue life to the stones so that they do not lie around, scattered as carcasses." [26].

After a fashion, this square mound echoes a distant pyramidal landmark of south Java, the celebrated Temple of Borobudur near Yogyakarta (Indonesia) [27] to which we will refer later on (Figure 6). Comparatively, there are fewer stone monuments on the southern hemisphere than on the northern one, perhaps due to a sparser distribution of the land masses.

**Figure 6.** View of the Borobudur Temple in JYogyakarta.

All in all considered, by acting in this seductive manner, Noguchi is able to incorporate important aspects of the aesthetics of Shintoism by revering nature as something supernatural and, at the same time, to beckon the essence of Zen Buddhism that pervades Nipponese idiosyncrasy [28], signalling the misty way of the occult and letting human imagination perform the task of accomplishing what cannot be entirely revealed or uttered [29].

Previously, he had pronounced a direct reflection onto Buddhist liturgy in his celebrated piece "Black Slide Mantra", a kind of helical toboggan for children cum carven steps. The playground, sculpted in polished sable stone [30], is located at Odori Park in central Sapporo (Figure 7).

**Figure 7.** Black slide mantra. Odori Park. Sapporo.

These references are clearly supplying the Japanese fields and cities with stone monuments of an imaginary archeology stemming from a culture that, in a sense, never was. However, it could have been this way if the crucible of History would have forged the events otherwise. Slowly but surely these new forms are enriching their idealized past for the future generations and so the creations have received wide acclaim everywhere [27].

It was chiefly thorugh such powerful design examples in mind and with an eye on Buddhist rites, that the planning of the Takino funereal park was to take a start in the 1980s.

## 2. Materials and Methods

In the first place we need to remind that Takino is not just a cemetery tucked in the outskirts of a town in the sense of Europe or the US, but rather and as the name Rei-en 霊園 suggests, a place of pilgrimage, in which we ascend to a veritable "Sanctuary of Souls". The departed are not merely buried and swiftly forgotten, instead they are comforted, accompanied, and ever-present in the pristine manner inherent to the Japanese ideas of afterlife and netherworld [31].

That said, in the general plan and from an architectural perspective, we can distinguish three main compounds around which the whole landscape sequence orbits (Figure 8).

**Figure 8.** General plan of the Takino Cemetery.

Relating to the sculpture of Buddha, the Stonehenge circle, and the Moai alignment, the methodology will consist of a thorough and systematic analysis of the origin, implications, and interplay of these three elements in order to improve the archeological complex and be able to assess the new realizations that have taken place at Takino, especially since 2017.

### *2.1. The Giant Buddha Statue*

As explained previously, the Daibutsu is situated in the first place. This Buddha modelled after the Kamakura statue, is completely central to the precinct of the cemetery (Figure 9), secluded by a hill, and from 2017 onwards it is preceded by a sizeable rectangular pond (in black in Figure 8) that impedes direct entrance trough the passageway and towards the massive sculpture weighting 1.500 tons and rising to a height of 13.5 m [29].

**Figure 9.** The Daibutsu statue at the Takino cemetery in its original configuration.

The imposing figure is flanked by two symmetrical mounds, which, in this way, define an important avenue leading directly to the giant statue. We must hereby stress that the sculpture is not directly perceptible from the entrance gates to the funereal park.

We need to outline that, in Japan, traditionally the Buddha image is not exposed to the outside but rather accommodated in a grand temple of timber frame as in Todai-ji at Nara. The only exception is the Kamakara Daibutsu and this is likely due to the arson that once occurred in the hosting precinct; it was of such magnitude that the high priests of the time decided to leave the bronze statue alone without reconstructing the precedent temple [32].

In addition, stone is not the preferred material for statures of the Buddha; instead we find mostly wood or gilded bronze in some special cases. This fact builds on the scarcity of permanent remnants in Japanese archeology as already stated. In the case of Takino, to settle these problems of conservation, it was planned that the sculpture would be composed of resistant granite stone and after careful polishing, left open to the visitors among the gentle slopes of the selected area of 800 hectares (See Figure 9, above). To house it in a modern construction would have been inviable.

However, soon enough, the towering image exerted an equal spiritual burden on the believers and by-passers. The stone Buddha was considered too solemn, imposing or even melancholic. The gloomy and cold weather of this area might have played a role for such ominous perception [33].

In this respect, we can find an important transitional case of an adumbrated but not actually visible Buddha at the Byodoin (Temple of Peace) of the town called Uji. The whole precinct is an example of Amitabha Buddhism in which the pilgrimage destination to be reached was the paradise of the West. The travellers were forced to accede to the complex by the river Uji and not by the current entrance from the narrow winding streets inside the town. It is remarkable that the name Uji in Japanese 宇治 means the Kingdom of Heaven or the Heavenly Paradise [32].

The central hall is called Hôôden 鳳凰殿, the Palace of the Emperor Phoenix. It is in fact crowned by two symmetric phoenixes, which, when nested over a temple, auspicated the flow of spiritual and material perfection for the entire realm. Their mere presence suggested a breeze of harmonious sounds echoed by the peculiar wind instrument called Shô. The Hôôden was considered an emblem of Japan to the point of being reproduced life-size as a Japanese Pavilion for the Chicago World-Fair of 1893 (Figure 10). The person in charge of that pavilion was the celebrated artist and writer Okakura Kakuzo, the author of The Book of Tea, written solely in English in 1906 [33].

**Figure 10.** A postcard showing the front of Hôôden in Chicago (no longer extant).

The relevant feature of this temple for the case of Takino is that the lattice work and oval opening of the main gates allowed to contemplate from the outside the countenance of the Buddha without need to penetrate inside the temple (Figures 11 and 12). These elaborate process of visualization of a deity or religious image is called darshan in Sanskrit दर्शन.

**Figure 11.** View of the lattice of the Main Gates at the Hôôden in Uji.

**Figure 12.** Three-dimensional reconstruction of the main hall of the Buddha with the elliptic opening to see the Holy Visage without entering the precinct. Uji.

The Byôdôin scenography and iconography are so meaningful that they have been reconstructed in places far away from Japan as in Oahu, Hawaii (US). In this careful reproduction, the access and view from a body of water are duly respected; it was dedicated in 1968 in memory of the Japanese Immigration (Figure 13) [34]. Besides the Buddha, historic characters like George Washington are revered in the surroundings under the category of kami (deity).

**Figure 13.** The Byôdôin at Oahu, Hawaii. Valley of the Temples.

For esoteric theology matters, the hidden Buddha or Hibutsu 秘仏 is an invariant of East Asian religions. Since the historical Buddha Shakyamuni ascended to Heavens after entering paranirvana 涅槃 (nehan in Japanese), it cannot be manifest or appear frequently in the earthen world. Those who could possibly be summoned are the bosatsu, impersonations of the Buddha who voluntarily elected to remain in this world in order to help human beings attain nirvana in their turn [35]. For that reason, it is rare in Japan to find a statue of the Buddha exposed to the outdoors and some of the images in the temples can be regarded only on special occasions, every fifty or even one hundred years. Even the shadow of the Buddha is considered a good omen and this belief has been incorporated into a common expression of everyday Japanese language [36].

A somewhat extreme case of the former attitude is to be found at the previously referred temple of Borobudur in the outskirts of Yogyakarta Indonesia [20]; in the third level of the pyramid we would find a retinue of curious bell-shaped stupas, providing some spiritual ring to believers (Figures 6 and 14). The stupas, the first Buddhist monuments, are hollow in this case, and just through rhomboid apertures we can surmise a finely carved Buddha statue kept within. It is somewhat astounding that the original builders had to take such effort to create the sculptures (Figure 15) only to subtract them later from open view as they are perpetually put to rest behind heavy stone blocks.

A less remote example of a covered Buddha is located at Gyeongju in Korea. The hill of Seokguram where it was erected, has been a consistent source of attraction for the local people and foreign travellers as well. The seemingly nondescript artificial cave indicated a novel concept of space [37] for Asia that was subsequently forsaken for centuries, until incidentally rediscovered by the Japanese, because it was mainly built in stone and since then is experiencing a revival to the point of having been declared as a World Heritage site (Figures 16–18).

**Figure 14.** The bell-shaped stupas on the third level of Borobudur.

**Figure 15.** A damaged stupa reveals the hidden Buddha, which could only be descried from the loopholes of the stone envelope. Candi Borobudur. Yogyakarta. Source: author.

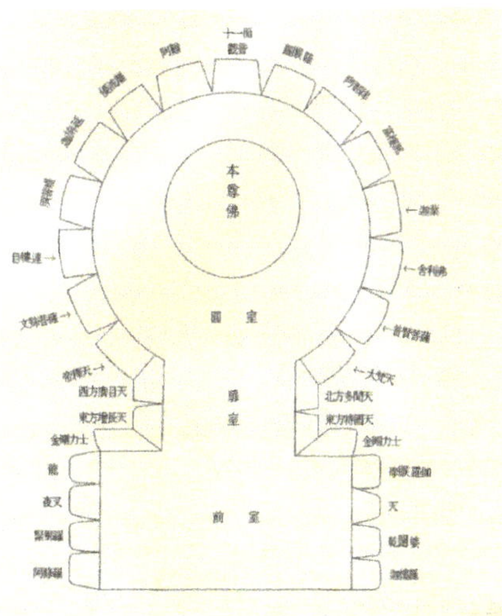


**Figure 16.** Plan of Seokguram showing (in Chinese characters) the many different guardians and protective figures (four kings and arahats) that surround the Buddha in carved reliefs.

**Figure 17.** Sectional view of the grotto including simulation of lighting levels by a circular opening no longer extant.

**Figure 18.** A view of inner harmony at the adytum of Seokguram.

Its main features are as described in the manuscript of Ko Yusoep [34]:

This great statue, beginning as a bloodless, passionless lump of granite, has been endowed with a strong pulse, breath, divinity, gentleness and dignity. When unveiled, the joy was surely not limited to the sculptor, but extended throughout Silla and echoed towards the cosmos

And in the book of Joon-Sik Choi [34] we can read these impressions:

You cannot help but wonder how on earth it was possible to carve such a soft and flowing appearance out of granite, a rock that is so hard that it is considered the most difficult stone to sculpt… Isn't it enough for the best works of art simply to exist?

The Seokguram enclave links with a cherished Buddhist tradition of hypogeal compounds [35,36] that goes back in time to thousands of years and originated mainly in India [17] in coenobiums known as chaitya. The climate of the Indian plateau and the geologic characteristics of its soil favoured the introduction of this peculiar system of construction [8] intended for religious or spiritual communities.

The tumuli that formed the stupa, as we said, the original Buddhist monument [15], subtly evolved to house inner space. It began by nominal diggings on the surface of the rock cliffs to serve as mere storage and shelter, until gradually becoming sumptuous hypogeal temples devoted to the Buddha and its grand ceremonies. The main reasons for such a procedure were both climatic and spiritual: the imperious necessity of building a covered space to conduct religious ceremonies or to offer prayers when the storms of monsoon weather prevented holy mission or any kind of outdoor activities for monks and disciples alike [7]. Such spaces also reduced the probabilities of incidental stamping or crashing onto swarms of insects that thrived in the damp of the stormy season. In the same manner, in Korea, the Seokguram cave, a celebrated Buddhist precinct (Figure 17), emerged from the antecedent of ritual tumuli [9].

Being a form of revolution, the tumulus or the stupa beckon a cosmic order that is revealed and displayed through its external features, aspects that were taken as a prototype in Japan, Korea, and China [15,16]. The tumuli had at the beginning a Daoist essence as we will see, that was slowly adopted and incorporated in Buddhism. The same happened in Japanese temples with Shintoism. Domes consisting of masses of stones have been used in Korea since ancient times to create a sort of ceremonial mound. Often, they are covered with vegetation, mosses, and grass. The first tumuli had no inner space, but they implied a geomantic meaning as a sort of altars that required to be encircled for rituals [9]. At a later stage, such Feng Shui of the land was made explicit by the designated sequence of images of the twelve astral beings which compose the Eastern Zodiac. These entities regulated the seasons and also the time within the day (Figures 19 and 20).

**Figure 19.** Tumulus of Hwangnam-ri. Gyeongju (Korea).

Ox 丑 Rat 子
Hare 卯 Tiger 寅
Serpent 巳 Dragon 辰
Ram 未 Horse 午
Bird 酉 Ape 申
Boar 亥 Dog 戌

**Figure 20.** The cosmological group disposition of the twelve animals found at diverse tumuli in Korea. The hour of the rodent coincides with midnight (north), and the hour of the horse stands for midday (southwards). The order segues from right to left and top to bottom in each inclined column, the duration of each hour was deemed 120 min, double of the current modern hour. The special Chinese characters for every entity are given.

The said astral creatures, provided with common animal heads on a human-like torso, appeared on the base of the tumuli at regular intervals from the solar directions [9], starting with the horse 午 (Wu in Chinese), which roughly translates as Meridian or Noon. This astral horse is aligned with due south as it signals midday (Figure 20). The normal sequence of the beings, clarified by the authors is Rodent (24 h), Ox (2 h), Tiger (4 h), Hare (6 h), Dragon (8 h) Serpent (10 h), Horse (12 h), Ram (14 h), Ape (16 h), Bird (18 h), Dog (20 h), and Boar (22 h). It is interesting to remark that unlike Japan or China, in Korea, the entrance to the tumuli has a deviation of five to ten degrees from the solar south–north axis.

In Japan, the twelve beings are frequently presided by four major entities, which include the dusk war of turtle and serpent to the north, the turquoise dragon to the east, the vermillion phoenix to the south, and the white tiger facing west. Mostly, the four Guardian Kings as they are called, appear on the inside of the tumulus or statue concerned, and, sometimes, they are formed outside of the building as in the famous temple of Shitenno-ji in Osaka [38].

We have delved on the tumuli since the first known large-scale Buddhist monument was the stupa स्तूप, and especially the Stupa of Sanchi (India), which is basically a mound with some relics inside it but never a tomb. Images of the Buddha were rare 2400 years ago and not technically feasible, as the first ones were made around 250 B.C. by Indo-Bactrian sculptors descending from Alexander's lasting Hellenistic influence on the area.

The tumulus is connected with the stupa in some subtle ways, particularly in Korea, where the mounds evolved to the point of being hollowed (remarkably without any knowledge of vaulting techniques) and harbour a sizeable statue, which was a constant source of inspiration not only in this country but in Japan, as it was discovered by Nipponese engineers. In this manner, the Seokguram Buddha is entirely covered but it is still visible from the outside through a glass panel (we cannot know about the view in the past).

In the case of Takino, the Buddha statue was at first conceived to stay in the outside, but since this was resented by the public, there was a decision to cover it partially, leaving only the head perceptible as in other temples of Japan and especially the Byodoin (Figures 10 and 11). This was particularly the project decision recognized by the team of the architect Tadao Ando, because the idea of holding the Buddha inside of some sort of makeshift modern building was rapidly discarded as unfeasible and impractical. In addition, it was still desirable to maintain a panoramic vision of such elaborate and magnificent statue, and it was spiritually necessary that the Buddha could grant a direct and unfiltered blessing over the land of the pilgrims that the hallowed presence encompassed.

The method of partial concealment of the statue had to be one fundamentally based on the creation of sustainable landmasses, resembling nature in all aspects, as is the truncated volcano (another famous symbol). The tumulus or forested mound comes then as a natural solution since it was also used for the entombments of ancient emperors of Japan in the historic period known as Kofun Jidai (Era of the Mounds). A conventional building with a façade, windows, pediments, or roofs, would not have served the purpose (being abhorrent among other things), and certainly could not rival with the Stonehenge circle of fathomless antiquity. In fact, both elements evoke proto-historical moments of history related to caves and megaliths, but to become an actual shrine an obscure cave is not enough; it must the sublimated by the light much in the same way as the Roman Pantheon does [39]. Such is both the wish and abstract style of architect Tadao Ando. The former discussion explains why the Takino Buddha cannot be just covered, it can only be framed or better "preserved".

Regarding the reaction of the authorities and the public, it has been thoroughly positive, because Japan is typically defined as a culture of shadows [40] where ambiguity and relativity are always preferred to stark contrast or straightforward expression. An illustrating example is the fact that the western method of chiaroschuro is substituted in Japanese art by "notan", literally "thick and thin" lines [41]. So far, we have described the historic evolution of design intentions for the section of the Buddha statue at Takino. Surprisingly, this superb sculpture was completed with the alignments of Stonehenge and similarly at the main side of the complex by a lengthy retinue of 42 Moai statues similar to those erected on Easter Island. Let us present both archeological milestones in the following chapters.

### 2.2. The Stonehenge Ring

Although much older in historical apparition, henges can compare to the tumuli described beforehand in the sense that they are both megalithic rings that suggest rituals of circumambulation usually following the solar cycles and cosmic rituals.

It might seem contradictory to display a reproduction of an ancient British stone circle in the design of the park, in coexistence with the Great Buddha; the reasons for that are variegated (Figure 21).

**Figure 21.** Stonehenge in Salisbury, England.

The mere existence of different replicas of Stonehenge can be understood under the mind frame of the megalithic structure as a universal symbol of pre-historic monumental attitude in Europe [42]. We need to remind also that these kinds of megaliths often linked to Celtic culture marked the Western edge or Finis Terrae of the ancient world (in Britanny, England, and Ireland). Their Maltese counterparts marked the Southern Edge in turn. The fact that we can find many references to Stonehenge in both pop culture and different replicas of the landmark, either for recreational use or for engineering and research purposes can point to two issues: one resides in the vestiges of eurocentrism along history and taking into account culture, with its historicity being more amply represented through media. The other can be understood from a more pragmatic stance. As widely known monuments, they attract tourism to other regions away from the original site, offering a closer or even more affordable version of the monument and an extra to the simple enjoyment of nature [43]. It is, after all, a question of re-naturalizing culture and re-culturing nature [44].

Customarily, in Japan, a Buddhist temple is completed or connected by virtue of syncretism with a Shinto sanctuary. This also happens in China and Korea with the singularity that instead of Shinto we might have a Daoist precinct or a Muist element. Some argue that the Shinto religion is a unique Nipponese evolution of Daoism [45].

Nevertheless, at a funeral necropolis, it would be inauspicious to bring in or invoke the kami or spirits of the Shinto, which eventually represent the everlasting life-forces of nature and favour transmigration or transformation, thus in perennial conflict with death conceived as cessation.

Shintoism is the traditional religion of Japan and the only one endorsed by the Imperial Household after the Meiji Restoration of 1868. One of its main postulates is remote antiquity (real or feigned). In an island recently inhabited by the Japanese as Hokkaido were, it has often been controversial to enforce the "Way of the Gods" [35].

The most representative element of Shinto shrines is the Torii 鳥居. Such element can be defined as an elaborate wooden lintel painted in vermillion (Figure 22). The Torii exerts the complex function of beckoning and attracting the Sun, which is the paramount deity of the Shinto's numerous pantheon (Amaterasu 天) and, therefore, an adequate symbol of Japan. This is reinforced by the ever-present sacred rope on display, called shimenawa しめ縄, with its zigzag paper fleece, representing sunrays, which decorates and blesses [46] the Torii or its annexed structures as the case may be [47]. Even the character of the word is formed with "bird" in the sense of the vermillion phoenix discussed at the Byodoin chapter. The underlying metaphor is that the sun would perch on the Torii much in the fashion the birds do.

**Figure 22.** The famous Torii of the ocean shrine of Miyajima.

The insular nature of Japan as an edge of the world, opposed to the great continent of China, has suffered on occasion the comparison with the British Isles. After all, both island nations retain the system of a religious ruler unlike France and China in which the celestial-line of predestined monarchs was abruptly ended after violent revolutions [48], whilst in Japan the dynastic line was just restored in the 19th century.

The designers of the Takino park dared to imagine that lintels of recycled stone from the sculpture of the Buddha could be converted into a circular Torii, an open tumulus of sorts that would coincide with a Stonehenge-like profile, suitably fitting their purposes of attracting the public less prone to the Buddhist faith and evocate at the same time primitive beliefs of cosmic and archeological resonance in an intriguing or derelict location (Figure 23) [49].

**Figure 23.** Stonehenge reproduction at Takino cemetery.

By invoking an improbable proto-historic past, perhaps inadvertently, the promoters of Takino created a unique opportunity of beholding an extraordinary landmark as pristine as in the same day of its erection, with no foreseeable problems of preservation unlike its Britannic counterpart. To the minds of the authors, the debate on authenticity and originality is somehow skipped since the original monument had no inscriptions, art, or reliefs of any sort. As we know, granite is an igneous rock as abundant in Japan as it is in England [19,43].

In this manner, intellectual curiosity for ancient monuments is completely satisfied in an idyllic surrounding and is hardly perturbed by the discrete interments of funeral ashes enshrined under the enge. Nature is most conveniently re-cultured [50]. It is interesting to remark that the orientation of the stone-blocks was somewhat altered (Figures 24 and 25), as in the replica, shattered fragments of the lintels were not left to scatter on the ground. This has shifted the possible original solar or cosmic alignments although this extreme has not been fully demonstrated [51]. We should also notice that Stonehenge's replica does not face the central axis formed by the Buddha statue but ends in a parallel path to that axis (See Figure 7).

At a given moment, the stone circle incorporated a sheltered altar (Figure 26) with three Buddha figures. This element was later removed as it produced mixed interpretations. Under the circle there is a glazed locked chamber accessible through stairs, which contains the ashes of the departed in some 400 air-tight metal niches. The number of urns can be expanded to a certain extent. Also, the sculpture of the Buddha was previously in direct visual connection with the stone circle. In the new configuration after 2017, this is no longer possible (Figures 9 and 27), wherein the importance of the Stonehenge circle is strengthened and counterpoints the whole picture.

**Figure 24.** Plan of Takino's Stonehenge replica. Due north is upside.

**Figure 25.** Plan of the original Stonehenge monument. It is slightly tilted with respect to Figure 23.

**Figure 26.** The Stonehenge replica with a former Buddhist altar.

**Figure 27.** The Stonehenge replica with the Buddha before the 2017 renovation.

### 2.3. *The Moai Alignment*

The third important nexus of the Takino complex is the row of 42 Moai. Moai are probably one of the most notorious remnants of the Rapa Nui People as well as the most recognizable icon of Easter Island. The standard Moai can be briefly defined as a colossal humanoid figure carved from the ashes of the Rano Raraku crater or any other volcanic material, with an oversized and elongated head composed of simplified and geometric lines [43].

Their bodies are minimalist, with arms resting at the sides or on the stomach. The legs are frequently imperceptible, although they suggest a crouched position (Figure 28). The torso rests on a ceremonial platform called ahu, which cannot be trodden under any circumstances, which replaces the lower limbs of the bodies. This sacred platform has not been respected at Takino cemetery, which could then be considered a sort of mockery of the original figures (Figure 29). The Moai could be detailed at times by a red crown or hat called Pukao and two rounded eyes made of coral or white conch.

**Figure 28.** Ahu Akivi. Easter Island. Source: Author.

**Figure 29.** The row of Moai at Takino cemetery. Source: Author.

Although the true meaning of the Moai is still under discussion, we are certain that they were a central piece of Rapa Nui society and are usually linked to the religious life of the islanders, often being understood as a representation of their deities or ancestors [44].

The Rapa Nui disappeared into the ashes of time, but their enigmatic giants can be compared to a variety of religious practices that exhibit similarities in numerous locations of Polynesia, the Lapita area of influence, and East Asia. We can name, in this group, the Tiki idols, the Filipino Anitos, and the Korean Dol Hareubang from Jeju Island, a territory where shamanism is still practiced today.

Korean shamanism, also known as Musok (무속/巫俗) or Mukyo (무교/巫教), has demonstrated unparalleled adaptability and has managed to integrate into contemporary society, crossing boundaries even beyond the Han Republic. The Dol Hareubang ("stone grandfather" in Jeju dialect) became an emblematic figure of the Korean pop culture but is also a valuable part of the Jejuan own identity, a friendly and merry character who can be found on souvenirs shops or bagatelles, but who is also related to ancient beliefs that are still alive [44].

The Dol Hareubang (돌하르방/偶石木) bear substantial similarities to the Moai, although these are smaller (about the size of an adult or child, but this can vary) and maintain slightly more realistic proportions. The grandfathers are sculpted into the porous basaltic rock of Jeju Island, and their contour and facial features are substantially rounder and caricatured than those of the Moai, with large eyes and a flat, bulbous nose. The headpiece is now integrated into their heads, whose headdress has become a bowler hat, giving them a silhouette oftentimes described as phallic, thus associating the stone idols of Jeju with fertility (Figure 30).

**Figure 30.** Miniature reproduction of the old man or grandfather of Jeju Island.

The arms typically rest on the tummy. Some Koreans identify this gesture with abundance or pregnancy, and it is said that touching the nose of a Dol Hareubang helps women have offspring, being Jeju Island one of the biggest tourist attractions for newlyweds on the peninsula. They feature integrated feet or bases, which were especially useful as they used to rest at thresholds or crossroads as guardian deities, in an analogous manner to which Haechi, protector and mascot of Seoul, did in mainland Korea [45].

The stone grandparents of Jeju Island are deeply connected with the local shamanism and are also referred as Beoksu Meori, "totem head" (벅수머리/法首). According to the Jeju tradition, the Dol Hareubangs serve as receptacles for the spirits, "grandfather" becoming synonymous with "ancestor".

Other analogous characteristics can be observed in antique megalithic sculptures of pre-Columbian America, such as the Olmec art or more clearly the Colombian stone idols of San Agustin, the "Andean Easter Island" [48]. These discoveries seemed to reinforce earlier theories of pre-Columbian transoceanic contact, such as those formulated by anthropologist Paul Rivet or Thor Heyerdahl which were later discarded by evidence based on genetic tests [48]. Another archeological site outisde the Pacific where huge stone heads have been identified is the hierothesion of Antiochus I at Nemrut Dağı (Turkey), originally the statues belonged to seated gods but the heads were removed at some point [46], and now they bear a curious resemblance with Atama Daibutsu of Takino as only the head of the Buddha is visible over the hill. Although other authors may consider these coincidences as superficial or extravagant, the Moai mark the Eastern Edges of the World, something by which the Japanese archipelago is commonly identified (land of the rising sun). Theretofore, ancestral spirituality of the Pacific seems to permeate and symbolically connect, through archeological vestiges, places like Jeju, Easter Island and Takino Cemetery, at least for the imagery of the people.

Turning back to the general plan of Takino, the Buddha was situated between two symmetrical mounds, which define an important axis leading directly to the giant statue. The Stonehenge compound is at the end of another axis running parallel to the Buddha's axis. Whereas the line of the Moai is segmented [52] and does not follow any of the principal directions related to the Buddha mound or Stonehenge's circle. They seem to pass as wraiths encompassing the said landmarks.

Apart from the three main elements analyzed, other milestones were considered for the design of Takino cemetery, among them the archaeological ruins of Tiwanaku (Bolivia) and Machu Picchu (Peru), Angkor Thom (Cambodia) and different kinds of pyramids. However, they were deemed unrecognizable or not sufficiently related with funereal culture and were consequently discarded [53].

## 3. Results

The former discussion on the three distinctive elements of Takino funereal park leads us to the situation in which the original design needed significant alterations, and these were mainly performed by a thoughtful intervention of architect Tadao Ando [54].

Ando had designed and constructed several Buddhist temples in the past, especially in the island of Awaji, although he is better known in the West for his projects of christian chapels (Mt. Rokko and Hotel Alpha in Tomamu).

It is said that working on the construction or embellishment of a Buddhist temple or shrine may provide atonement for even the gravest sin and in any case it is considered a powerful boon towards indivual afterlife or reincarnation [55]. There are withheld rumours that due to his advanced age, the architect desired to be granted with these spiritual benefits and so accepted the controversial commission. Unfortunately, these confidential matters remain beyond the sphere of public domain, since personal religion is hardly ever discussed in the media in Japan. Be that as is may, this fact should not dissuade us to perform a thorough description the main intentions of the realized project in what they relate to the core of Japanese culture.

The first one could be perhaps something often forgotten but very important for Buddhism, while it is not directly visual, the power of thurification.

Aroma or perfume have in themselves the capacity of converting people to the true faith without being seen, which is why incense it is used consistently at all Buddhist temples. But in the sacred literature, we find examples of mountains or hills called incense burner, fragrant or perfumed; this was achieved in the past by the predominat vegetation joined to favourable breezes [56].

Ando pursued the creation of one of such fragrant hills on the selected spot. To that aim, he resorted to lavenders. They are not native to the island of Hokkaido but have become fashionable after some exports from Provence and given the availability of cultivable land on this region.

To ensure a persistent fragance, he employed an artifical plantation of 150,000 lavenders at regular intervals on the hemispherical hollow hill that he was to create to surround the Buddha.

When the plants are in full bloom, in this case by the end of June, they give the scent of the typical violet or purple flowers, which is quite unusual especially for the Japanese landscape. This strategy has become a dominant feature of the landmark (Figure 31).

**Figure 31.** Tadao Ando, Hill of the Buddha (2017), with Buddha's forehead emerging from a field of lavender shrubs.

In a second important movement, Ando decides that beholding just the forehead of the Buddha is suitable for the visitors of the cemetery much in the same way as in the formerly described temple of Byodoin; therefore, he would create a kind of truncated dome, open to the sky, like the conical interior of a volcano, from which only the head will emerge. The dome, which encloses the statue had already been used at Seokguram (see above) and other rock-hewn monasteries in India and Sri Lanka (Figures 32 and 33), but in that case it was mostly devoid of light.

**Figure 32.** Tadao Ando, front view of the hill of the Buddha from the entrance arcosolium.

**Figure 33.** Side view of the new realization with the head protruding.

The intervention of Tadao Ando subtly alterates the topography of the site until the body of the statue is no longer visible [57]. Still, the people desirous to pray or pay homage to the Great Buddha have accesended to the this inner massive chamber, which is conical in shape, by an elaborate passageway through the newly created hill. Such corridor is conceived as a half cylindrical tunnel but reinforced with concrete folded plates on the inner vault.

The passageway is frontal to the sculpture, which has been rotated and relocated for the new project. However, progression to it is not straighforward because of the hidden mechanisms that the architect has devised (Figure 34).

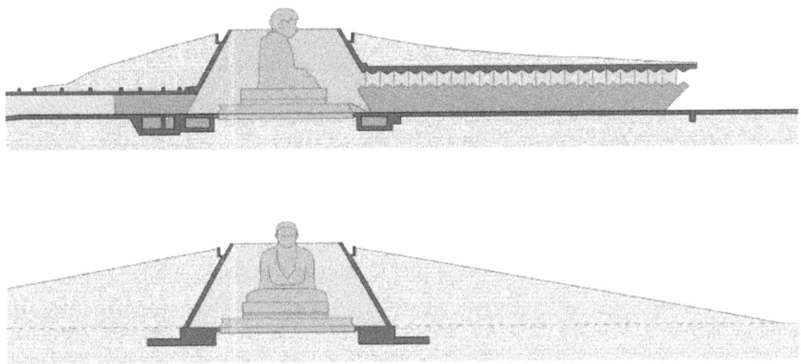

**Figure 34.** Sectional scheme of the Buddha Hill.

In the new project, we arrive to a ceremonial pond, which cuts the main axis and forces the visitor to take a pronounced detour around it, almost a circumambulation. The basin's surface is concave and prevents water splashes regardless of the wind condition, which can be steady around this area [58]. The floor of the stank is made of black marble and that gives the water a swarthy appearance, which, added to the myriad of ripples that form in the agitated surface, resembles the mythological Styx river, which the deceased were forced to ferry on Charon's boat in order to disembark on the Hades (Figure 35).

**Figure 35.** Side view of the black pool with aisles and concrete walls.

Thus, we are confronted with an eerie entrance to the underworld of the necropolis. A recreation of sorts of Kyoto's karesansui precincts, in this occasion not through a vital albeit inorganic garden but with a sullen and lifeless courtyard, it is as Hölderlin described in his famous verse [59]: "the Walls stay speechless and cold".

For the device of the oblong basin, it is coupled with a panelled concrete wall of reduced height. Lofty enough to impede the view from the outside while at the same time beckoning the grey sky to a stifled threnody.

With the so-created chicane of the lateral aisles of the pool, which reminds us of the sinuous corridor entrances of Egyptian temples in the ideas of Giedion and Le Corbusier [60], Tadao Ando forces us all but to crouch to regard the dark flow of water, and our gaze inevitably meets the reflection of the obscure cloudy sky on the grey diffuse walls and the liquid surface. We instantly known that we are close to being admitted to the presence of something supernatural.

At this point, we could agree with the Italian artist Edoardo Persico, who, praising the racing course at the roof of Fiat Automobile factory of Torino, wrote: "Under the skies and before God only the pond (track) is free" [61].

The extraordinary feature of low-height dense and opaque walls produces a secluded space, which is present at some Zen gardens and yards throughout Japan, although it is perhaps more clearly represented in the Suimui Utaki Garden Temple, at the entrance of Shurijo Palace in Naha (Okinawa, Rikyu Islands). It is known that when the king of the Rikyus visited a temple or shrine outside the Shurijo, he would open the gates and offer prayers at this diminutive shrine, and the priestesses would perform exquisite ceremonies (Figure 36). The plants within the rare masonry wall are banyan trees and Japanese laurels [62].

**Figure 36.** The Suimui Utaki garden temple. Shurijo (Okinawa).

After traversing the black pool (Figure 37), we get through the vaulted arcosolium (an arch under which interments are performed), a sort of tunnel 40 metres long at the end of which stands the surmised body of the statue, and a whitish marble light shines at the extreme [63]. Nonetheless, only when we enter the chamber that is open to the sky are we allowed to contemplate the long-sought hallowed face (Figure 38).

Finally, we have reached the sacred hall or garbagriha in Sanksrit terms. The space in which the deity manifests and where enlightment should take place for those who keep a pure vision (Figure 39). This huge conical space is somehow otherworldly, to walk around an inclined cone wall of huge dimensions is by no means comfortable but then we realize of the small chapels that surround the great Buddha; in them, different bosatsu (boddhisattva) and arahats (distinguished disciples) are worshipped. It is again a sublimated space in which the physical body must be transformed into a spirit, only at that moment we raise our heads and look to the zenith, from the Arabic word "semt" سَمْت, which means the path.

**Figure 37.** Side cylindrical pavillion at the extreme of the black pool.

**Figure 38.** The Great Buddha inside an open conical dome.

**Figure 39.** Zenithal vision of the conical dome.

## 4. Discussion

Founded on the previous results, the authors can demonstrate a tendency in Buddhist religious architecture, which reveals specific structures colligated to a sort of all-pervading cosmic harmony. This pertains to a well-rooted Asian tradition that should be increasingly recognizable in the years to come. In the authors' theory, such designs based on archeology show a clear direction for the recreating spaces of spirituality in Eastern Asia [63].

For instance, as we have just seen in the hill of the Buddha at Takino, in fact a concrete shell, with pleated or folded panes, the protruding head of the Buddha is now oriented to the northwest, exactly the sunrise position of the summer solstice, known in Japanese as Geshi 夏至.

In this magnificent roofless structure, Tadao Ando is able to reconcile his main two design paradigms and interests. The first one is the sunlit volume revolving around a centre that he exemplifies in the Roman Pantheon (Figure 40). He has frequently pointed out that such space cannot exist in nature, but it is naturalized by the due course of the sun and the seasons. That is the conceptual reason why the Buddha statue cannot be completely covered for instance with a glass dome, let alone technical hindrances, as it would have prevented climatic events from happening.

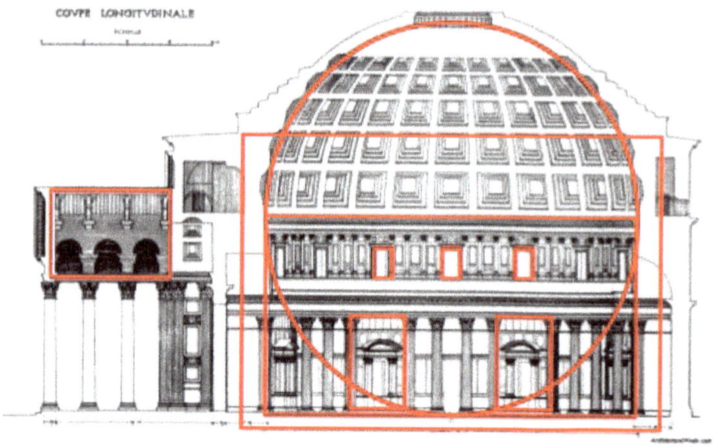


**Figure 40.** Roman Pantheon. Scheme of proportions that reveals the perfect sphere enclosed by its structure.

The second one, reinforcing the first, would be a yearning for total integration with nature and specifically with wind and water, the ingredients that enhance the growth of vegetation and the well-known fundaments of Feng Shui 風水. For instance, with the plantation of blossoming violet lavenders, the architect is capable to ensure a distinctive colour in the project for each one of the four seasons in accordance with the Daoist tetralogies, similar to Heidegger's Geviert (or the Fourfold, earth, sky, mortals, and immortals) [64], for winter it would be the white in the snow, in spring the lush green of the new leaves, fragrant purple flowers of lavender to the summer, and finally red brown for the fall.

A thorough cosmic reconstruction of the Universe is performed in this open dome shrine. The philosophical ideas underlying Buddhist order, such as the concept of Garan 伽藍 are sustained. This term originally meant a monastery yard but evolved to signify the harmonic realm created by the Buddha known as Konpon Daitō 根本大塔 [64,65] or axis mundi. The secluded space conceived for the Buddha along with his entourage, radiate and resonate in unison as a celestial harp through the land (Figure 41).

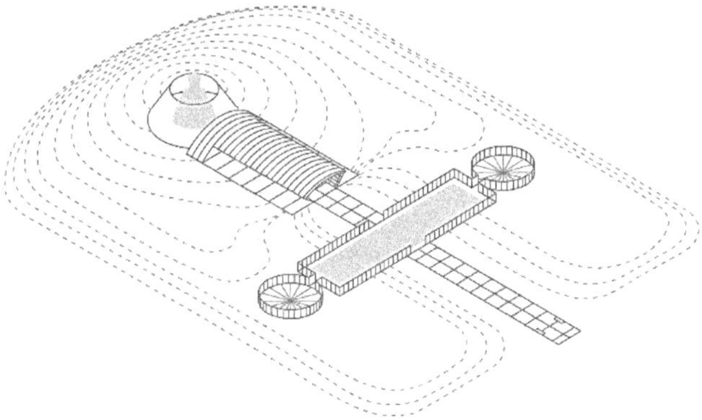

**Figure 41.** Perspective Scheme of the Buddha Hill and the traverse black pond.

But we must be advised that reducing or concealing the importance of the Buddha, with the hill operation described, also implies increasing the effect of the other heathen temples and statues, particularly the reconstruction of Stonehenge. There seems to be only a slight connection between them although we believe that deep inside they both honour the ancestral Japanese tradition of Mukaebi 迎え火.

Mukaebi, roughly translates as the welcome back fire. In ancient Japan, the ashes of respected ancestors and especially of young children were not interred in communal cemeteries, instead they were deposited near the threshold or entrance fence of a household. By treading over the cinders, the parents pined for the rebirth of their departed infants, especially in the wombs of fertile wives. Such ceremony has been maintained to our days and is performed each year to initiate the Obon season in July or August, depending on the region [66]. When the embers of the homecoming ritual fire are extinguished and turn to ashes they are buried near the small entrance garden conducing to a house and all the members of the family need to walk once over them, pleading for eternity.

In Takino, all the cinders, by the thousand, are deposited near the entrance and in the circular pavilions of the shrine of the great Buddha or the threshold of the Stonehenge in special sealed chambers; therefore, such ancestral ritual is re-enacted almost inadvertently every single time a visitor enters the said monuments [32,67,68]. The notable work of Ando with deft organization of an underground sustainable architecture has inspired other artists and colleagues inside and out of Japan to explore the concepts of hypogeal spaces like a catacomb or necropolis. Some of them are James Turrell (Figures 42 and 43) and Inmaculada Rodriguez Cunill (Figure 44). In the Naoshima art garden, Turrell has collaborated with Tadao Ando [69] to produce elliptical voids intended to contemplate the lights from the sun and the sky combined, especially at twilight. One of them has derived into a hotel for visitors (Figure 42).

In Spain, Turrell's work at the Montenmedio Hacienda (Figure 43), a natural open-air space of international resonance (supported by the contemporary art museum CARS), shows echoes of the cases examined above. The concealment of the central motif and its occasional appearance thanks to the solar cycle, point to an asceticism that characterizes most of Turrell's work [70,71].

On the other hand, the work known as Wolf Pack City (Figure 44) uses light through recycled materials that reflect luminous flux as if they were gold or silver (but they are not—it is actually refuse) and keeps the remains of despondent citizens through the names of those who could not achieve peace. Far from being a space to rest (it may seem so by the denomination of "necropolis", by the apparent noble materials, etc.) it is rather a space of rage, where we find a "corridor of the putrefaction" of the bodies, or a "corridor of the whitened sepulchres" (alluding to the biblical phrase of Christ). Such peculiar necropolis represents the utmost extreme from a space of tranquillity and transcendence. It is more of a scream of anguish than the silence conveyed by the halls of the Buddha discussed previously. The darkness of the catacombs is replaced by the (false) ostentation of the materials that conduce to light through reflection [72]. Although we might be critical with the eventual commercial use of one of the short symmetric cylindrical pavilions at the side of the entrance pond (Figure 45), in the end we recognize that Tadao Ando's formidable proposal evolves in the following sense: at first just intends to conceal by stealth the towering Buddha, which was found intimidating by the visitors. Later he arrives to a very keen solution of surrounding the statue with an artificial hill and an inner dormant volcano. Finally, upon the so-created cone, once it erupts, flowers of Mandala will rain as specified in the Lotus Sutra 雨曼陀羅華. To perfect the metaphor, these flowers should have been wysterias (Fuji 藤) but resulting impractical at this place, he has substituted them with more resistant lavenders [73,74]. It is a regal echo of mount Fuji (富士山) and a suitable crown for the Majesty of Death, whose virtue solely is to be allowed immortality.

**Figure 42.** James Turrell and Tadao Ando. Eco-Hotel at Naoshima.

**Figure 43.** James Turrell. Second Wind. Vejer de la Frontera. Spain.

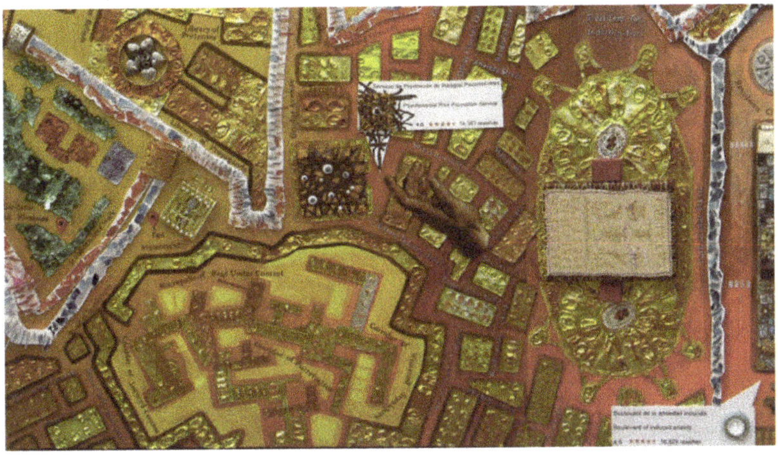

**Figure 44.** Excerpt from the "Wolf Pack City" showing the Necropolis of rage under control. Some of the wings of the building suggest broken swastikas. Work by artist I. Rodriguez Cunill (2021).

**Figure 45.** Side view of the black pool with short cylindrical pavilion for part of the funeral urns.

## 5. Conclusions

With these results and discussion, we have brought to the surface how, in a somewhat remote region of Japan, imagined archeological vestiges have been put to work to create an attractive park, privately owned, and managed, which serves as an expansive green area and at the same time satisfies the afterlife aspirations of the community, although with a considerable cost. An artificial earthen shell that generates an open dome shrine has been executed. The renovated complex is provided with ceremonial access and other facilities intended for religious services. The intervention completed by Tadao Ando is successful as much as it is able to colligate purposes of architectural excellence with religious and teleological yearnings. The power of archeology to provide significance to funereal monuments that are in this way transformed into necropolises and cities of the dead is remarkable, and it can even revalorize the land use. For instance, we know that Varanasi in India is a city to die in, to avoid reincarnation and the tiresome wheel of samsara. Still, we may wonder if the transition from life to death is operated smoothly in the cities in which we inhabit or rather is the other way round. At any rate, it is an enigmatic process that has inspired contemporary artists like Noguchi, Mishima, Turrell, and Rodriguez-Cunill.

The latest realizations at Takino introduce a new dimension of Buddhism ascetics, probably inaugurated by the Ajanta Caves in India, that is, the hidden grotto or monastery. In this manuscript we have detailed how the evolution from the ceremonial tumuli to the Buddhist chaitya must have been a strenuous process involving years of planning and painstaking labour.

It was raised when the adepts of the Buddha, especially of the Lesser Vehicle prevailing branches, at the times of monsoon weather for instance, could not reach out of their monasteries to preach or collect alms and, therefore, decided to work in the underground to spruce their halls of prayer and meditation. The tradition of showing the face of the Buddha unlike the adytum of Indian temples or the concealed Buddha represents the transition from Hinayana or esoteric Buddhism to the more social religiosity of the Mahayana, which prevails now in Japan and Korea and at a former time in China. In Japan, it appeared earliest at the Byôdoin Temple discussed in the article.

A revealing example of how this transition has operated in some cosmological aspects of the project of the Hill, is that the original position of the Buddha at Takino, which was faced as usual to the south, has turned to face the northwest to mark the sunrise on the summer solstice.

At the time of completing the Hill of the Buddha project, the promoters of Takino Park faced another key decision as if to give a new entity and reinforced interest to the Stonehenge ring by having veiled the importance of the Buddha.

There was, in Japan, a historic period referred to as Sakoku 鎖国, or the "enchained country", in which no one was permitted outside the archipelago for any reason, and that included the Japanese inhabitants themselves. Such extreme insularity has resulted to our minds in a necessity of marking the edges of the world with monuments, and in this way establish a bond with other isles no matter how remote. This would explain why the Moai from Ester Island and the Stonehenge circle have been brought forth to the collective imaginary at this cemetery in Sapporo.

The insight that the authors propose after studying in deep Takino funereal park is that invoking archeology provides us with a powerful and effective tool to establish a sure-footed path of creation towards a more serene landscape architecture, which, in turn, renews the much desired link between human constructions, culture and nature or, in other words, an alternate road to endurance through sustainability. The answer that it offers to the eternal question of how are we to live is as such: be boundless like the Sky and the four Directions, sustain all the creatures in your love.

**Author Contributions:** Conceptualization, J.C.-L. and I.R.-C.; methodology, J.C.-L.; software, C.P.-B. and M.G.-V.; formal analysis, M.G.-V. and C.P.-B.; investigation, J.C.-L. and I.R.-C.; writing—original draft preparation, J.C.-L. and I.R.-C.; writing—review and editing, V.M.-S.; visualization, M.G.-V.; supervision J.C.-L. and I.R.-C. All authors have read and agreed to the published version of the manuscript.

**Funding:** This research received no external funding.

**Data Availability Statement:** The original contributions presented in this study are included in the article. Further inquiries can be directed to the corresponding author.

**Acknowledgments:** J. Cabeza-Lainez wants to praise the creative genius of Mishima Yukio. The forgotten sons are remembered. Many thanks to Mattarena Tuki from Hangaroa and Antonio Rodríguez de las Heras for his kind help. I. R.-C. wishes to pay homage to Eva Guill-Walls, also recognizes Maria Luisa Rodriguez Cunill and Antonio Maestre Cernadas. V. Marquet-Saget expresses deep gratitude towards his grandfather, affectionately called 'Boby', who first introduced him to the wonders of architecture. César Puchol wants to thank Andrea Kalff and her family for offering sustenance in Hawaiian travels. Miguel Gutierrez-Villarubia would like to thank Iulia Sanchez Martin for contributing with photographs and exploration of Japan.

**Conflicts of Interest:** The authors declare no conflicts of interest.

## References

1. Hudson, M.J. *Ruins of Identity. Ethno-Genesis in the Japanese Islands*; University of Hawaii Press: Honolulu, HI, USA, 1999; ISBN 0824821564.
2. Choi, J.-S. *Gyeongju, the Heart of Korean Culture: A UNESCO World Heritage*; Hanul Publishing Group: Paju, Republic of Korea, 2011.
3. Umesao, T. *An Ecological View of History. Japanese Civilization in the World Context*; Trans Pacific Press: Melbourne, Australia, 2003; ISBN 9781876843892.
4. Umesao, T.; Befu, H.; Kreiner, J. *Japanese Civilization in the Modern World: Life and Society*; Senri Ethnological Studies: Osaka, Japan, 1984.
5. Umesao, T.; Jennison, R. Japan as Viewed from an Eco-Historical Perspective. *Rev. Jpn. Cult. Soc.* **1986**, *1*, 25–31.
6. Watsuji, T. *Climate and Culture: A Philosophical Study*; Greenwood Press: Westport, CT, USA, 1961.
7. Yu, H.-J. *Smiles of the Baby Buddha: Appreciating the Cultural Heritage of Kyongju*; Changbi Publishers, Inc.: Paju-si, Republic of Korea, 1999.
8. Chatzichristodoulou, M.; Jefferies, J.; Zerihan, R. (Eds.) *Interfaces of Performance*; Ashgate Publishing, Ltd.: Farnham, UK, 2009.
9. Salguero-Andujar, F.; Cabeza-Lainez, J. Comparative, geometric and structural analysis of the Sokkuram Temple at Kyongju (Korea). In Proceedings of the Second National Symposium on Cultural Heritage of Korea, Gyeongju, Republic of Korea, 11 November–3 December 2017.
10. Cabeza Lainez, J.M. *La Visión y La Voz: Arte, Ciudad y Cultura En Asia Oriental*; UCOPress Ediciones Universidad de Córdoba: Córdoba, Spain, 2017.
11. Olcina Cantos, J.; Martin Vide, J. *Influencia del Clima en la Historia*; Cuadernos de Historia; Arco Libros—La Muralla, S.L.: Madrid, Spain, 1999.
12. Cabeza Lainez, J.M. Desde Sri Lanka hasta Japón: Ideas acerca de la evolución del Stupa. In *La investigación sobre Asia Pacífico en España*; San Ginés Aguilar, P., Ed.; Colección Española de Investigación sobre Asia Pacífico 1; Universidad de Granada: Granada, Spain, 2007; pp. 553–570.
13. Salguero Andujar, F.; Rodriguez-Cunill, I.; Cabeza-Lainez, J.M. The Problem of Lighting in Underground Domes, Vaults, and Tunnel-Like Structures of Antiquity; An Application to the Sustainability of Prominent Asian Heritage (India, Korea, China). *Sustainability* **2019**, *11*, 5865. [CrossRef]
14. Ebersole, G.L. *Ritual Poetry and the Politics of Death in Early Japan*; Princeton University Press: Princeton, NJ, USA, 1989.
15. Rodriguez-Cunill, I.; Gutierrez-Villarrubia, M.; Salguero-Andujar, F.; Cabeza-Lainez, J. Sustainability in Early Modern China through the Evolution of the Jesuit Accommodation Method. *Sustainability* **2021**, *13*, 11729. [CrossRef]
16. Wertheim, M. *The Pearly Gates of CyberSpace*; W. W. Norton: New York, NY, USA, 2000.
17. Jones, L. *Hermeneutics of Sacred Architecture*; Harvard University Press: Cambridge, MA, USA, 2000; 2 Volumes.
18. Blake, W. *Milton a Poem, Copy B Object 2*; Eaves, M., Essick, R.N., Viscomi, J., Eds.; The William Blake Archive; British National Library: London, UK, 2019.
19. Matsuie, M. *The Summer House*; Kindle ed.; Libros del Asteroide: Barcelona, Spain, 2024.
20. Bollnow, O. *Human Space*; Hyphen Press: London, UK, 2011.
21. Barrett, J.C.; Boyd, M.J. *From Stonehenge to Mycenae: The Challenges of Archaeological Interpretation*; Bloomsbury Academic: London, UK, 2021.
22. Parker Pearson, M. *Stonehenge: A Brief History*; Bloomsbury Academic: London, UK, 2023.
23. Ramswamy, S.G. *Design and Construction of Concrete Shell Roofs*; CBS Publishers & Distributors: New Delhi, India, 2005.
24. Darvill, T.; Marshall, P.; Pearson, M.P.; Wainwright, G. Stonehenge Remodelled. *Antiquity* **2012**, *86*, 1021–1040. [CrossRef]
25. Anghel, A.A.; Cabeza-Lainez, J.; Xu, Y. Unknown Suns: László Hudec, Antonin Raymond and the Rising of a Modern Architecture for Eastern Asia. *Buildings* **2022**, *12*, 93. [CrossRef]
26. Castleden, R. *The Making of Stonehenge*; Routledge: London, UK, 1993. [CrossRef]
27. Hill, R. *Stonehenge*; Harvard University Press: Boston, MS, USA, 2021. [CrossRef]
28. Cleal, R.M.J.; Walker, K.E.; Montague, R. *Stonehenge in Its Landscape: Twentieth Century Excavations*; Archaeological Report 10; English Heritage: London, UK, 1995.
29. Cabeza-Lainez, J.M.; Salguero-Andújar, F.; Rodríguez-Cunill, I. Prevention of Hazards Induced by a Radiation Fireball through Computational Geometry and Parametric Design. *Mathematics* **2022**, *10*, 387. [CrossRef]

30. Millán-Millán, P.M.; Cabeza-Lainez, J. HBIM Methodology to Achieve a Balance between Protection and Habitability: The Case Study of the Monastery of Santa Clara in Belalcazar, Spain. *Buildings* **2022**, *12*, 510. [CrossRef]
31. Saraiva, L.; Martzloff, J.C.; Baldini, J.; Dhombres, J. *História Das Ciências Matemáticas: Portugal e o Oriente—History of Mathematical Sciences: Portugal and East Asia*; Fundação Oriente: Lisboa, Portugal, 2000.
32. Grayson, J.H. *Korea: A Religious History*; RoutledgeCurzon: London, UK, 2002; ISBN 0-7007-1605-X.
33. Kerr, A.; Sokol, A.K. *Another Kyoto*; Penguin Press: New York, NY, USA, 2018; ISBN 978-0141988337.
34. Bunko, T. *Memoirs of the Research Department of the Toyo Bunko (the Oriental Library)*; Publications—Tōyō Bunko. Ser. B; The Toyo Bunko: Tokyo, Japan, 1926.
35. Salguero-Andújar, F.; Cabeza-Lainez, J.-M. New Computational Geometry Methods Applied to Solve Complex Problems of Radiative Transfer. *Mathematics* **2020**, *8*, 2176. [CrossRef]
36. Cabeza-Lainez, J.M.; Pulido-Arcas, J.A. New configuration factors for curved surfaces. *J. Quant. Spectrosc. Radiat. Transf.* **2013**, *117*, 71–80. [CrossRef]
37. Cabeza-Lainez, J.M.; Xu, Y. Empire, Science and Geometry at the origins of early modern architecture. *Symmetry Cult. Sci.* **2020**, *31*, 321–351. [CrossRef]
38. Doganoff, I. *Berechnung Von Kugelschalen Über Rechteckigem Grundriss: Mit Berechnungsdiagrammen*; Mitteilungen aus dem Institut für Massivbau Hannover; Werner: Düsseldorf, Germany, 1962.
39. Cabeza-Lainez, J. Innovative Tool to Determine Radiative Heat Transfer Inside Spherical Segments. *Appl. Sci.* **2023**, *13*, 8251. [CrossRef]
40. Doganoff, I.; Hoffmann, C.; Rühle, H. *Shell and Fold Structure Roofs Out of Precast, Pre-Stressed Reinforced Concrete Elements*; Bulletin of the International Association for Shell Structures: Düsseldorf, Germany, 1964.
41. Cross, H. The Column Analogy: Analysis of Elastic Arches and Frames. Available online: https://www.ideals.illinois.edu/items/5207 (accessed on 10 January 2025).
42. Cabeza-Lainez, J.M. Las bóvedas de cerámica armada en la obra de Eladio Dieste. Análisis y posibilidades de adaptación a las condiciones constructivas españolas. In *Actas del Tercer Congreso Nacional de Historia de la Construcción*; Santiago, H., Ed.; CEHOPU: Madrid, Spain, 1996.
43. Cabeza-Lainez, J. Architectural characteristics of different configurations based on new geometric determinations for the conoid. *Buildings* **2022**, *12*, 10. [CrossRef]
44. Pierre, C. Dome Architecture in Ancient Asian Art and Seokguram Grotto: In comparison with Western Dome Constructions. *Korean J. Art Hist.* **2022**, *313*, 155–170. [CrossRef]
45. Dessi, U. Religious Diversity in Japan. In *Religious Diversity in Asia*; Borup, J., Fibiger, M.Q., Kühle, L., Eds.; BRILL. ProQuest Ebook Central: Leiden, The Netherlands, 2019; pp. 48–66. ISBN 9789004415812.
46. Harvey, P. *An Introduction to Buddhism: Teachings, History and Practices*, 2nd ed.; Cambridge University Press: Cambridge, UK, 2012; ISBN 978-0521676748.
47. Uchio, T.; Harrison, B. Japanese and Easter Island—Ecotourism and the Relationship with Local Residents. *Jpn. J. Policy Cult.* **2024**, *32*, 77–106.
48. Lipo, C.P.; Hunt, T.L.; Haoa, S.R. The 'walking' megalithic statues (moai) of Easter Island. *J. Archaeol. Sci.* **2013**, *40*, 2859–2866. [CrossRef]
49. UNESCO. Seokguram Grotto and Bulguksa Temple. UNESCO World Heritage Convention. Available online: https://whc.unesco.org/en/list/736/ (accessed on 27 February 2025).
50. Puchol, C.; Chon, S.-S. Overseas propagation of Korean Shamanism through Diaspora movement and the linage 'New York Musok', 실천민속학연구 제40호. Available online: https://www.kci.go.kr/kciportal/ci/sereArticleSearch/ciSereArtiView.kci?sereArticleSearchBean.artiId=ART002877487 (accessed on 10 January 2025).
51. Zucker, P. Ruins. An Aesthetic Hybrid. *J. Aesthet. Art Crit.* **1961**, *20*, 119–130. [CrossRef]
52. Lipo, C.P.; Hunt, T.L. Mapping prehistoric statue roads on Easter Island. *Antiquity* **2005**, *79*, 309–317. [CrossRef]
53. Cabeza-Lainez, J. Finding the Exact Radiative Field of Triangular Sources: Application for More Effective Shading Devices and Windows. *Appl. Sci.* **2023**, *13*, 11318. [CrossRef]
54. Hinz, C.; Wilhelm, J.L. An environmental aesthetic approach to ruins: The situativity, relationality and emergence of experiences in a derelict sanatorium. *Soc. Cult. Geogr.* **2024**, *26*, 201–217. [CrossRef]
55. Rodriguez-Cunill, I.; Cabeza-Lainez, J.; Lopez-Cabrales, M.d.M. Art and the City Fiction in Japanese American Internment Camps: Sequels for Resiliency. *Arts* **2023**, *12*, 195. [CrossRef]
56. Cabeza-Lainez, J. A New Principle for Building Simulation of Radiative Heat Transfer in the Presence of Spherical Surfaces. *Buildings* **2023**, *13*, 1447. [CrossRef]
57. Salguero-Andújar, F.; Prat-Hurtado, F.; Rodriguez-Cunill, I.; Cabeza-Lainez, J. Architectural Significance of the Seokguram Buddhist Grotto in Gyeongju (Korea). *Buildings* **2022**, *12*, 3. [CrossRef]
58. Rodríguez Cunill, M.I. Algunas precisiones en torno al marco en la obra transitable. *Bol. Arte* **2018**, 551–572. [CrossRef]

59. Cabeza-Lainez, J. New Geometric Theorems Derived from Integral Equations Applied to Radiative Transfer in Spherical Sectors and Circular Segments. *Mathematics* **2024**, *12*, 2875. [CrossRef]
60. Frazer, J.G. *La Rama Dorada: Magia y Religión*; Fondo de Cultura Económica: Ciudad de México, Mexico, 2021; ISBN 9786071606464.
61. Gutierrez-Villarrubia, M.; Cabeza Lainez, J. La majestad de la muerte. In *Postrimerías, Valdés Leal desde la Facultad de Bellas Artes de Sevilla*; Infante del Rosal, F., Arjonilla Alvarez, M., Bilbao Peña, D., Eds.; Colección Catálogos, n.° 1; Editorial Universidad de Sevilla: Sevilla, Spain, 2022; ISBN 9788447224043.
62. Wittgenstein, L. The Mythology in Our Language. Remarks on Frazer's Golden Bough. HAU Books: Chicago, IL, USA, 2018; ISBN 9781912808403.
63. Mishima, Y. *The Temple of Dawn*; Shinchosa: Tokyo, Japan, 1973.
64. Gudmunsen, C. *Wittgenstein and Buddhism*; Palgrave Macmillan: London, UK, 1997; ISBN 9781349031306.
65. Hu, H. 明朝帝王陵 *(增订版) [Mausoleum of the Emperor of the Ming Dinasty Ming (Revised Edition)]*; Xueyuan Editorial: Beijing, China, 2013; ISBN 9787507742718.
66. Liu, Y. 中华文明史视角下的陵墓考古 (Archaeology of the mausoleum from the perspective of the History of Chinese Civilization). *People's Trib.* **2021**, *32*, 102–106.
67. Quesada-García, S.; Valero-Flores, P.; Mendoza-Alvarez, D.; Cabeza-Lainez, J. Cognitive Accessibility in Rural Heritage: A New Proposal for the Archaeological Landscape of Castulo. *Sustainability* **2022**, *14*, 11581. [CrossRef]
68. Lee, O.Y. *Smaller is Better. Japan's Mastery of the Miniature*; Kodansha USA Publishers: New York, NY, USA, 1984.
69. Mengke (s.f. [c. 300 a.C.]). 孟子·告子章句上 *(Mencius Gaozi, Chapter 1)*.
70. Wang, Z.; Guo, W.; Zuo, X. 明十三陵整体空间营造手法探讨 (Dissertation on general techniques of spatial construction of the Ming Tomb). In *Space Governance for High-quality Development, Proceedings of the Annual National Planning Conference of China (ANPCC), Chengdu, China, 25–30 September 2021*; Urban Planning Society of China: Beijing, China, 2021.
71. You, W.B. 中国"儒、道、佛"生死伦理观比较 (Comparison of the Ethics of Destiny between Chinese philosophy of confucianism, daoism and buddhism). 中国医学伦理学 *Chin. Med. Ethics* **2008**, *21*, 116–117+125.
72. Xunzi (s.f. [3rd Century b.C.]. 《禮 論》 *(Theory of the Rites)*.
73. Peña-García, A.; Cabeza-Lainez, J. Daylighting of road tunnels through external ground-based light-pipes and complex reflective geometry. *Tunn. Undergr. Space Technol.* **2023**, *131*, 104788. [CrossRef]
74. Rodríguez Cunill, M.I. Medio ambiente, videoinstalaciones y otras propuestas artísticas audiovisuales. In *Medios de Comunicación y Medio Ambiente*; Gutiérrez-SanMiguel, B., Ed.; University of Salamanca: Salamanca, Spain, 2002. Available online: https://dialnet.unirioja.es/servlet/articulo?codigo=294636 (accessed on 19 December 2024).

*land*

MDPI

*Article*

# Earthquake Destruction and Resilience in Ancient Helike, Gulf of Corinth, Greece: A Case Study of Past Human–Environment Relationship

**Dora Katsonopoulou [1], Ioannis Koukouvelas [2,*] and Mariza Kormann [3]**

[1] The Helike Society and the Helike Project, 10682 Athens, Greece; eliki@otenet.gr
[2] Department of Geology, University of Patras, 26504 Patras, Greece
[3] Independent Researcher, Oxfordshire OX16 9AE, UK; mariza.kormann@kellogg.oxon.org
* Correspondence: iannis@upatras.gr

**Abstract**

Geoarchaeological work and excavations of the Helike Project over the last 30 years in the Helike coastal plain, Gulf of Corinth, have yielded abundant evidence on ancient settlements, as well as the surrounding landscape and environmental changes that resulted from geological phenomena and catastrophic events. The research methods applied by the Helike Project followed a multidisciplinary approach, including combined archaeological excavations and palaeoseismological trenching, geophysical prospection, archaeometric, environmental, and soil micromorphology analyses, and computer-based landscape modelling. A wealth of settlement remains that were unearthed across the plain, ranging in date from the Early Helladic period (3rd millennium BC) to the Late Antiquity (5th century AD), indicates that the ancient inhabitants of the area chose to always resettle in the area by adjusting their ways of living to the geomorphology and natural hazards, prevailing each time. Our results show that disasters in the area increased between the Geometric and Roman times due to severe earthquakes that occurred approximately every 300 years. In particular, archaeological and geological finds recovered from the Late Classical–Hellenistic Helike settlement, which was revived in the western part of the plain shortly after the disastrous 373 BC earthquake, have enriched our knowledge regarding the historical seismicity of the region and past human–environment relationships.

**Keywords:** Ancient Helike; 373 BC earthquake; ancient settlement and mobility; environmental changes; resilience; Helike Fault earthquakes; Gulf of Corinth; Greece

## 1. Introduction

Helike, the principal city of ancient Achaea, located on the southwest shore of the Gulf of Corinth, was founded in the Bronze Age and was severely damaged by an earthquake and tsunami in 373 BC. This remarkable catastrophic event made a great impression on the ancient world. It was reported by many ancient authors, contemporary and of later dates, down to the middle Byzantine times [1–3]. The area of Helike is one of the most seismically active regions in Greece [4–7], bound to the south by the Helike Fault and to the north by the Aigion Fault.

Because Helike was reportedly submerged by the 373 BC earthquake, previous efforts to locate its site were concentrated in the sea, and their results were inconclusive [8]. To resolve whether the ancient city lay in the sea or on land, the Helike Project launched a long-term multidisciplinary research program in 1988. First, a systematic underwater survey



https://doi.org/10.3390/land14071392

was conducted southeast of Aigion using a side-scan and sub-bottom sonar (Figure 1a). The results showed no ruins of an ancient city underwater, and thus, research was redirected to the adjacent broad, coastal plain between the Selinous (west) and Vouraikos (east) Rivers using extensive borehole drilling and geophysical exploration [9,10] (Figure 1a). Since 2000, the systematic excavations of the Helike Project have brought to light architectural remains dating from the Early Bronze Age (EBA), the Mycenaean, Geometric, Classical, Hellenistic, Roman, and Late Byzantine periods [11–13].

Georarchaeological studies of the finds recovered from the excavations suggested evidence of a destructive mudflow that followed the 373 BC earthquake [7] and a lagoon formation in the Early Helladic and Classical Helike sites [9,14]. Further, applications of soil micromorphology applied to the Late ClassicalHellenistic Helike site showed evidence of earthquakes and other geological hazards in soil and sediment in a direct stratigraphical association with the architecture of the site [15].

Digital elevation models (DEMs) have been used in the Helike region to model the terrain's characteristics, focusing on landscape models of the Early Helladic, Classical, Hellenistic, and Roman periods. The continuous occupation of the landscape is connected to a relationship between the sea and land, manifested through marine transgressions, earthquakes, tsunamis, tectonic uplift, sediment deposition, and delta subsidence [16]. These dynamic models allow us to enter the story of the Helike landscape at different temporal levels and scales, ranging from micro-faunal identification and absolute dating to large-scale spatial considerations, thus strengthening our archaeological interpretation of long-term trends, sequences, and changes. Archaeological evidence from the Helike Project excavations provides a web of interdisciplinary data that shows how past societies interacted with the environment, thus contributing to our understanding of the destruction/resilience system that can be useful for a sustainable future [17]. The collected data cover habitation in the area over a long chronological timeframe, ranging from the Early Bronze Age to the Late Antiquity. Resilience is the intrinsic ability of an ecosystem or a society to adjust its functioning prior to, during, or following environmental changes or natural disasters so that it can maintain its status [18].

In this paper, we used the results of the Helike Project's geoarchaeological work and excavations to produce an accurate timeline, which can reflect data of seismic history along the Helike Fault spanning almost three millennia. These data also offer reliable calibration dates for palaeoseismologically defined earthquakes. Finally, we provide tangible evidence of societal reactions to natural disasters, people's adjustments to environmental changes, and habitation patterns in earthquake-prone areas.

## 2. Geology and Seismology of the Helike Area

The Gulf of Corinth, located in central Greece, represents an active intra-continental rift zone that runs nearly perpendicular to the Greek orogenic belt. The rift's current geological activity is shown by prominently exposed fault scarps, faulted colluvial wedges, raised shorelines, and earthquake environmental effects, including landslides on land and underwater [7,19–23] (Figure 1).

Dating of uplifted syn-rift sediments in northern Peloponnese indicates that the rift began developing approximately 4 million years ago, progressing from east to west and from south to north [21,24–28]. From the lower Pleistocene period (1.8–1.5 Ma) onward, the deformation and sediment deposition within the rift have shifted northward alongside an increased extension rate [20–29].

The rift experiences an almost north/south-oriented extension, as indicated by both geodetic and fault-slip analyses [27,28,30,31]. Throughout the rift, the orientation of the fault array changes progressively from ENE to WNW as one moves from east to west [7,32–35].

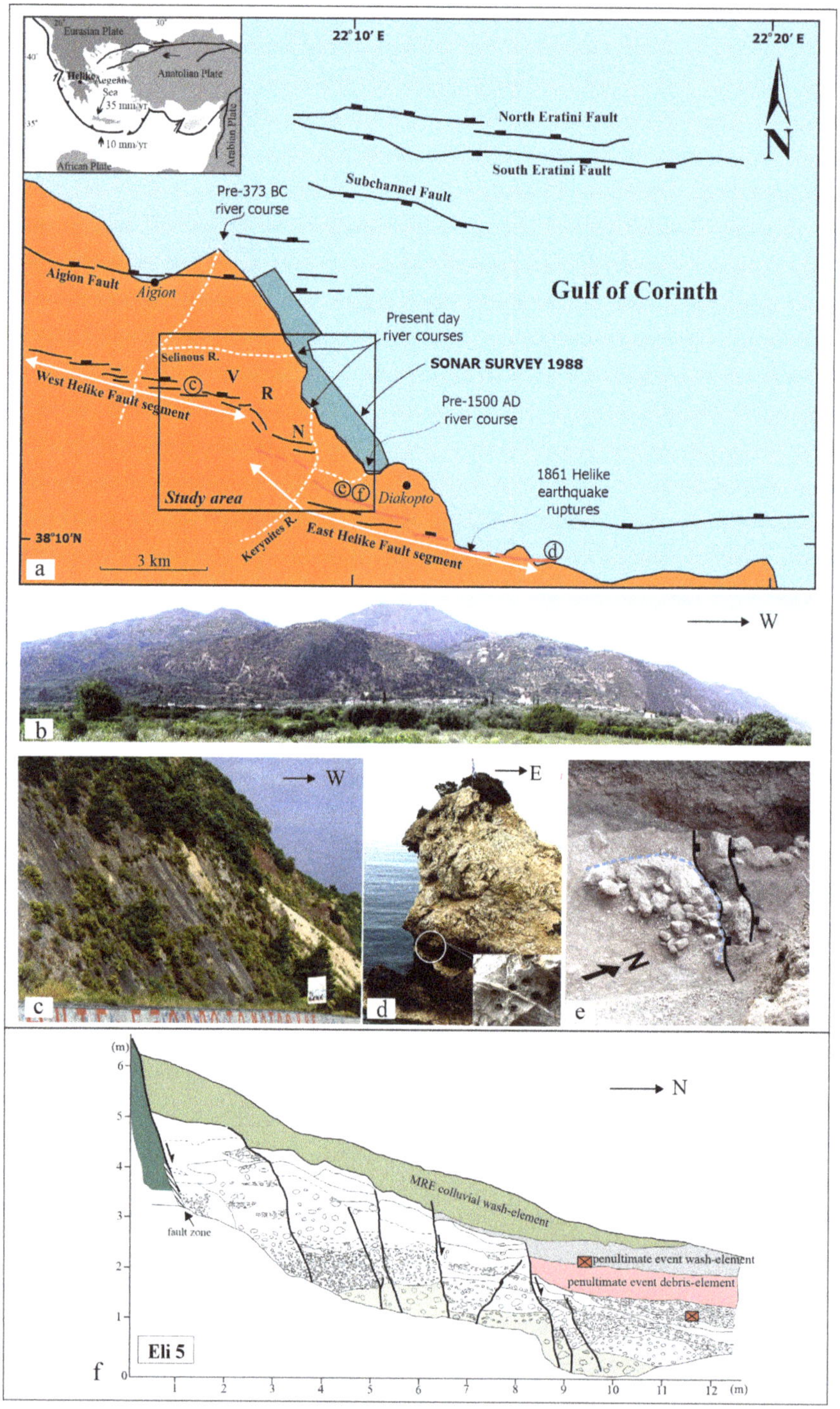


**Figure 1.** (**a**) Simplified structural map of the broader study area, presenting the major active faults in proximity with the Helike site, shown in the black rectangle (the map is modified after [7,33]. The inset

shows the geotectonic setting of the study area in the Eastern Mediterranean and Helike's location. Note: R = Rizomylos, N = Nikolaiika, and V = Valimitika indicate the main archaeological sites in the Helike region. (**b**) Panoramic view of the central part of the Helike Fault scarp. This photo is taken looking south, and the width of the photo is approximately 7.5 km. (**c**) Details of the exposed fault plane of the HF. This photo is taken looking west, and the height of the outcrop is 30 m. For the location of the photo, see (**a**). (**d**) Overview (view to the east) showing notched Mesozoic limestones that formed along the Helike Fault plane; for location, see (**a**). The inset shows the L. lithophaga hole entrances in limestone bedrock. For the location of the photo, see (**a**). (**e**) Photograph showing a semicircular pile of stones, outlined by a dashed blue line, excavated within the T13 trench. The thick black lines mark surface ruptures of the ancient remains during the 5th century AD earthquake (photo is modified after [36]). For the location of the photo, see (**a**). (**f**) Eli5 trench log and its interpretation in terms of palaeoseismic stratigraphic units and events across the HF (the log is modified after [37]). The red boxes represent charcoal-dated samples; for dates, see Figure 2 and Table 1. MRE = most recent event. For the location of the photo, see (**a**).

Historical earthquakes have been documented in the Gulf of Corinth rift [4,5,38,39] (Figure 1a). In particular, the seismic activity in the western Corinth rift, where the ancient city of Helike lies, is characterized by significant historical earthquakes and short-in-time swarms of clustered seismic events that lasted for one to two years [40–42]. The region between the Diakopton and Rion-Antirrion straits, according to seismic reflection studies and onshore surveys, is traversed by a complicated network of both offshore and onshore faults [7,33,43–47] (see Figure 1a). These investigations have revealed a complex basin structure with notable along-strike variations that are linked to alterations in the underlying basement fabric. Specifically, the deformation concentrates on normal faults dipping to the north, which deform the southern coast and cut through the offshore area of the rift. Evaluation of data from several palaeoseismological trenches has shown that the coastal faults of the northern Peloponnese exhibit high slip rates, varying from 0.3 to 11 mm per year [7,33,48,49]. Figure 1f shows an example of a palaeoseismological trench excavated in the Helike area.

### 2.1. The Seismic History of the Helike Fault

Understanding a fault's seismic history and its connectivity with nearby faults is important for defining two parameters of forthcoming earthquakes, i.e., their magnitude and time-frequency. In our case study, these two parameters are also critical for understanding the history of the settlement, which evolved in this central part of the Gulf of Corinth over a long period. To define these critical parameters, we have applied trench excavation using combined techniques of archaeology and palaeoseismology. The data acquired concern the dating and the magnitude of earthquakes and their environmental impact.

The Helike Fault (HF) controls the evolution of the Helike delta plain and the environmental changes that occurred in the area, e.g., a change in river course through time (Figure 1a). This fault has a clear surface trace (Figure 1b,c) that crosscuts a series of pre-earthquake landforms, and its steep fault plane offsets recent deposits. It is 23 km long, divided into two prominent segments (Figure 1a). Its trace is slightly irregular and separated from adjacent normal faults (Figure 1a). The HF is a crustal-penetrating fault, has a 600 m minimum cumulative throw, and separates a mountain range footwall block from an adjacent alluvial plain where three coalescence deltas have formed (Figure 1b). The HF dips consistently from 55–60° north (Figure 1c), and on a mesoscopic scale, it is accompanied by secondary synthetic normal faults. Similarly, like many normal faults, it decreases its dip with increasing depth (listric geometry) and sole into a north-dipping low-angle fault [40,50]. This low-angle fault at depth coalesces with many higher-angle normal faults in the Gulf of Corinth [40].

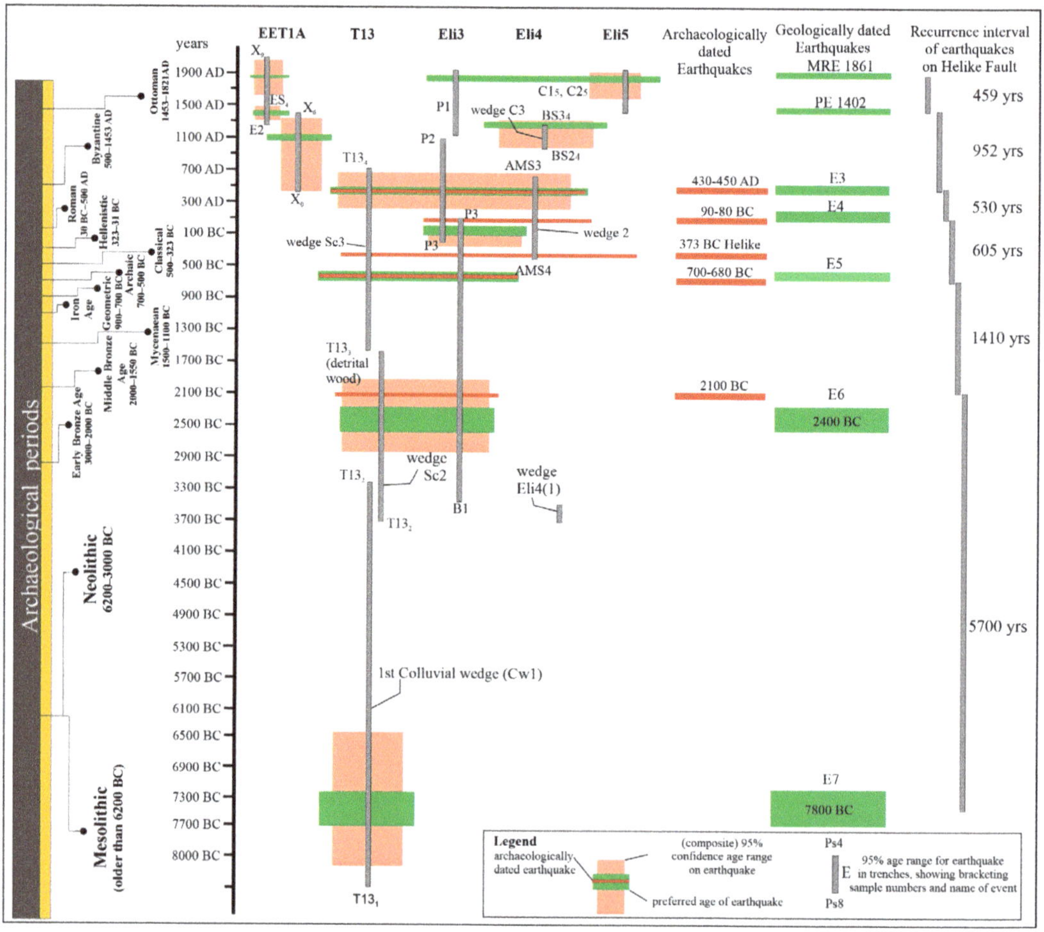


**Figure 2.** Chronogram of palaeoearthquakes, recognized in the EET1A [44], Eli3 and T13 [33,36], and Eli4 and Eli5 [37] trenches excavated across the HF, based on $^{14}C$ and archaeological dating. For details, see the text. The recurrence interval of the earthquakes is defined based on the historical data for MRE (most recent event) and PE (penultimate event), and the palaeoearthquakes correlated with the archaeological evidence for E3–E6 events.

### 2.2. *Palaeoseismological Studies of the Helike Fault (HF)*

Since the HF controls the landscape and the shorelines of the western Gulf of Corinth, it naturally affects ancient settlements developed on the Helike Delta (Figure 1). The slip rate of the fault ranges from 0.3 to 13.7 mm/yr, depending on the time scale and method of observation [39,40,51–57]. For this study, the slip rate of the fault is recalculated based on a series of palaeoseismological trenches [33,36,37]. In total, 13 single-slot California-type trenches were excavated across the fault scarp, which is east of the Kerynites River. Their stratigraphy is dominated by coarse-grain clastic materials (Figure 1f), which cover a period of approximately the entire Holocene (Figure 2). In particular, the stratigraphy of the trenches is controlled by three splays of the HF, and within their tectonostratigraphy, seven fault-controlled colluvial wedges were noted [33,36,37] (Figures 1f and 2).

Among the 13 trenches excavated, are reviewed in this study the EET1A [54–57], Eli3 and T13 [33,36], and Eli4 and Eli5 [37]. These five trenches include all previously mentioned colluvial wedges, dated by 20 $^{14}C$-calibrated samples with two standard deviations of uncertainty. Another criterion for their selection is based on the archaeological finds unearthed within their limits (Figure 1e). T13 reached a maximum depth of 5.5 m and contained a Roman tomb offset by a past earthquake surface rupture. In addition, a tile-covered Roman grave was discovered in Eli3. Notably, the trench T13 stratigraphy expands approximately over the entire Holocene, and Eli3 shows the best-preserved colluvial wedge of the 1861 earthquake [33]. In this colluvial wedge, pottery sherds of the Ottoman period were found. Eli4, Eli5, and EET1A preserved a clear tectonostratigraphy of the last four events hosted along the Helike Fault. These events are annotated as MRE, PE, E3, and E4 (Figure 2).

To estimate the age of the earlier earthquake events (E4–E7), dating was based on palaeoseismology, although with significant uncertainty (Figure 2, Table 1). In particular, the E7 event is recognised in only one trench. In the T13 trench, E7 is constrained based on the dates of two samples, T13-1 and T13-2 (Table 1). Lacking further palaeoseismological data, we considered that the probable age of the E7 is close to the middle of this period, at about 7.8 $\pm$ 2.4 ka BC. Event six (E6) is again poorly dated at about 2.4 $\pm$ 1.4 ka BC, based on two $^{14}C$-dated samples from trench T13 [36] (Figure 2). However, this event was dated ca. 2100 BC during archaeological excavations at the Early Bronze Age Helike settlement (see next Section 3.1). Thus, we adopted the archaeologically determined age for this event. Similarly, the E5–E3 events are identified based on $^{14}C$-calibrated data with archaeological dating (Figures 2 and 3, Table 1). Therefore, seismic events dated via archaeological data obtain accurate age estimates in contrast with ages constrained only by $^{14}C$-calibrated data. The archaeometric error usually ranges between 300–615 years (Figure 3). Accordingly, we can constrain the recurrence interval and the true slip rate (SRT) between events (Figures 2 and 3).

Finally, based on the palaeoseismological and archaeological data of the excavation of these five trenches and their dating, we compiled a new complete seismic history of the HF. This timeline resolves parameters such as the slip per event, the apparent slip rate (SRA), and the true slip rate (SRT), defined by six closed seismic cycles (Figure 3). A series of historical earthquakes (1861, 1402, 1147) have been documented and could correspond particularly with the penultimate event (PE) [4,5,38,39]. Among them, we have more than one candidate event, but we lack the exact geographical limits, and sometimes we have only contradicting age determinations. Two notably harmful and thoroughly recorded earthquakes occurred in 1147 and 1402 AD, respectively. Both had devastating consequences for the town of Galaxidi, located opposite our study area, as well as for the entire region around Helike between Xylokastron (east) and Aigion (west). We suggest that the Helike Fault most probably hosted the 1402 earthquake. The MRE is a well-documented event in our study area, which occurred in 1861 in the western Gulf of Corinth [4,38–40,58].

Based on the new data, we recalculated the apparent slip rate of the fault as appeared in the published papers by [33,36] at an increased value, from 0.5 mm/yr to 0.61 mm/yr. In addition, the SRT between events ranges between 0.17 to 2.6 mm/yr. However, as the SRT = 0.17 mm/yr for the event E7 (Figure 3) was recognised in only one trench, and because archaeological data for this period are missing, we can exclude E7 from our calculations because its time constraint is uncertain. Then, the range of the SRTs for all other events shows a bimodal distribution, with a low SRT ranging from 0.5 to 0.7 mm/yr. Moreover, a high SRT ranged from 2.2 to 2.6 mm/yr. Accordingly, a time-predictable recurrence model for the Helike Fault is proposed based on the last five seismic cycles. The maximum anticipated earthquake magnitudes reach approximately 6.7 [7,33,36].

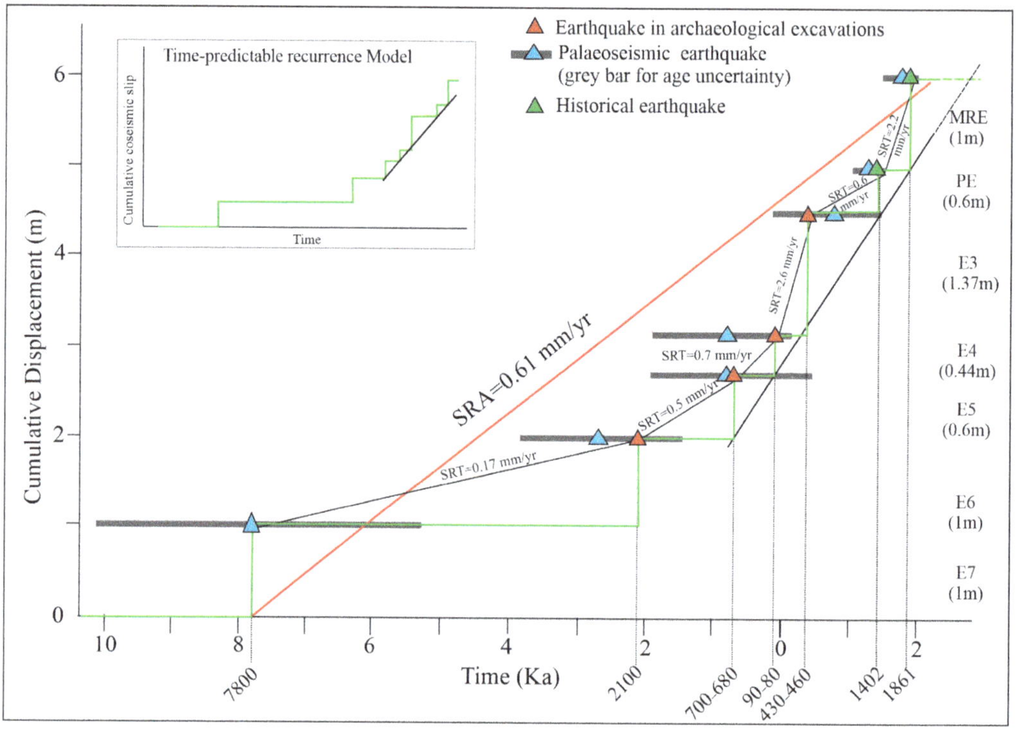


**Figure 3.** Slip history diagram of the HF. Geological records extend from today to 9.8 ka. The red line indicates the apparent slip rate (SRA), and the black lines indicate the true slip rate (SRT) defined by closed seismic cycles. The inset shows the proposed earthquake predicted model adopted from the published literature (for discussion, see text).

**Table 1.** Radiocarbon ($^{14}C$) and thermoluminescence (TL) ages of samples recovered from Eli3, T13, Eli4, Eli5 and EEFT1A palaeoseismological trenches.

| | Ref.no./ Laboratory | Layer | Dated Material | Measured Age TL or $^{14}C$ Age | Calibrated Age |
|---|---|---|---|---|---|
| Eli3 | P1 ** | C3 | Pottery | TL 745 ± 56 | |
| | P2 ** | Top of C1 | Pottery | TL 1079 ± 85 | |
| | P3 ** | Base of C1 | Pottery | TL 2049 ± 150 | |
| | B1 * | | Bone | $^{14}C$ 3020 ± 880 | 3600 BC–700 AD |
| | W1 * | F4 | Wood fragment | $^{14}C$ 300 ± 120 | 1440–1810 |
| T13 | T13-1 * | Top of SC3 | Soil | $^{14}C$ 8850 ± 110 | 8300–7600 BC |
| | T13-2 * | Base of L | Soil | $^{14}C$ 4820 ± 90 | 3790–3370 BC |
| | T13-3 * | Top of SC1 | Charcoal | $^{14}C$ 3470 ± 80 | 1980–1600 BC |
| | T13-4 * | Top of BO1 | Charcoal | $^{14}C$ 1450 ± 80 | 420–720 AD |
| Eli4 | $BS3_4$ * | Base of C3 | Soil | $^{14}C$ 870 ± 40 | 1030–1260 AD |
| | $BS2_4$ * | Top of F3 | Charcoal | $^{14}C$ 1000 ± 40 | 970–1160 AD |
| | $C1_4$ * | Base of F3 | Charcoal | $^{14}C$ 2170 ± 40 | 380–90 BC |
| | $BS1_4$ * | Top of F2 | Soil | $^{14}C$ 4840 ± 40 | 3710–3520 BC |
| Eli5 | $C1_5$ * | Top of C3 | Charcoal | $^{14}C$ 130 ± 40 | 1670–1950 AD |
| | $C2_5$ * | Top of F3 | Charcoal | $^{14}C$ 340 ± 40 | 1400–1950 AD |

* University of Georgia, Center for Applied Isotope Studies. ** Institute of Geology and Sate Seismological Bureau Beijing, China. Based on Oxcal 3.5, with 2$\sigma$ of 0.92–0.95.

## 3. Settlement and Mobility in Helike: The Archaeological Evidence

### 3.1. Occupation in Prehistoric Times: The Early Helladic and the Mycenaean Sites

Archaeological data from the Helike Project excavations, so far, indicate that the earliest Helike settlement, dated to the Early Bronze Age (EBA), was built in the central part of the Helike Plain between the Selinous and Kerynites Rivers (Figure 1a, R). At this location, in the area of modern Rizomylos, excavations brought to light the impressive remains of a coastal proto-urban settlement built only a few hundred metres from the sea on the seaward side of an intermittent lagoon that has existed in this part of the plain since the Late Neolithic period [9]. The settlement shows organised town planning, consisting of large rectilinear buildings with spacious rooms built across cobbled streets and open paved areas. Of particular interest among the buildings is a construction of the type known as a "corridor house", suggesting the presence of a central administration in control of the settlement [11].

The enormous amounts of pottery collected from the excavated buildings include many complete vases, a notable number of fully restored vessels, and a significant number of intact pots corresponding to the Early Helladic (EH) II and III periods, thus dating the settlement to ca. 2500–2200/2100 BC [11,59]. An even earlier occupation of the site, going back possibly to the Neolithic period, is suggested by the $^{14}C$ dating and recovered ceramics from the boreholes drilled to greater depths underlying the excavated EH remains [11].

Archaeological evidence and palaeoenvironmental studies of the sediment samples from the EH buildings suggest that the settlement was destroyed ca. 2100 BC by an earthquake, most probably accompanied by extensive fire, and submerged in a coastal lagoon under sediments containing a mixture of freshwater, brackish, and marine microfossils [60]. The lagoon later silted over, thus keeping the settlement protected from any subsequent human interventions. Today, the settlement is buried 3–5.5 m below the surface and 1 km from the contemporary shoreline.

In the following Late Bronze Age (LBA), the Mycenaean habitation site of Helike is found about 1.5 km to the southeast of the EBA settlement, in the eastern part of the Helike Plain near the Kerynites River in the modern village of Nikolaiika [61] (Figure 1a, N).

### 3.2. Occupation in Historical Times: The Geometric/Archaic Sites

In the same general area of modern Nikolaiika, our excavations uncovered at a depth of 2–3 m below the surface, architectural remains of well-built walls of a large building dated on the basis of rich associated pottery to the Geometric period.

The geological data acquired from the excavations at the Geometric Helike site are of particular interest. Along the southern part of the excavated Geometric walls, a segment of the active HF was revealed, crossing the excavation and shaping its stratigraphy into a footwall and hanging wall blocks. In the tectonostratigraphy of the excavation (Figure 4), it was possible to identify the occurrence of an entirely unknown Geometric earthquake, ca. 700–680 BC, of a magnitude estimated at 6.3R. The catastrophic event coincided with the destruction and abandonment of the site, which was consequently buried under a succession of torrential and river-borne sediment.

It is noted that, at this location, a mudbrick superstructure based on a tall stone footing was chosen by ancient builders, a technique particularly effective in areas susceptible to mudflows and floods. Also, the use of the natural terrace of solidified conglomerate substratum on the HF as a foundation for the construction suggests that they possessed the technical know-how in securing the structure in terms of statics.

**Figure 4.** Archaeological excavation showing Geometric period architectural remains and destruction layer offset by the Helike Fault (HF). The dashed line shows the proposed continuation of the fault under the destruction layer. The footwall block includes well-cemented conglomerates, and the hanging wall block is composed of successive colluvial material.

### 3.3. The Classical and Hellenistic Sites

Based on the archaeological evidence from our excavations in the Helike Plain so far, it appears that the Classical Helike settlement occupied the area between Nikolaiika (east) and Rizomylos-Eliki (west) (Figure 1, N-R). In other words, following the Geometric/Archaic period, the settlement expanded toward the central and western parts of the plain. In particular, the discovery of ruined Classical walls and a Classical destruction layer in Rizomylos is noticeable. Remains of destroyed Classical walls of the 4th century BC, found buried under lagoonal sediments at a depth of 3 m below the surface in a location about 150 m southwest of the earlier EH settlement in Rizomylos, show evidence of a possible wave action (tsunami), perhaps related to the 373 BC earthquake [2].

In the second location, about 250 m south–southeast of the EH settlement, a thick Classical destruction layer of big cobblestones from destroyed stone walls, large clay roof tiles, and a good deal of pottery was excavated at a depth of about 2 m below the ground's surface. The strong presence of Bithynia fossils in the strata covering the excavated destruction layer suggests the formation of a lacustrine environment at this location, postdating its destruction. Geological data of a purely terrestrial environment from a borehole, drilled a short distance from the archaeological excavation, suggest that this location lies at the transitory margin from an ephemeral lake or its coastal marshy section to an inland environment. These data seem to agree with the reported location of Helike at a distance of 12 stadia (about 2 km) from the sea [2].

Only a few decades after the 373 BC seismic event, a new settlement re-emerged in the western part of the Helike Plain, in modern Eliki-Valimitika (Figure 1a, V). Excavations by our team in this region brought to light a good number of Late Classical–Hellenistic buildings, including the well-preserved remains of a large-scale textile workshop, suggesting that Helike soon recovered from the earthquake's fatal blow and became a flourishing town in the Hellenistic period [12,13]. The advantageous location of the settlement on the southwest shore of the Gulf of Corinth favoured maritime travelling and trade since prehistoric times, as the excavation finds from the Helike EBA settlement suggest [11]. In particular, the Hellenistic textile dye works in Helike, the only workshop positioned near the sea among its contemporaries around the Gulf of Corinth, played a key role in the commerce and economy of the region, thus demonstrating the importance of coastal locations in textile manufacturing and cultural exchange.

It is worth mentioning that the preference shown by the Helike builders in the earliest phase of the construction, dated to the second half of the 4th century BC, for a polygonal mode of masonry, may well be related to a desire for a more solid and sturdy structure following the catastrophic experience of the 373 BC earthquake a few decades ago.

### *3.4. Occupation in Later Times: The Roman Site*

Following the destruction of the Hellenistic settlement by another earthquake ca. 90/80 BC, the people of Helike moved further back east, where the pre-373 BC earthquake habitation sites were located (Figure 1a, R and N). The Roman settlement, the immediate successor to the Hellenistic one, soon expanded and developed toward the east along and to the south of the central Roman road (Figure 5a), crossing the Helike Plain near the coastline [10].

**Figure 5.** (**a**) Segment of the Roman road and Hellenistic architectural remains of walls excavated in the western Helike Plain. (**b**) Skeletal remains of a man in an extended position, found under the collapsed ruins of a Roman building in Nikolaiika.

The settlement enjoyed a long life from the late 1st century BC to about the middle of the 5th century AD. Here, a successor installation of the abandoned Hellenistic dyeing workshops was found, which continued to operate into the Roman period. Thus, the skill for crafts seems to be a diachronic feature of the Helike culture, not interrupted by a natural catastrophe that occurred, but transferred and continued within the new settlement site by the surviving population.

Similarly to the previous Helike settlements, the Roman site was devastated by a strong earthquake in the first half of the 5th century AD, as indicated by our excavation data [62]. In this case, however, the disaster left us with tragic evidence of the violent death of a man and his animal (Figure 5b), who were found buried under the ruins of one of the destroyed buildings [62,63].

## 4. The Dynamic Landscape of Helike

A school of thought, advocated by Ingold [64] and others, argues that a landscape is cultural and temporal and that temporality is imprinted in the patterns of dwellings throughout the ages. People's engagement with the landscape draws its unique significance through an array of related features that only have meaning in people's minds. The archaeologist's task is to uncover such meanings, where each object in the landscape becomes a clue to meaning, which is fundamentally temporal.

This view is not dissimilar to and, we believe, is inspired by key thinkers across different philosophical traditions. For example, Heraκleitos (c. 535–475 BC) put forward the idea that "you cannot step into the same river twice", which highlights how landscapes are always in flux and shaped by time, as discussed by Patrick [65]. More recently, Lock [66] emphasised how humans are embedded in time and place. He defined 'dwelling' as living authentically in harmony with, and shaping, the landscape rather than simply as something to exploit.

Ingold [64] argues for a rigorous landscape theory that can be investigated and demonstrated through GIS technologies. Because GIS is essentially quantitative, we need to understand the mismatch between the study of people and landscapes, which is essentially qualitative. The central argument is how this relationship can be modelled within GIS; what can we hope to achieve through the digital representation of place, space, and their temporal meaning?

Polla and Verhangen [67] focused on theories about the way and reason why people moved in the landscape. They used computer-based modelling approaches, experimenting with the various parameters involved, such as in Lock et al. [68], where all assumptions are laid out in detail, allowing for different scenarios to be explored and compared to real-world outcomes. Computer modelling is, thus, a heuristic approach that aids the development of theory and interpretation through hypothesis testing.

This section focuses on movement in the landscape in the region of Ancient Helike, where landscapes are not just physical spaces but ones that hold deep emotional, cultural, and historical meanings. As Heidegger [69] has emphasised, dwelling is a fundamental way humans exist. People do not just inhabit spaces; they develop a sense of belonging so that, even after disasters, their identity remains strongly tied to that place [70–73]. Returning may indicate restoration of identity and continuity, as there is a sense of home even if the physical environment has changed. As such, the Helike region was constantly reoccupied after each disaster from the Early Helladic to the Late Antiquity.

### *Computer-Based Modelling of the Helike Landscape*

Digital elevation models (DEMs) are widely used in geographic information systems (GIS), remote sensing, and topographic analysis to model terrain characteristics. A methodology for modelling ancient landscapes has been developed by one of us (MK), whereby a representation of the current landscape is dynamically modified according to the following five variables: the rates of sediment deposition, sea level rise, and tectonic uplift, together with earthquake-related events of pulse tectonic uplift and subsidence. The simulation runs backwards in time from a present landscape DEM.

Focusing on the green height profile in Figure 6, there are repeated patterns where all elevations increase at a certain rate (tectonic uplift, sea level rise, and sediment deposition), followed by inflexion points caused by strong earthquakes with combined pulse tectonic uplift and subsidence (Figure 6). Strong earthquakes are repeated six times, as shown in the Early Helladic (2100 BC), Geometric (700–680 BC), Classical (373 BC), Late Hellenistic (90–80 BC), and Roman (mid-5th AD) periods, and more recently, in the 19th century (1861). These events have also been verified by geological and archaeological evidence, with some reported by historical records.

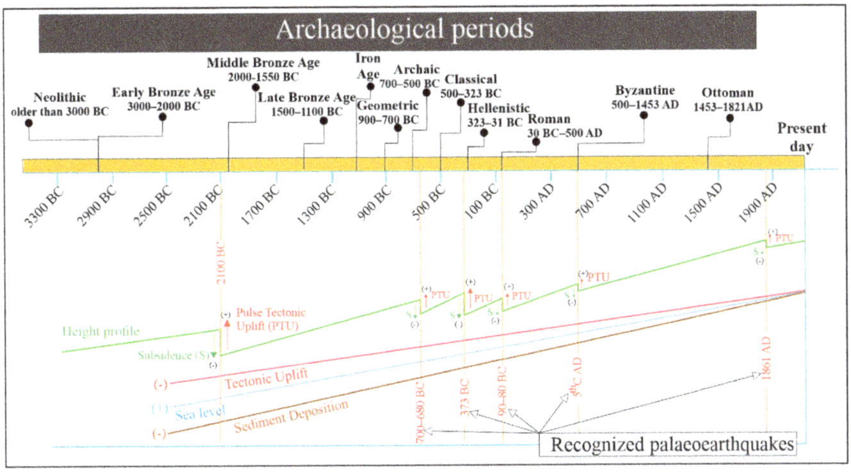


**Figure 6.** Depiction of repeated patterns of elevation trajectories in the Helike landscape, in a qualitative way.

The simulations reported in this section are focused on the events of 2100 BC and 373 BC, as Soter [52] estimates an average sediment deposition of 16.8 m over the past 7000 years. The deposition process is related to fluctuations in sea level rise and tectonic uplift, with periods of accelerated sediment accumulation followed by steady isostatic adjustments. Furthermore, variations in deposition rates across the Helike Plain are probably largely driven by differences in terrain characteristics and local elevation. Specifically, archaeological data from the Early Helladic site suggest a deposition rate of approximately 1 m per thousand years (m/kyr) over the last 4100 years.

Concerning the sea level rise, Soter [52]'s findings reveal an average rate of 0.5 m per thousand years (m/kyr) over the past 5000 years for the periods under consideration. For example, for a simulation dated to 373 BC, the estimated sea level in the Gulf of Corinth would have been 1.19 m lower than present, based on the calculation $0.5 \times 2373 = 1.19$ m. Concerning tectonic uplift and subsidence, these must be estimated based on the sea level, which depends on the accurate dating of marine organisms. Radiocarbon dating requires calibration due to the reservoir effect, where deep-water mixing causes marine samples to appear older than they are. In the Gulf of Corinth, Soter [52] estimated a local correction of 380 years based on radiocarbon dating for marine samples in the Helike Delta to account for carbon depletion from surrounding limestone.

By plotting the calibrated shoreline ages against elevation, Soter [52] demonstrated that the shoreline was approximately 20 m lower around 9000 BP than its present level. Additionally, two significant discontinuities were identified. The first discontinuity suggests a pulse uplift of 6–7 m in the earthquake of 2100 BC, while the second may correspond to the 373 BC earthquake, during which an uplift of approximately 2 m occurred. Furthermore, the data provide preliminary evidence that the 1861 AD earthquake resulted in an uplift of approximately 1 m.

Taking these factors into account, the analysis estimates subsidence in the range of 4 to 9 m for the earthquake of 373 BC, and the assumed co-seismic subsidence is 4.5 m [52]. This aligns with the suggestion of at least 3 m of subsidence and is further supported by evidence of significant shoreline submergence, likely caused by plain tilting, soil liquefaction, and submarine landslides, as observed in the 1861 AD earthquake.

The quantification discussed above can be incorporated into a general equation to calculate time-dependent heights of every single point in a DEM model as follows:

$$H_{t1} = H_{t0} + (S_{t1} - U_{t1}) + t * (R_S - R_U - R_D) \quad (1)$$

where $H_{t1}$ is the DEM height at time t1, $H_{t0}$ is the height at the start time, $S_{t1}$ is the pulse subsidence, $U_{t1}$ is pulse tectonic uplift, t is the time, and $R_S$, $R_U$, and $R_D$ are the rates of sea level rise, tectonic uplift, and sediment deposition.

The DEM is mapped to a raster file of the Helike landscape, where pixels representing heights at or below zero are coloured blue, indicating that the area is below sea level. This clearly shows the sea's progression and regression over time and the changes in the landscape. Figure 7 presents a model of the current landscape, with the left side showing the present-day terrain and the right side displaying the DEM map overlay.

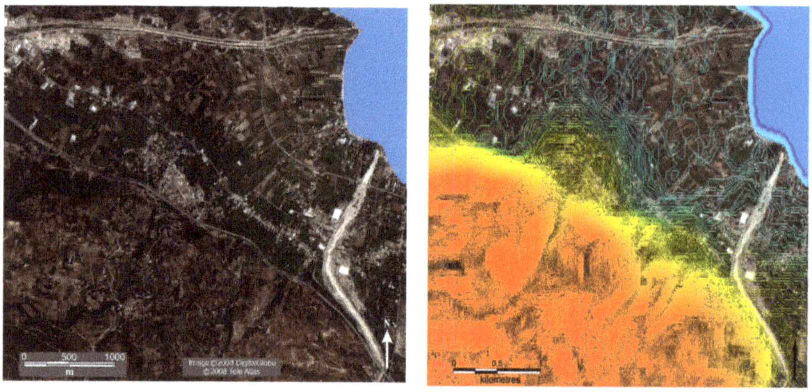


**Figure 7.** (**Left**) Map of the current Helike Landscape. (**Right**) The DEM map overlay. The ranges of the colour code are only qualitative and as follows: 0–50 m elevation: cyan, 50–200 m: yellow, 200–997 m: orange.

The general shape of the shoreline before the earthquake of ca. 2100 BC, as shown in Figure 8 (left), aligns with the archaeological findings from excavations by the Helike Project.

These excavations uncovered an Early Helladic II-III settlement featuring a "corridor house" and gridded streets [9,11]. Evidence from boreholes further supports the conclusion that this settlement was situated near the shoreline. Figure 8 (right) illustrates the landscape immediately after the earthquake. The marine transgression appears to have led to three lagoons, with the largest one centrally positioned in the bay and oriented northwest–southeast. It seems that, following the earthquake, the sea formed an enclosed lagoon in this orientation, located below the modern railway line.

Over time, the lagoon was gradually silted over, and by the Classical period, the shoreline had shifted close to the present-day railway line. Figure 9 (left) illustrates the landscape before the 373 BC earthquake. A small blue area in the centre, near excavated trench H8, suggests the presence of a lagoon, though it was more likely a swamp in 373 BC.

**Figure 8.** (**Left**) The Helike landscape in the Early Helladic period before the earthquake of 2100 BC. (**Right**) The landscape after the earthquake of 2100 BC.

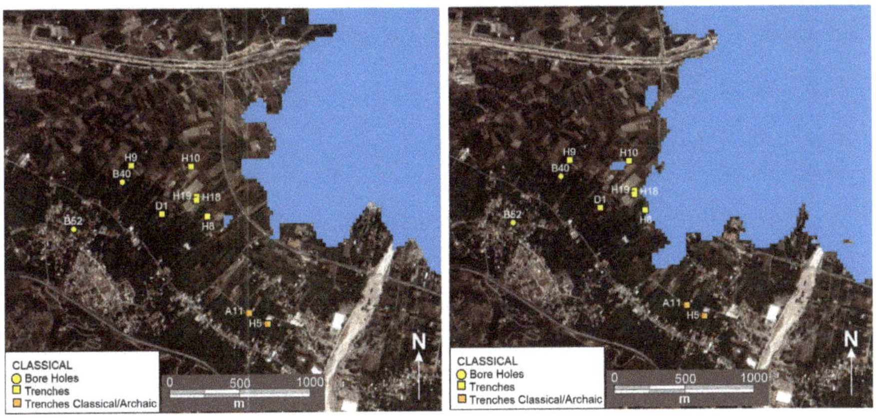


**Figure 9.** (**Left**) The Helike landscape in the Classical period before the earthquake of 373 BC. (**Right**) The landscape after the earthquake of 373 BC.

Even today, part of the central Helike area is named "Valtos", meaning "the swamps", likely a remnant of an ancient lagoon that was initially open to the sea but later cut off by sediment deposition. According to a study on the Middle and Late Holocene paleogeography of the Helike Delta [14], in the 3rd/2nd millennium BC, the lagoon was small, stretching in the central/western part of the plain. After 1000 BC, it gradually grew to the northeast and by about 400 BC to the west. Following the 373 BC period, the lagoon occupied the central part of the plain, leaving on either side dry land for resettling.

This interpretation aligns with evidence of the Classical settlement in the area. Archaeological trench H8, the closest to the swamp, contained significant amounts of Archaic/Classical pottery at depths of 2.45–3.00 m, indicating that the area had silted over by that time. Trenches H18 and H19, located nearby, revealed Classical walls at a depth of 3 m, while H19 also contained ostracods [60] at 3.20 m beneath the walls, suggesting a previously brackish environment. These findings reinforce the presence of earlier marine, lagoonal, and marshy conditions at the site.

Figure 9 (right) illustrates the model following the 373 BC earthquake, highlighting the formation of two lagoons: one close to the centre of the picture and another further north, close to the Selinous River. Archaeological trench H10 is next to the lagoon at the centre.

The Classical horizon was found at a depth of 3.70–3.90 m. Above it, there is evidence of a freshwater environment with freshwater ostracods, green and black clay, and up to 2 m, a black clay marshy environment appears. The model shows large areas being submerged near the shore after the earthquake.

The model shown in Figure 10 (left) simulates a 150-year progression from 373 BC to the Hellenistic period, corresponding to Eratosthenes' account (3rd century BC) [2].

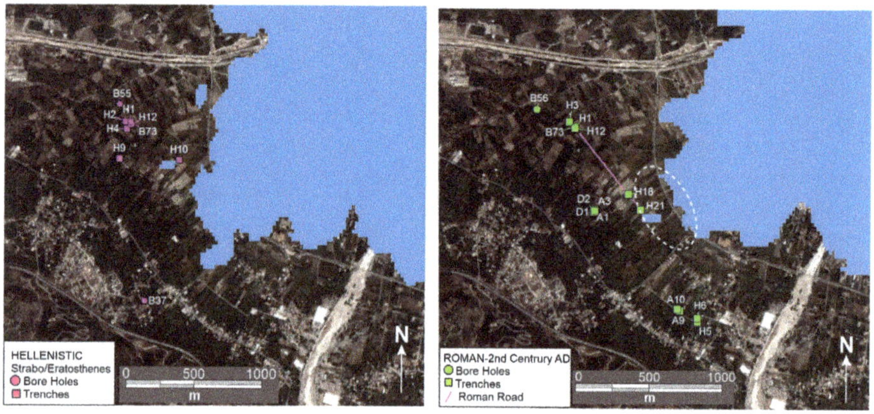


**Figure 10.** (**Left**) The Helike landscape in the Hellenistic period. (**Right**) The landscape in the Roman period, at the time of Pausanias (2nd century AD).

Two key observations emerge: first, the Hellenistic settlement pattern shifted northwest compared to the Classical period. Second, the shoreline remained relatively stable, while the central shore area experienced prolonged submergence.

The interpretation is that the area was submerged under the sea for a considerable time and that the environment changed from marine to brackish. Furthermore, Eratosthenes reported that the bronze statue of Poseidon was in the "poros", which has been re-interpreted as an inland lagoon [2]. The model showing a possible scenario in the 2nd century AD is presented in Figure 10 (right). By the 2nd century AD, the area was extensively silted over, as shown by the discovery of a major segment of an excavated Roman road, which runs through the site [10] (see magenta line in Figure 10, right).

Pausanias may have travelled along this Roman road and reported that the ruins of Helike were still visible underwater but heavily deteriorated. The models indicate that the lagoon is probably situated at the location annotated in Figure 10 (right), oriented NW–SE in the same location as the ancient lagoon. This view is further reinforced by extensive gravity measurements in the area by Dimitropoulos and Noutsis [74], whose results closely match the location and orientation of the lagoon as annotated.

This section has presented and demonstrated a method for landscape analysis through raster-based digital elevation models of the Helike landscape. DEMs are generated from dynamic models of sea level rises, coastal and tectonic pulse uplifts, subsidence, and sediment deposition. The present-day DEM was used to compile landscape models in the Early Helladic, Classical, Hellenistic, and Roman periods. Each model is validated by borehole and excavation data and interpreted in the light of ancient written sources. The models make several contributions to understanding the implications of the geomorphism of the Helike landscape to the continuity and discontinuity of human occupation.

## 5. Discussion

### 5.1. Seismicity and Settlement

Past earthquakes provide useful information on seismic risk assessment in a region. However, the scientific records offer quantifiable data on prior earthquakes from the beginning of the twentieth century. This period represents but a small record of the Holocene seismicity [4,5,38,39]. To expand this database, archaeological records are substantial in determining the damage inflicted on human structures revealed at archaeological sites [75–82]. Specifically, archaeology unravels the effects of ancient earthquakes on human communities by examining damage patterns and identifying seismic signs imprinted in human constructions and artefacts [7,11–13,15]. Archaeology, palaeoseismology, epigraphy, and historical texts are the primary sources of earthquake evidence. However, only the most severe seismic events would be recognised in all four sources, whereas smaller magnitude events might occur in some and rarely in all. In the current study, we calibrated the age of past earthquake events using dates resulting from combined $^{14}$C dating and archaeological evidence. The age of younger earthquakes is determined based on historical sources and $^{14}$C dating.

The timing of a disaster, its geographical location, the society affected, and the nature of the hazard are crucial elements that interact in various ways to differentiate a hazard from a disaster [83,84]. In the case of a series of earthquakes spanning over the Holocene, like in the Helike area, the pivotal factors are as follows. The first factor refers to the rate of onset and the duration of the incident. Devastating earthquakes, including the main event and aftershocks, may persist for months in the inflicted area. The second factor represents the extent of the affected region related to the earthquake magnitude or the proportion of the population struck. The third factor concerns the nature of the society inflicted. A society may be adaptable and mobile, like nomadic groups, or rigid, as, for instance, a settlement administered by a strong hierarchy or with a deep connection to specific geographical features. In our case, since Helike was the capital city in the region from the Early Bronze Age until the 4th century BC, i.e., for more than 20 centuries, it possessed a certain core societal value. The fourth factor involves the nature of the hazard and its recurrence, along with the accumulation of various types of hazards. In the case of Helike, the recurrence interval is typically in the order of more than four centuries, and in some cases, it is more than a millennium (Figure 11).

Additionally, successive hazards of different kinds may have the same cumulative effect characterised as a cascade disaster, also described as a ratchet effect by [7,85–87]. The 373 BC earthquake is not directly related to the HF and, therefore, could be considered a cascade event [7].

Frequent earthquakes will challenge the resilience of a community. Archaeological data acquired from the Helike Project excavations suggest that settlement locations changed positions over the centuries but always remained in the coastal plain, and primarily in the area bounded between the rivers Selinous (west) and Kerynites (east). It is worth noting that each habitation site, after abandonment due to earthquake destruction, bears no evidence of any subsequent occupation. On the contrary, revived settlements usually unfold in a new location, within the major Helike area, as set each time by current geological conditions. Thus, the Helike settlement sites since the Early Bronze Age compose a nucleus each time, which preserves the special features of the particular period of the settlement untouched. Figure 12 presents their mobility across the plain in time based on the excavation data.

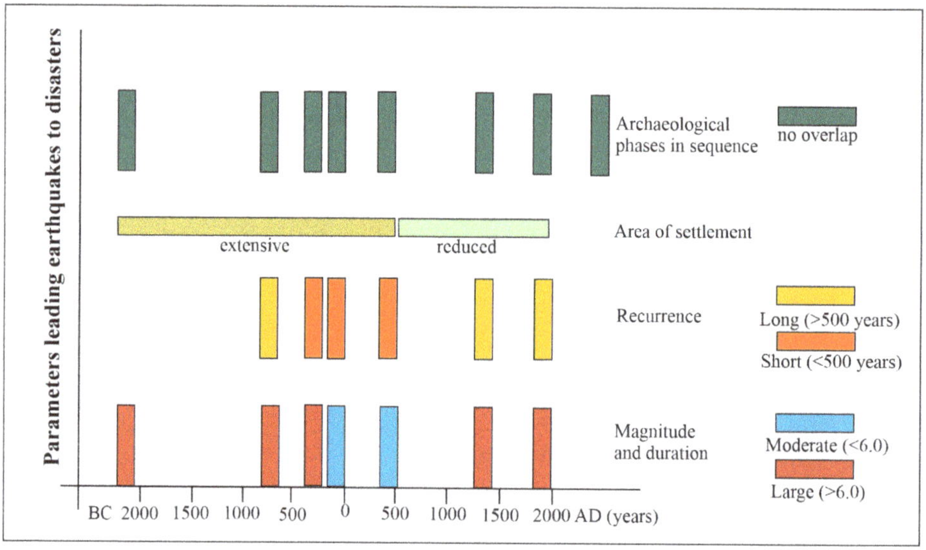


**Figure 11.** Chart showing timing and magnitude of earthquake disasters in the Helike Plain, their geographical distribution, and the impact on the various chronological settlements (for discussion see text).

**Figure 12.** Map of the Helike Plain showing movements and settlements from the 3rd millennium BC to the 1st millennium AD (for discussion see text).

### 5.2. Earthquake Destruction and Resilience

Beginning with the earliest proto-urban town of the 3rd millennium BC, exhibiting an organised building plan, including among its buildings a most outstanding monumental "corridor house", Helike featured as an Early Bronze Age coastal centre associated with the rise of a new socio-economic structure at the time. In addition to advanced characteristics at the level of the settlement organisation, overseas links and contacts are evidenced through the presence of an array of exotic, prestigious items and commodities, including a unique depas cup of Anatolian type, ornaments of gold and silver, Melian obsidian tools, and the adoption of 'international' fashions of life such as the 'serve and drink' culture and 'potter's marks' [11]. This cosmopolitan settlement came to an end ca. 2100 BC, struck by a strong earthquake accompanied by fire and subsequently buried under the sediments of a lagoon [60]. The collapse of the environmental and socio-economic systems resulted in the absence of any habitation sign in the central part of the plain for a very long period, as indicated by the stratigraphic excavation data. Human presence in this area is found again in the Archaic/Classical period down to the Hellenistic/Roman times, i.e., after a period of over a thousand years.

The facing of a serious environmental threat, ca. 2100 BC, and the burial of the destroyed settlement under lagoon deposits, seems to have driven people away from the watery lower lands of the mid-plain. Thus, in the following period from the Late Bronze Age (15th century BC) through to the Late Geometric/Archaic period (7th century BC), the new settlement was built on a safer location that occupied the lower plateau and higher sloping of the hill above the modern village of Nikolaiika (Figure 1a, N) at a distance of 1.5 km to the southeast of the abandoned Early Bronze Age habitation site (Figure 12). The settlement, dated from the 14th to the 11th century BC, occupied an area of about 20 acres, and was organised in groups of houses, including complexes formed of smaller rooms [61,88]. Mycenaean 'broad' Helike, as named in the Homeric epics, was part of the western territory of the Mycenaean kingdom of Agamemnon and participated—providing ships and crew—in the Trojan expedition under Agamemnon and his forces.

In the same general area, at the foot-mountain of the same hill, evolved the successor town of Geometric Helike, which founded the first and most powerful colony of the Achaeans in South Italy before the end of the 8th century BC (ca. 730 BC) [89]. Habitation in the proto-Geometric period, preceding the fully developed settlement of the Geometric metropolis of the 8th century BC, is evidenced via the Derveni grave pottery group, including the most popular Achaean drinking cup, the kantharos [90]. The influx of Corinthian products that reached the hinterland of Achaea since the second half of the 8th century BC [91] and exceptional objects from the eastern Mediterranean, such as the metal Phoenician models of clay vases dated to the early 7th century BC from a number of Achaean sites [92], suggest that Helike remained the commercial centre through which the area had fostered contacts with the East and the rest of the ancient world since the Early Bronze Age.

Previous experience of seismic disturbance may account for the solutions implemented by the Geometric society for the adjustment of architecture to the geomorphology, as we may infer from the foundation of the large Geometric building on the solidified conglomerate of the Helike Fault at this location (see Section 3.2). More importantly so, since the dimensions, the form and spaciousness of the construction—rather rare by the standards of the Geometric era—in conjunction with the abundance of drinking vessels recovered, suggest that the building hosted communal dining with many participants and most probably belonged to a local ruler or chieftain. The constant care and concern of the people in obtaining safety and stability for their structures is again evidenced in the use of polygonal masonry in the constructions of the post-373 BC earthquake settlement (see Section 3.3).

This flourishing culture was disrupted once again by another strong earthquake that occurred ca. 700–680 BC, and caused the abandonment of the site and its burial under thick river-borne fluvial deposits. However, recovery was attained shortly, as suggested by the archaeological finds dated to the 6th and 5th centuries BC from this area [12,93]. Further, the discovery of destroyed Classical walls by the 373 BC earthquake in more than one location to the west (see Section 3.3) indicates the expansion of the Classical settlement in a considerable length in the plain, coinciding with Helike's elevation at the position of the capital city of the

Achaean Dodekapolis, a place already held before during the predominance of the Ionians in the area. The foremost position of the Classical polis as both the political and religious seat of the Confederation, is witnessed, among other things, by the realisation of major projects in its territory, as was the construction of a sizeable marble temple and its sculptural decoration, all made of the most celebrated and costly marble of the ancient world, the Parian. The unique presence of the Parian marble in this region, the only site in the entire western Peloponnese before Olympia, places Helike along the maritime routes network for the export and diffusion of the valuable Parian marble to the West [94].

At the peak of its culture, Helike was smitten once again in 373 BC by another natural disaster, this time by a combined natural force of an earthquake and a sea wave [7]. Although believed in the past that there were no survivors and that the city disappeared from the face of the earth, the archaeological record has shown that, a few decades later, Helike was reborn, and a new settlement was built in the western part of the plain. The settlement soon grew in size and flourished, including in its premises, a large-scale textile industry, thus retaining its commercial and economic dynamics in the region and beyond (see Section 3.3). But yet, once more, the Hellenistic society's attainments were forcibly discontinued by the strong earthquake of 90–80 BC [95] in another return of destruction over resilience. The latter, though, was soon restored within the same century through the rebuilding of the successor settlement of the Roman period in a new location to the east of the abandoned Hellenistic site. In keeping with the manifestations of a long tradition as transmitted by their ancestors, survivors continued to practice the inherited skills for crafts in the new settlement, as indicated by the excavation finds (see Section 3.4) [62].

The archaeological record so far suggests that throughout the centuries, in a repeated cycle of shortened natural disasters alternating with long periods of cultural and societal stability and growth, the Helike society consistently demonstrated resilience against destruction. It could be said that having confronted for generations with major environmental risks, society learned from past experience and formulated efficacious solutions to enhance resilience (Figure 13).

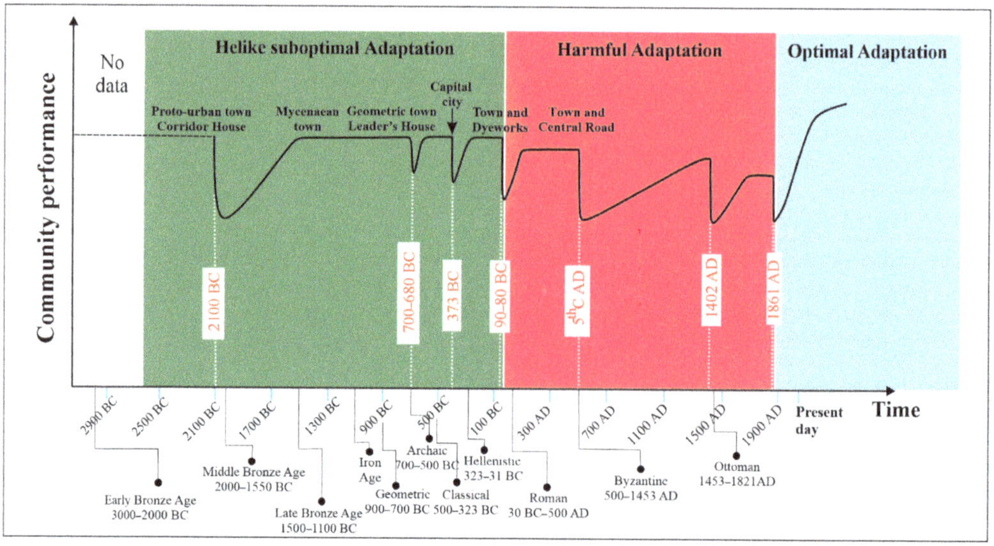


**Figure 13.** Schematic diagram of the recovery process (resilience curve) following earthquake events in the Helike area. The diagram illustrates archaeological periods, the age of earthquakes, and the recovery

time related to the pre-earthquake level. The green-shaded area indicates suboptimal society adaptation after earthquake damage, allowing the community to maintain a social level similar to the prior one; the red-shaded area highlights harmful society adaptation following the closely occurring earthquakes, every 300 years, between the Geometric and Hellenistic times. The light blue-shaded area shows the modern optimal adaptation following the 1861 earthquake. Adaptation nomenclature follows Klasa et al. [96].

Aware of this repeated system moving between destruction and resilience, the ancients worshipped Poseidon, the god of waters and earthquakes. In Helike, a land of high seismicity, Poseidon was the dominant deity, worshipped under the epithet of Helikonios [97]. In his sanctuary, mentioned in the Homeric epics, bull sacrifices took place to appease the god and secure safety for Helike's land and people's lives.

## 6. Conclusions

Helike serves as a striking example of a true palimpsest. Originally established during the Bronze Age (3rd millennium BC), it was repeatedly resettled during its long-lasting life down to Late Antiquity (1st millennium AD) as a result of natural phenomena. This study provides, for the first time, a complete catalogue of the historical earthquakes of the HF, a most on-land normal fault, which affected an area of continuous habitation over the last ten millennia. Furthermore, for almost all these earthquakes, we furnish their co-seismic and post-seismic phases based on (a) accurate earthquake recognition, (b) landscape description, and (c) human settlement and movement in the region (Figure 12). More specifically:

1. We extended our documentation of earthquake phenomena into the past ten millennia based on palaeoseismology and archaeology data. Furthermore, we contributed to a better understanding of the recurrence interval of earthquakes along the Helike Fault as a causative structure that resulted in ground motions and physical impacts on settlements in the study area, thus filling in existing knowledge gaps. In particular, we focused on a cluster of four strong earthquake events (700–680 BC, 373 BC, 90–80 BC, and 420–450 AD), which devastated the Helike area at recurrence intervals ranging between 530 and 288 years. From a geological point of view, these earthquakes caused significant environmental changes in the area between the Selinous and Kerynites Rivers [7,37,95].
2. It has been shown that the diachronic course of the Selinous and Kerynites Rivers, which drain the Helike Delta and border the permanently inhabited Helike area, was repeatedly altered by surface-rupturing earthquakes on the Helike Fault and a powerful off-fault earthquake event in 373 BC. However, the mechanisms associated with these earthquake events have not yet been thoroughly understood.
3. Shifts of the ancient shoreline, landwards and seawards, in relation to the current seashore, have significantly contributed to our understanding of the diachronic evolution of the Helike settlements and the landscape over the Holocene.
4. Living in a high-seismicity area and often experiencing natural disasters, the ancient people of Helike developed a certain adaptation of 'seismically' living, as characteristically shown by the building modes employed in the architecture to secure their constructions in terms of statics and endurance.
5. Regarding the people's predilection for continuous resettling in a seismically active plain, the landscape has played a crucial role. The Helike Plain is an alluvial, fertile land surrounded to the south by mountains, where three rivers originate, providing an abundance of water. To the north, the place is open to the sea, allowing easy access to sea voyages and trade routes. Its location as a commercial coastal centre in this part of northwest Peloponnese and the Gulf of Corinth favoured contacts and exchanges within the Mediterranean basin toward the East since the 3rd millennium BC, and to

the West and its markets since the early historical times, when Helike founded the most powerful Achaean colony in South Italy, the city of Sybaris.

**Author Contributions:** D.K.: Conceptualization, excavation, supervision, writing—original draft, and overall editing. D.K. and I.K.: Methodology, writing—review and editing, and data analysis. I.K.: Geological analysis and palaeoseismological interpretation. M.K.: Landscape analysis, computer modelling, and writing—review and editing. All authors have read and agreed to the published version of the manuscript.

**Funding:** This research received no external funding.

**Data Availability Statement:** The data collected during the field work are available on request from the authors.

**Acknowledgments:** We thank the Institute for Aegean Prehistory in Philadelphia, USA, and the local Municipalities of Diakopton and Aigion for their support and contributions to the Helike Project excavations. We also thank the 3E company of Athens for supporting the earlier Helike Project work of borehole drilling.

**Conflicts of Interest:** The authors declare no conflicts of interest.

## References

1. Guidoboni, E.; Comastri, A.; Traina, G. *Catalogue of Ancient Earthquakes in the Mediterranean Area up to the Tenth Century*; Istituto Nazionale di Geofisica: Rome, Italy, 1994; p. 504.
2. Katsonopoulou, D. The earthquake of 373 BC. Literary and archaeological evidence. In *Helike III, Archaeological Sites in Geologically Active Regions*; Katsonopoulou, D., Soter, S., Koukouvelas, I., Eds.; The Helike Society: Athens, Greece, 2005; pp. 15–32.
3. Partida, E. The Disaster and the Experience of 373 BC Followed through the Architecture and Topography of Delphi. In *POSEIDON—God of Earthquakes and Waters Cult and Sanctuaries*; Katsonopoulou, D., Ed.; The Helike Society: Athens, Greece, 2017; pp. 227–258.
4. Papazachos, B.; Papazachou, C. *The Earthquakes of Greece*; Ziti Publications: Thessaloniki, Greece, 2003; p. 304.
5. Papadopoulos, G.A. *Historical Earthquakes and Tsunamis in the Corinth Rift, Central Greece, Publication 12, National Observatory of Athens*; Institute of Geodynamics: Athens, Greece, 2000; p. 128.
6. Soter, S. Macroscopic seismic anomalies and submarine pockmarks in the Corinth–Patras rift, Greece. *Tectonophysics* **1999**, *308*, 275–290. [CrossRef]
7. Koukouvelas, I.K.; Piper, D.J.W.; Katsonopoulou, D.; Kontopoulos, N.; Verroios, S.; Nikolakopoulos, K.; Zygouri, V. Earthquake-triggered landslides and mudflows: Was this the wave that engulfed Ancient Helike? *Holocene* **2020**, *30*, 1653–1668. [CrossRef]
8. Marinatos, S.N. A submerged town of classical Greece. *Archaeology* **1960**, *113*, 186–193.
9. Soter, S.; Katsonopoulou, D. Submergence and uplift of settlements in the area of Helike, from the Early Bronze Age to Late Antiquity. *Geoarchaeology* **2011**, *26*, 584–610. [CrossRef]
10. Tsokas, G.N.; Tsourlos, P.I.; Stampolidis, A.; Katsonopoulou, D.; Soter, S. Tracing a Major Roman Road in the Area of Ancient Helike by Resistivity Tomography. *Archaeol. Prospect.* **2009**, *16*, 251–266. [CrossRef]
11. Katsonopoulou, D.; Katsarou, S. Mainland cosmopolitanism and the rise of personal prestige: New evidence from the coastal Early Helladic town of Helike, northwest Peloponnese. *BSA* **2017**, *112*, 1–32. [CrossRef]
12. Katsonopoulou, D.; Partida, E. Filling a Gap in the History of an Achaean Leader City: Post-373 B.C. Helike. *NAC* **2023**, *52*, 169–199.
13. Weir, R. Was Hellenistic Helike a Typical Polis? A Comparative Study of Numismatic Contexts. In *Helike After 373 BC—An Interdisciplinary Approach*; Katsonopoulou, D., Ed.; The Helike Society: Athens, Greece, 2024; pp. 123–135.
14. Koutsios, A. Middle to Late Holocene Paleogeography of Helike Delta Plain. Application in Archaeological Research. Ph.D. Thesis, University of Patras, Patras, Greece, 2009; p. 497.
15. Gaggioli, A. The geoarchaeology of seismically triggered soft sediment deformation structures (SSDS). *J. Archaeol. Sci.* **2024**, *165*, 105961. [CrossRef]
16. Kontopoulos, N.; Avramidis, P. Geological Reconstruction of Ancient Coastal Landforms in North Peloponnesos, Greece. An Example from Aliki Lagoon. In *Ancient Helike and Aigialeia—Archaeological Sites in Geologically Active Regions*; Katsonopoulou, D., Soter, S., Koukouvelas, I., Eds.; The Helike Society: Athens, Greece, 2005; pp. 293–301.
17. Redman, C.L. Resilience Theory in Archaeology. *Am. Anthropol.* **2005**, *107*, 70–77. [CrossRef]

18. Lonergan, K.E.; Greco, S.F.; Sansavini, G. Ensuring/insuring resilient energy system infrastructure. *Environ. Syst. Decis.* **2023**, *43*, 625–638. [CrossRef]
19. Nixon, C.W.; McNeill, L.C.; Bull, J.M.; Bell, R.E.; Gawthorpe, R.L.; Henstock, T.J.; Christodoulou, D.; Ford, M.; Taylor, B.; Sakellariou, D.; et al. Rapid spatiotemporal variations in rift structure during development of the Corinth Rift, central Greece. *Tectonics* **2016**, *35*, 1225–1248. [CrossRef]
20. Doutsos, T.; Kontopoulos, N.; Poulimenos, G. The Corinth-Patras Rift as the initial stage of continental fragmentation behind an active island arc (Greece). *Basin Res.* **1988**, *1*, 177–190. [CrossRef]
21. Maffione, M.; Herrero-Bervera, E. A relative Paleointensity (RPI)-calibrated age model for the Corinth syn-rift sequence at IODP hole M0079A (gulf of Corinth, Greece). *Front. Earth Sci.* **2022**, *10*, 813958. [CrossRef]
22. Rusch, K.; Stümpel, H.; Gauß, W.; Müth, S.; Sokolicek, A.; Kissas, K.; Rabbel, W. Geological Challenges of Archaeological Prospecting: The Northern Peloponnese as a Type Location of Populated Syn-Rift Settings. *Remote Sens.* **2020**, *12*, 2450. [CrossRef]
23. de Vals, M.; Moretti, I. Geology and construction: Survey of archaeological sites and their natural environments (Gulf of Corinth, Greece). *Comptes Rendus. Geosci.* **2022**, *354*, 51–73. [CrossRef]
24. Pechlivanidou, S.; Cowie, P.A.; Duclaux, G.; Nixon, C.W.; Gawthorpe, R.L.; Salles, T. Tipping the balance: Shifts in sediment production in an active rift setting. *Geology* **2019**, *47*, 259–262. [CrossRef]
25. Gawthorpe, R.L.; Leeder, M.R. Tectono-sedimentary evolution of active extensional basins. *Basin Res.* **2000**, *12*, 195–218. [CrossRef]
26. Perissoratis, C.; Piper, D.J.W.; Lykousis, V. Alternating marine and lacustrine sedimentation during late Quaternary in the Gulf of Corinth Rift basin, central Greece. *Mar. Geol.* **2000**, *167*, 391–411. [CrossRef]
27. Leeder, M.R.; Mack, G.H.; Brasier, A.T.; Parrish, R.R.; McIntosh, W.C.; Andrews, J.E.; Duermeijer, C.E. Late-Pliocene timing of Corinth (Greece) rift-margin fault migration. *Earth Planet. Sci. Lett.* **2008**, *274*, 132–141. [CrossRef]
28. Ford, M.; Hemelsdaël, R.; Mancini, M.; Palyvos, N. Rift Migration and Lateral Propagation: Evolution of Normal Faults and Sediment-Routing Systems of the Western Corinth Rift (Greece). *Geol. Soc. Spec. Publ.* **2017**, *493*, 131–168. [CrossRef]
29. Ori, G.G.; Roveri, M.; Nichols, G. Architectural patterns in large-scale Gilbert-type delta complexes, Pleistocene, Gulf of Corinth, Greece. In *The Three Dimensional Facies Architecture in Terrigenous Clastic Sediments and its Implications for Hydrocarbon Discovery and Recovery*; Miall, A.D., Tyler, N., Eds.; SEPM, Concept in Sedimentology and Paleontology: Tulsa, OK, USA, 1991; pp. 207–216.
30. Floyd, M.A.; Billiris, H.; Paradissis, D.; Veis, G.; Avallone, A.; Briole, P.; McClusky, A.S.; Nocquet, J.M.; Palamartchouk, K.; Parsons, B.; et al. A new velocity field for Greece: Implications for the kinematics and dynamics of the Aegean. *J. Geophys Res.* **2010**, *115*, B10403. [CrossRef]
31. Nixon, C.W.; McNeill, L.C.; Gawthorpe, R.L.; Shillington, D.J.; Michas, G.; Bell, R.E.; Moyle, A.; Ford, M.; Zakharova, N.V.; Bull, J.M.; et al. Increasing fault slip rates within the Corinth rift, Greece: A rapidly localising active rift fault network. *Earth Planet. Sci. Lett.* **2024**, *636*, 118716. [CrossRef]
32. Gawthorpe, R.L.; Fabregas, N.; Pechlivanidou, S.; Ford, M.; Collier, R.E.L.; Carter, G.D.O.; McNeill, L.C.; Shillington, D.J. Late Quaternary mud-dominated, basin-floor sedimentation of the Gulf of Corinth, Greece: Implications for deep-water depositional processes and controls on syn-rift sedimentation. *Basin Res.* **2022**, *34*, 1567–1600. [CrossRef]
33. Koukouvelas, I.; Stamatopoulos, L.; Katsonopoulou, D.; Pavlides, S. A palaeoseismological and geoarchaeological investigation of the Eliki Fault, Gulf of Corinth, Greece. *J. Struct. Geology* **2001**, *23*, 531–543. [CrossRef]
34. Doutsos, T.; Piper, D.J.W. Listric Faulting, Sedimentation, and Morphological Evolution of the Quaternary Eastern Corinth Rift, Greece: First Stages of Continental Rifting. *Bull. Geol. Soc. Am.* **1990**, *102*, 812–829. [CrossRef]
35. Verroios, S.; Zygouri, V. Geomorphological Analysis of Xilokastro Fault, Central Gulf of Corinth, Greece. *Geosciences* **2021**, *11*, 516. [CrossRef]
36. Koukouvelas, I.; Katsonopoulou, D.; Soter, S.; Xypolias, P. Slip rates on the Helike Fault, Gulf of Corinth, Greece: New evidence from geoarchaeology. *Terra Nova* **2005**, *17*, 158–164. [CrossRef]
37. Pavlides, S.; Koukouvelas, I.; Kokkalas, S.; Stamatopoulos, L.; Keramydas, D.; Tsodoulos, I. Late Holocene Evolution of the East Eliki fault, Gulf of Corinth (Central Greece). *J. Quat. Int.* **2004**, *115–116*, 139–154. [CrossRef]
38. Ambraseys, N. *Earthquakes in the Mediterranean and Middle East: A Multidisciplinary Study of Seismicity up to 1900*; Cambridge University Press: Cambridge, UK, 2009; p. 968.
39. Mouyaris, N.; Papastamatiou, D.; Vita-Finzi, C. The Helice fault? *Terra Nova* **1992**, *4*, 124–129. [CrossRef]
40. Doutsos, T.; Poulimenos, G. Geometry and kinematics of active faults and their seismotectonic significance in the western Corinth–Patras rift (Greece). *J. Struct. Geol.* **1992**, *14*, 689–699. [CrossRef]
41. Albini, P.; Rovida, A.; Scotti, O.; Lyon-Caen, H. Large Eighteenth–Nineteenth Century Earthquakes in Western Gulf of Corinth with Reappraised Size and Location. *Bull. Seismol. Soc. Am.* **2017**, *107*, 1663–1687. [CrossRef]
42. Kaviris, G.; Elias, P.; Kapetanidis, V.; Serpetsidaki, A.; Karakonstantis, A.; Plicka, V.; De Barros, L.; Sokos, E.; Kassaras, I.; Sakkas, V.; et al. The Western Gulf of Corinth (Greece) 2020–2021 Seismic Crisis and Cascading Events: First Results from the Corinth Rift Laboratory Network. *Seism. Rec.* **2021**, *1*, 85–95. [CrossRef]

43. Moretti, I.; Sakellariou, D.; Lykousis, V.; Micarelli, L. The Gulf of Corinth: An active half graben? *J. Geodyn.* **2003**, *36*, 323–340. [CrossRef]
44. McNeill, L.C.; Collier, R.; Pantosti, D.; Marco de Martini, P.; D'Addezio, G. Recent history of the Eastern Helike Fault: Geomorphology, paleoseismology and impact on paleoenviornments. In *Helike III, Archaeological Sites in Geologically Active Regions*; Katsonopoulou, D., Soter, S., Koukouvelas, I., Eds.; The Helike Society: Athens, Greece, 2005; pp. 243–265.
45. Bell, R.E.; McNeill, L.C.; Bull, J.M.; Henstock, T.J.; Collier, R.E.L.; Leeder, M.R. Fault architecture, basin structure and evolution of the Gulf of Corinth Rift, central Greece. *Basin Res.* **2009**, *21*, 824–855. [CrossRef]
46. Taylor, B.; Weiss, J.R.; Goodliffe, A.M.; Sachpazi, M.; Laigle, M.; Hirn, A. The structures, stratigraphy and evolution of the Gulf of Corinth Rift, Greece. *Geophys. J. Int.* **2011**, *185*, 1189–1219. [CrossRef]
47. Charalampakis, M.; Lykousis, V.; Sakellariou, D.; Papatheodorou, G.; Ferentinos, G. The Tectono-Sedimentary Evolution of the Lechaion Gulf, the South Eastern Branch of the Corinth Graben, Greece. *Mar. Geol.* **2014**, *351*, 58–75. [CrossRef]
48. Palyvos, N.; Pantosti, D.; DeMartini, P.M.; Lemeille, F.; Sorel, D.; Pavlopoulos, K. The Aigion-Neos Erineos normal fault system (Western Corinth Gulf Rift, Greece): Geomorphological signature, recent earthquake history and evolution. *J. Geophys. Res.* **2005**, *110*, B09302. [CrossRef]
49. Sorel, D. Pleistocene and still-active detachment fault and the origin of the Corinth-Patras rift [Greece]. *Geology* **2000**, *28*, 83–86. [CrossRef]
50. Rigo, A. Etude Sismotectonique et Géodésique du Golfe de Corinthe (Grèce). Ph.D. Thesis, Paris VI, Paris, France, 1994; p. 281.
51. Armijo, R.; Meyer, B.; King, G.C.P.; Rigo, A.; Papanastassiou, D. Quaternary evolution of the Corinth Rift and its implications for the Late Cenozoic evolution of the Aegean. *Geophys. J. Int.* **1996**, *126*, 11–53. [CrossRef]
52. Soter, S. Holocene uplift and subsidence of the Helike Delta, Gulf of Corinth, Greece. In *Coastal Tectonics*; Stewart, I., Vita-Finzi, C., Eds.; Geological Society, London, Special Publications: London, UK, 1998; Volume 146, pp. 41–56.
53. Briole, P.; Rigo, A.; Lyon-Caen, H.; Ruegg, J.; Papazissi, K.; Mistakaki, C.; Balodimou, A.; Veis, G.; Hatzfeld, D.; Deschamps, A. Active deformation, of the gulf of Korinthos, Greece: Results from repeated GPS surveys between 1990 and 1995. *J. Geophys. Res.* **2000**, *105*, 25605–25625. [CrossRef]
54. Stewart, I. Holocene uplift and palaeoseismicity on the Eliki Fault, western Gulf of Corinth, Greece. *Ann. Geof.* **1996**, *39*, 575–588. [CrossRef]
55. De Martini, P.; Pantosti, D.; Palyvos, N.; Lemeille, F.; McNeill, L.; Collier, R. Slip rates of the Aigion and Eliki faults from uplifted marine terraces, Corinth Gulf, Greece. *Comptes Rendus Geosci.* **2004**, *336*, 325–334. [CrossRef]
56. McNeill, L.C.; Collier, R.E.L.L. Uplift and slip rates of the eastern Eliki fault segment, Gulf of Corinth, Greece, inferred from Holocene and Pleistocene terraces. *J. Geol. Soc.* **2004**, *161*, 81–92. [CrossRef]
57. McNeill, L.C.; Collier, R.E.L.; de Martini, P.M.; Pantosti, D.; D'Addezio, G. Recent history of the eastern Eliki fault, Gulf of Corinth: Geomorphology, palaeoseismology and impact on palaeoenvironments. *Geophys. J. Int.* **2005**, *161*, 154–166. [CrossRef]
58. Schmidt, J. *Studien Über Erdbeben*; Carl Schottze: Leipzig, Germany, 1879; pp. 68–83.
59. Katsarou-Tzeveleki, S. Morphology and Distribution of Pottery at the Early Helladic Settlement of Helike. In *Helike IV, Protohelladika—The Southern and Central Greek Mainland*; Katsonopoulou, D., Ed.; The Helike Society: Athens, Greece, 2011; pp. 89–126.
60. Alvarez-Zarikian, C.A.; Soter, S.; Katsonopoulou, D. Recurrent submergence and uplift in the area of Ancient Helike, Gulf of Corinth, Greece: Microfaunal and archaeological evidence. *J. Coast. Res.* **2008**, *24*, 110–125. [CrossRef]
61. Tsokas, G.N.; Tsourlos, P.I.; Stambolidis, A.; Katsonopoulou, D.; Tsokas, A.N.G. Geophysical prospection at the Mycenaean cemetery in the area of ancient Helike, Achaea, Greece. In *Helike V: Ancient Helike and Aigialeia*; Katsonopoulou, D., Ed.; The Helike Society: Athens, Greece, 2017; pp. 167–180.
62. Katsonopoulou, D. An Unknown Earthquake Revealed in Roman Helike. In *HELIKE VI, Helike After 373 BC—An Interdisciplinary Approach*; Katsonopoulou, D., Ed.; The Helike Society: Athens, Greece, 2024; pp. 349–364.
63. Karali, L. Death by Earthquake in Ancient Helike. In *Helike VI, Helike after 373 BC—An Interdisciplinary Approach*; Katsonopoulou, D., Ed.; The Helike Society: Athens, Greece, 2024; pp. 365–372.
64. Ingold, T. The Temporality of the Landscape. *World Archaeol.* **1993**, *25*, 152–174. [CrossRef]
65. Patrick, G.T.W. *The Fragments of the work of Heraclitus of Ephesus on Nature*; Press of Isaac Friedenwald: Baltimore, MD, USA, 1889.
66. Lock, G. *Using Computers in Archaeology: Towards Virtual Pasts*; Routledge: London, UK, 2003.
67. Polla, S.; Verhagen, P. Introduction. In *Computational Approaches to the Study of Movement in Archaeology*; Polla, S., Verhagen, P., Eds.; Walter de Gruyter GmbH: Berlin, Germany; Boston, MA, USA, 2014; pp. 1–4.
68. Lock, G.; Kormann, M.; Pouncett, J. Visibility and Movement: Towards a GIS-Based integrated approach. In *Computational Approaches to the Study of Movement in Archaeology*; Polla, S., Verhagen, P., Eds.; Walter de Gruyter GmbH: Berlin, Germany; Boston, MA, USA, 2014; pp. 23–42.
69. Heidegger, M. *Building, Dwelling, Thinking*; Krell, D.F., Ed.; Martin Heidegger: Basic Writings; Hofstadter, A., Translator; Routledge: London, UK, 1993; pp. 217–265.

70. Wagstaff, M. The Landscape of the Selinous/Kerynites Delta in AD 1700. In *Ancient Helike and Aigialeia-Archaeological Sites in Geologically Active Regions*; Katsonopoulou, D., Soter, S., Koukouvelas, I., Eds.; The Helike Society: Athens, Greece, 2005; pp. 67–74.
71. Sutton, S.B. Settlement patterns, settlement perceptions: Rethinking the Greek village. In *Beyond the Site: Regional Studies in the Aegean Area*; Kardulias, P.N., Ed.; University Press of America: Lanham, MD, USA; New York, NY, USA; London, UK, 1994; pp. 313–335.
72. Weiberg, E. Topography and Settlement: Perception of the Bounded Space. In *Protohelladika—The Southern and Central Greek Mainland*; Katsonopoulou, D., Ed.; The Helike Society: Athens, Greece, 2011; pp. 45–61.
73. Bradley, R. *An Archaeology of Natural Places*; Routledge: London, UK, 2000; p. 192.
74. Dimitropoulos, K.G.; Noutsis, G. Gravity Evidence on the Location of the Lost Ancient City of Helike. In *Helike III Ancient Helike and Aigialeia, Archaeological Sites in Geologically Active Regions*; Katsonopoulou, D., Soter, S., Koukouvelas, I., Eds.; The Helike Society: Athens, Greece, 2005; pp. 197–200.
75. Galadini, F.; Ceccaroni, E.; Dixit Dominus, G.; Falcucci, E.; Gori, S.; Maceroni, D.; Bonasera, M.; Di Giulio, G.; Moro, M.; Saroli, M.; et al. Combining earth sciences with archaeology to investigate natural risks related to the cultural heritage of the Marsica region (central Apennines, Italy). *Mediterr. Geosci. Rev.* **2022**, *4*, 287–318. [CrossRef]
76. Sintubin, M.; Stewart, I.S.; Niemi, T.; Altunel, E. (Eds.) Ancient earthquakes. *Geol. Soc. Am. Spec. Pap.* **2010**, *471*, 280.
77. Bottari, C.; Capizzi, P.; Sortino, F. Unravelling the Seismic Source in Archaeoseismology: A Combined Approach on Local Site Effects and Geochemical Data Integration. *Heritage* **2024**, *7*, 427–447. [CrossRef]
78. Rapp, G. Assessing archaeological evidence for seismic catastrophes. *Geoarchaeology* **1986**, *1*, 365–379. [CrossRef]
79. Nikonov, A.A. Reconstruction of the main parameters of old large earthquakes in Soviet Central Asia using the palaeoseismological method. *Tectonophysics* **1988**, *147*, 297–312. [CrossRef]
80. Rucker, J.D.; Niemi, T.N. Historical earthquake catalogues and archaeological data: Achieving synthesis without circular reasoning. In *Ancient Earthquakes*; Sintubin, M., Stewart, I.S., Niemi, T.N., Altunel, E., Eds.; Geological Society of America: Boulder, CO, USA, 2010; Volume 471, pp. 97–106.
81. Marco, S. Recognition of earthquake-related damage in archaeological sites: Examples from the Dead Sea fault zone. *Tectonophysics* **2008**, *453*, 148–156. [CrossRef]
82. Stewart, I.S. Advancing disaster risk communications. *Earth Sci. Rev.* **2024**, *249*, 104677. [CrossRef]
83. Rodríguez-Pascua, M.A.; Pérez-López, R.; Giner-Robles, J.L.; Silva, P.G.; Garduño-Monroy, V.H.; Reicherter, K. A comprehensive classification of earthquake archaeological effects (EAE) in archaeoseismology: Application to ancient remains of Roman and Mesoamerican cultures. *Quat. Int.* **2011**, *242*, 20–30. [CrossRef]
84. Leroy, S.A.G. *Natural Hazards, Landscapes, and Civilizations. Reference Module in Earth Systems and Environmental Sciences*; Elsevier: Amsterdam, The Netherlands, 2020. [CrossRef] [PubMed Central]
85. Wood, E.; Miller, S.K. Cognitive dissonance and disaster risk communication. *J. Emerg. Manag. Disaster Commun.* **2021**, *2*, 39–56. [CrossRef]
86. Ford, M.; Williams, E.; Malartre, F.; Popescu, S.-M. Stratigraphic architecture, sedimentology and structure of the Vouraikos Gilbert-type fan delta, Gulf of Corinth, Greece. In *Sedimentary Processes, Environments and Basins. A Tribute to Peter Friend*; Nichols, G., Williams, E., Paola, C., Eds.; International Association of Sedimentologists Special Publication: Huelva, Spain, 2009; Volume 39, pp. 49–90.
87. Chambers, R. Vulnerability. Editorial introduction. *IDS Bull.* **1989**, *20*, 1–7. [CrossRef]
88. Kolia, E. Excavation research in eastern Achaea (in Greek). *Arxaiologia* **2015**, *119*, 62–73.
89. Luppino, S. La Produzione Artistica di Sibari Dalla Fondazione alla Distruzione del 510 a.C. In *Ancient Helike and Aigialeia*; Katsonopoulou, D., Soter, S., Schilardi, D., Eds.; The Helike Society: Athens, Greece, 1998; pp. 361–369.
90. Coldstream, J.N. *Greek Geometric Pottery: A Survey of Ten Local styles and Their Chronology*; Methuen: London, UK, 1968; 528p.
91. Bozana-Kourou, N. Grave finds from the area of Aigion (in Greek). In *Memory of N. Kontoleon*; Schefold, K., Pouilloux, J., Eds.; Somateio ton filon Nikolaou Kontoleontos: Athens, Greece, 1980; pp. 303–320.
92. Papadopoulos, J.K. Magna Achaea: Akhaian Late Geometric and Archaic pottery in South Italy and Sicily. *Hesperia* **2001**, *70*, 373–460. [CrossRef]
93. Kolia, E. Archaic Terracotta Reliefs from Ancient Helike. *Hesperia* **2014**, *83*, 409–445. [CrossRef]
94. Katsonopoulou, D. Recent Evidence of Sculptures in Parian Marble. In *Paros and Its Colonies*; Katsonopoulou, D., Ed.; The Institute for Archaeology of Paros and the Cyclades: Athens, Greece, 2018; pp. 101–113.
95. Katsonopoulou, D.; Koukouvelas, I. Earthquake Environmental Effects: The Case of the Late Classical-Hellenistic Helike, Gulf of Corinth, Greece. *Geosciences* **2024**, *14*, 311. [CrossRef]

96. Klasa, K.; Trump, B.D.; Dulin, S.; Smith, M.; Jarman, H.; Linkov, I. A Resilience-Augmented Approach to Compound Threats and Risk Governance: A Systems Perspective on Navigating Complex Crises. *Environments* **2025**, *12*, 64. [CrossRef]
97. Petropoulos, I. Poseidon's Homeric Epithets and Titles. In *POSEIDON—God of Earthquakes and Waters. Cult and Sanctuaries*; Katsonopoulou, D., Ed.; The Helike Society: Athens, Greece, 2017; pp. 99–110.

land

Article

# A New Bronze Age Productive Site on the Margin of the Venice Lagoon: Preliminary Data and Considerations

Cecilia Rossi [1], Rita Deiana [2,3,*], Gaia Alessandra Garosi [4], Alessandro de Leo [4], Stefano Di Stefano [4], Sandra Primon [5], Luca Peruzzo [6], Ilaria Barone [3,6], Samuele Rampin [6], Pietro Maniero [6] and Paolo Mozzi [3,6]

1 Soprintendenza Archeologia, Belle Arti e Paesaggio per il Comune di Venezia e Laguna, Palazzo Ducale, Piazza San Marco 1, 30124 Venezia, Italy; cecilia.rossi@cultura.gov.it
2 Department of Cultural Heritage, University of Padova, Piazza Capitaniato 7, 35139 Padova, Italy
3 Interdepartmental Research Center for Cultural Heritage, University of Padova, Piazza Capitaniato 7, 35139 Padova, Italy; ilaria.barone@unipd.it (I.B.); paolo.mozzi@unipd.it (P.M.)
4 Se.Arch. Srl, Via Copernico, 8, 39100 Bolzano, Italy; gaia.garosi@live.it (G.A.G.); deleo@serviziarcheologia.com (A.d.L.); distefano@serviziarcheologia.com (S.D.S.)
5 Freelance Geologist, 30124 Venezia, Italy; sandra.primon@gmail.com
6 Department of Geosciences, University of Padova, Via Gradenigo 6, 35129 Padova, Italy; luca.peruzzo.1@unipd.it (L.P.); samuele.rampin@unipd.it (S.R.); pietro.maniero@studenti.unipd.it (P.M.)
* Correspondence: rita.deiana@unipd.it

## Abstract

The possibility of collecting new archaeological elements useful in reconstructing the dynamics of population, production and commercial activities in the Bronze Age at the edge of the central-southern Venice Lagoon was provided between 2023 and 2024 thanks to an intervention of rescue archaeology planned during some water restoration works in the Giare–Mira area. Three small excavations revealed, approximately one meter below the current surface and covered by alluvial sediments, a rather complex palimpsest dated to the late Recent and the early Final Bronze Age. Three large circular pits containing exclusively purified grey/blue clay and very rare inclusions of vegetable fibres, and many large, fired clay vessels' bases, walls and rims clustered in concentrated assemblages and random deposits point to potential on-site production. Two pyro-technological structures, one characterised by a sub-circular combustion chamber and a long inlet channel/praefurnium, and the second one with a sub-rectangular shape with arched niches along its southern side, complete the exceptional context here discovered. To analyse the relationship between the site and the natural sedimentary succession and to evaluate the possible extension of this site, three electrical resistivity tomography (ERT) and low-frequency electromagnetic (FDEM) measurements were collected. Several manual core drillings associated with remote sensing integrated the geophysical data in the analysis of the geomorphological evolution of this area, clearly related to different phases of fluvial activity, in a framework of continuous relative sea level rise. The typology and chronology of the archaeological structures and materials, currently undergoing further analyses, support the interpretation of the site as a late Recent/early Final Bronze Age productive site. Geophysical and geomorphological data provide information on the palaeoenvironmental setting, suggesting that the site was located on a fine-grained, stable alluvial plain at a distance of a few kilometres from the lagoon shore to the south-east and the course of the Brenta River to the north. The archaeological site was buried by fine-grained floodplain deposits attributed to the Brenta River. The good preservation of the archaeological structures buried by fluvial sediments suggests that the site was abandoned soon before sedimentation started.


https://doi.org/10.3390/land14071452

**Keywords:** Bronze Age; Venice lagoon; pyro-technological structures; fired clay vessels; electrical resistivity tomography (ERT); Frequency Domain Electromagnetic Measurements (FDEM)

## 1. Introduction

The Recent Bronze Age (1350/1300-1150 BCE) in the central-eastern Veneto region (Italy) witnessed a significant and accelerating population increase, notably along the major rivers (Bacchiglione, Brenta, Sile, Piave, Lemene and Livenza), as well as at their estuaries, indicating control over the coastal zone. While the precise drivers of this growth are still under investigation, it is consistent with the broader pattern of expanding settlement documented across the whole Veneto region during this period [1].

Observing the coastal strip of the Adriatic Sea from south to north, with a focus on the Venice lagoon's edge and its immediate hinterland, the later phase of the Recent Bronze Age saw the emergence of the embanked site of Cona–Cantarana, strategically positioned to control a key point between the Po River delta and the Adriatic coast, and the embanked site of Boion, located in proximity to a probable former sea outlet of the Brenta River. Moving further north, the most notable population clusters were situated during this same period within a peri-lagoonal zone, specifically including the territory of Mestre and the municipalities of Meolo and Quarto d'Altino. A limited presence is also attested within the lagoon itself, as evidenced by the archaeological record at Barena del Vigno. Eastern Veneto appears to follow a similar trend, showing a consistent pattern of settlement along the peri-lagoonal fringe, especially near the mouth systems of the major rivers and within the lagoon areas themselves, as illustrated by the sites of Cittanova–Eraclia and Caorle–San Gaetano.

Mirroring the broader Veneto region, the Final Bronze Age 1–2 (1175/50-1000 BCE) in the Venetian area marks a period of significant population reduction, with the settlement at Caorle–San Gaetano representing the sole continuous occupation controlling the para-coastal routes. Conversely, the Final Bronze Age 3 is characterised by a resurgence in territorial occupation, albeit one that was not as widespread or ubiquitous as in the recent Bronze Age but rather confined to specific areas. It is during this later phase that settlement was re-established in the vicinity of future *Altinum*, strategically positioned to control the lower course of the Sile River and its associated peri-lagoonal zone. Caorle–San Gaetano was, in all probability, succeeded in its role by the area of present-day Concordia, which was already active during the recent Bronze Age [2] (pp. 300–304), [3–7].

Within this established context of protohistoric settlement along the Venice lagoon's littoral and immediate hinterland, the recently identified site at Mira–Giare constitutes a novel and somewhat unexpected data point emerging from preventive archaeological investigations. The site appears to fill a perceived settlement lacuna within a central segment of the lagoon's littoral. The prior absence of archaeological evidence may be ascribed to a documentary deficit, potentially resulting from a lack of previous systematic archaeological investigations and/or issues concerning the subsurface visibility of the buried deposit.

These preliminary considerations prompted an in situ intensification of site investigation during the excavation phase, deploying a suite of supplementary analyses designed to maximise data acquisition for comprehensive archaeological and geomorphological reconstruction.

The main scientific questions we tried to answer with the in-depth investigations conducted while the archaeological excavation was still underway were as follows: what was the context of the structures, and how far did the production site extend? Answering these questions would allow for the discoveries to be framed within a broader context, both geographically and historically–culturally. It would also help in estimating the importance and complexity of the productive activity documented by the pyro-technological structures brought to light.

## 2. Setting

The archaeological site is located some hundred metres from the current inner margin of the lagoon of Venice (Figure 1), which is here bordered by a several-metre-high artificial level. The Venice lagoon is characterised by geological subsidence that brings about a continuous rise in relative sea level, estimated at around 1.2–1.3 mm $yr^{-1}$ for the last two thousand years [8,9]. The study area is now at an elevation around $-1$ m Mean Sea Level (MSL).

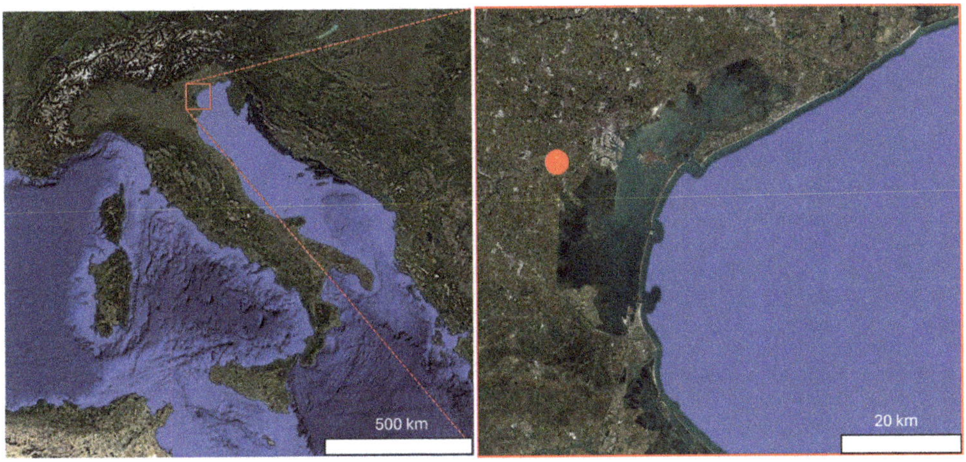


**Figure 1.** Position of Mira-Giare site (red circle) in the lagoon of Venice (Google Earth image modified by authors).

The site falls within the deltaic system of a mediaeval branch of the Brenta River, which has been debouching in the lagoon starting from the 12th century AD [10,11] (Figure 2). Between the mediaeval fluvial ridges, large stretches of the alluvial plain built by the Brenta River during the Last Glacial Maximum (LGM) crop out. These latter alluvial deposits belong to the Brenta megafan and are characterised by a typical calcic soil that is also found buried under the post-LGM lagoonal and fluvial deposits [12,13].

A Holocene palaeochannel of the Brenta River has been traced in the area through remote sensing, and it was radiocarbon-dated to the late Bronze Age [11]. Due to relative sea level rise [5], the Bronze Age inner margin of the lagoon is now below MSL. Geological investigations evidence the presence of buried deltaic deposits of the 2nd millennium BC in the underground area of the present lagoon SE of the Giare area [14]. This allows us to hypothesise the location of the Bronze Age lagoon margin as shown in Figure 2.

The new archaeological evidence of the Giare site consists of a productive context characterised by the presence of pyro-technological structures and provisionally dates to the transition from the late Recent Bronze Age to the early Final Bronze Age, based on a preliminary typological assessment of the recovered artefacts. Bronze Age chronological references throughout this paper are based on the frameworks proposed by [1,15,16].

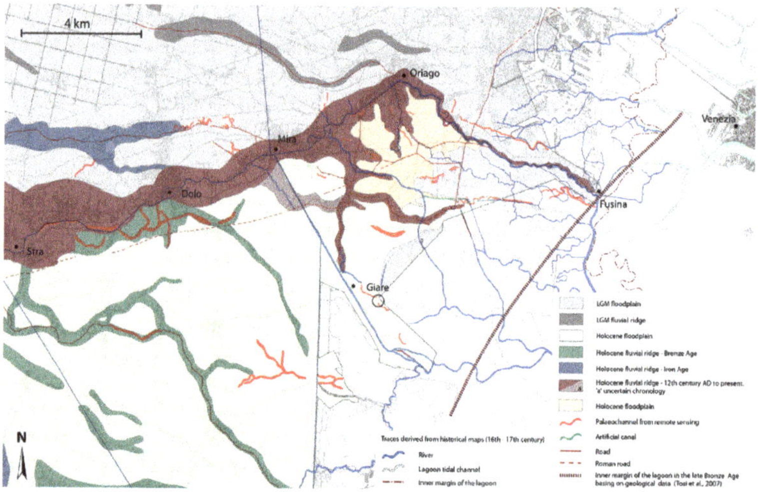


**Figure 2.** Geomorphological sketch map of the area around the Giare site [5,11,14].

## 3. Materials and Methods

Investigations at the Giare archaeological site, located in the municipality of Mira (VE), near the current lagoon margin, began in 2023. This was in connection with a major public works project aimed at the hydraulic arrangement, environmental restoration and landscape redevelopment of a portion of the drainage water network in the central-southern Venice lagoon (Figure 3).

Developed by the Consorzio di Bonifica Acque Risorgive, the project underwent preliminary Preventive Archaeological Interest Verification (VPIA) in 2020, as required by Article 25 of Legislative Decree 50/2016 (now Article 1, paragraphs 2 and 3, Annex I.8 of Legislative Decree 36/2023). The preliminary documentation assessed the archaeological risk as moderate, within a context characterised by massive areal deposits resulting from recent agricultural exploitation. As a result, the project was approved, conditioned upon continuous archaeological monitoring for soil disturbance operations reaching greater depths beyond the ground surface.

Archaeological excavations began in March 2023, carried out by Se.Arch. Srl, under the scientific direction of Soprintendenza Archeologia, Belle Arti e Paesaggio per il Comune di Venezia e Laguna. In April of the same year, the first anthropogenic deposits were identified, characterised by abundant pre-Roman ceramic material. This discovery, surprising as it was not predicted by the existing literature, yielded the opening of a trench and three excavation test units in distinct areas of the site, with the aim of clarifying the nature and chronology of the findings (Figure 4).

Stratigraphic investigations were carried out between April 2023 and July 2024, with numerous interruptions due to adverse weather conditions and rising groundwater. From the early stages of excavation, the exceptional nature of the discovery became evident: the recovered ceramic materials and the associated pyro-technological structural evidence pointed to a production site, dating between the late Recent Bronze Age 2 (first half of the 12th century BCE) and the early Final Bronze Age 1–2 (1175/50-1000 BCE), a chronological phase not previously recorded in this locality [17].

**Figure 3.** Territory under the jurisdiction of the Soprintendenza Archeologia, Belle Arti e Paesaggio per il Comune di Venezia e Laguna (red line indicates the perimeter of the municipalities). The location of the site is highlighted with a yellow circle (Google Earth image modified by authors).

Non-invasive geophysical measurements in the analysis of production sites of the Bronze Age, and more generally in this type of context, are typically taken by magnetic measurements [18–22], and in a few cases, this method appears to be coupled with Ground-Penetrating Radar (GPR) acquisitions [23,24] for the same purposes. In general, the absence of actual remains of structures of a specific size, with the presence of distributed traces of thin layers, pits and small-in-size remains instead, which can be related to pyrotechnological activities within shallow depths, allows the fruitful use of these methods, efficient in shallow systems. Furnaces and fired materials represent the ideal target for magnetic methods, while more sporadically, GPR provides high resolution in searching for the presence of shallow and thin structures, which are generally present in this type of site [25]. Electrical resistivity tomography (ERT), well-known for its use in searching built buried structures [26,27], is well documented in investigations involving geomorphologically complex sites, such as those often encountered in Bronze Age contexts [28–31]. The presence of palaeochannels and systems related to water circulation that may involve depths greater than the actual, shallower sediment of the archaeological settlement provides an easy target for this prospecting method. Therefore, ERT measurements may prove helpful in studying the relationship between settlement and water palaeo-system circulation in these contexts. Much more recent, but of extreme interest for a more detailed analysis of the more superficial layers of the subsurface, particularly in the context of palaeochannels and lagoon systems, is the use of low-frequency electromagnetic (FDEM) measurements [32–36]. This fast acquisition technique makes it possible to reconstruct maps in the shallowest part of the investigated subsoil, at least up to the depths typically reached by magnetic methods (about 1.5 m from ground level). The variations in soil conductivity and the correlation of these variations with the possible magnetic response of the investigated deposits are measured by this method. The EM field's quadrature component indicates the system's electrical conductivity. The in-phase component of the EM field expresses the correlation with magnetic susceptibility, providing important indications for the study of sites, such as those of the Bronze Age, which were placed in wet contexts or demonstrated a strong presence of fluvial/water palaeo-systems.

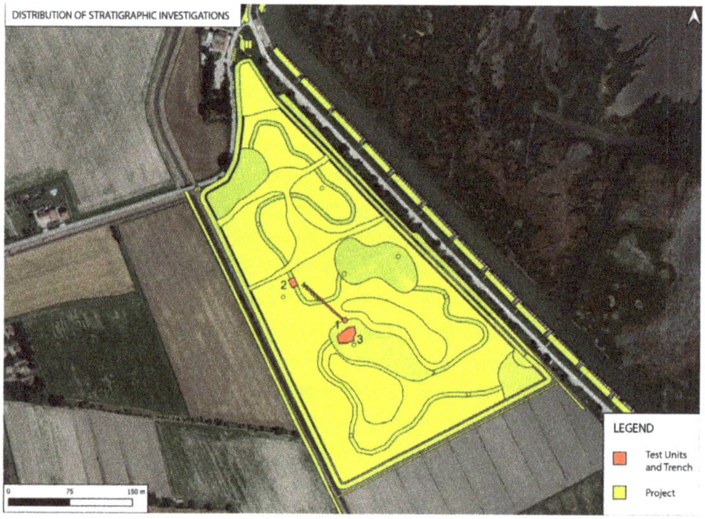


**Figure 4.** Mira (Venice), site plan of Parco delle Giare with location of archaeological stratigraphic tests (1, 2, 3 in the figure) (Google Earth image modified by authors).

Given the geomorphological context and proximity to Venice's lagoon of the Giare site, FDEM measurements in two different areas around the archaeological excavations and three different ERT lines in these areas were collected (Figure 5). Geophysical measurements supported research into possible additional buried structures, such as those that emerged through the excavation, attempting to identify the extent of the archaeological site, having evaluated the characteristics of the system and the archaeological remains.

FDEM measurements were collected using a GF Instruments multi-depth Mini Explorer conductivity meter. The instrument is equipped with three-depth probes with depth ranges (high/low) of 0.5/0.25 m, 1.0/0.5 m and 1.8/0.9 m that allow the simultaneous acquisition of the quadrature component data of the EM field, related to electrical conductivity variations in the investigated system, and the in-phase component data of the EM field, associated with the magnetic susceptibility of the system. The high-depth mode was used in this context, resulting in final depth ranges of 0.5 m, 1 m and 1.8 m to the probe position, at about 0.15–0.2 m above ground level throughout this acquisition. The acquisition was georeferenced using the integrated GPS mounted on the Mini Explorer controller. The data thus acquired produced three maps of the change in electrical conductivity, expressed in mS/m and p.p.m, and the system's susceptibility in three different depth ranges. FDEM maps were obtained by interpolating the data using Golden Software Surfer 15.

ERT measurements were collected using an IRIS Syscal Pro 48 ch resistivity meter, arranging for each line 48 electrodes spaced 0.5 m apart for a total length of 23.5 m and a depth of investigation of about 5 m. A dipole–dipole skip 4 array provided excellent lateral and vertical resolution to focus on detail in the shallow part of the system analysed here. The optimised dataset included direct and reciprocal measurements for each quadrupole. We set a time window of 250 ms, 3–6 stacks with 5% tolerance and variable current injection to target potential values of at least 50 mV. The inversion processing of the ERT sections was performed using an Occam inversion approach [37], as embedded in ProfileR2D freeware software (version 2.5 Oct. 2003) by A. Binley. Data with reciprocal errors exceeding 5% have been removed as outliers. The same resistivity range (0–25 Ohm*m) was finally applied to enable the timely comparison of detected anomalies and consistent interpretation of 2D ERT vertical sections obtained with Golden Software Surfer 15.

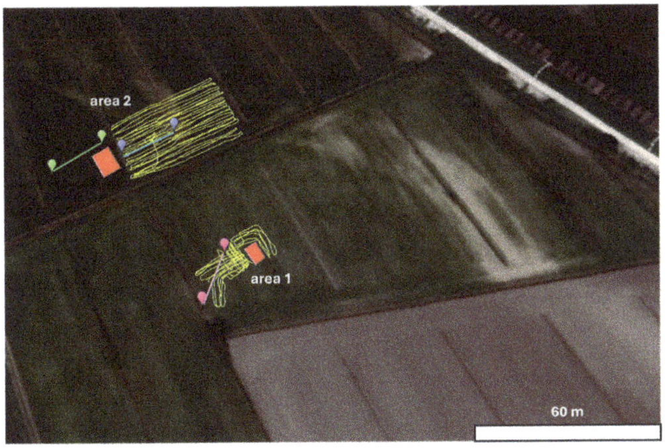


**Figure 5.** Locations of the FDEM measurements (yellow lines) and ERT lines (green, blue and pink) in the two excavated areas in red.

Coring was carried out using a composite Eijkelkamp manual auger with Edelman and gauge samplers, which allow for the collection of undisturbed sediment samples. Cores were aligned along a transect starting from the archaeological site, about perpendicular to the palaeochannel observed at the surface in satellite images (Figure 6).

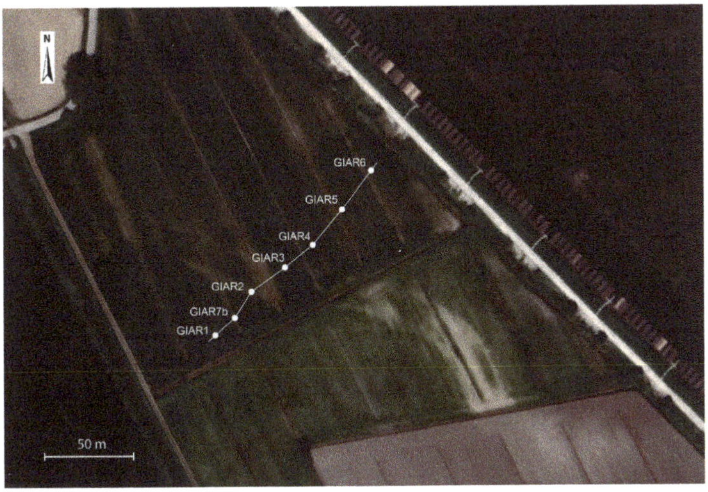


**Figure 6.** Location of geological cores.

## 4. Results

### *4.1. Archaeological Evidence*

The archaeological deposits were identified approximately 75 cm below ground level, which had been previously lowered by about 30 cm to remove the superficial arable layer across the entire work area. In a cross-section, the stratigraphy reveals an upper sub-agricultural layer (Stratigraphic Unit 2, abbreviated as SU 2; this abbreviation will be used for all stratigraphic units hereafter) approximately 40 cm thick, characterised by a compact and well-preserved silty loam matrix with a light-grey colour. Below this layer, a peat deposit (SU 3) follows, varying in thickness between 3 and 10 cm, composed primarily of undecomposed vegetal material. Further down, an alluvial deposit (SU 4) is present, approximately 25 cm thick, characterised by a rather compact clay–silt composition and a light-grey colour with diffuse iron–manganese mottling of orange hue (Figure 7). Likely originating from a natural sedimentation event, this alluvial layer seals the archaeological context, probably immediately or shortly after its abandonment.

**Figure 7.** Mira (Venice), Parco delle Giare. Southern section of the in-depth transect in the channel/basin identified in Test Unit 2 (Photo: Se.Arch. Srl—Courtesy of SABAP VE-Lag).

Test Unit 2 (Figure 4) presented a rather complex archaeological palimpsest. Excavations revealed three large circular pits (SU 8; SU-14; SU 9; SU-17; SU 20; SU-21) averaging 40 cm deep and approximately 125 cm in diameter, containing exclusively purified grey/blue clay and very rare inclusions of vegetable fibre. The associated discovery of a large quantity of bases, walls and rims of large fired clay vessels, clustered in concentrated assemblages and random deposits, points to potential on-site production of these artefacts. The analysis of these fired clay vessels and the few associated pieces of pottery began during the excavation activities, as part of a teaching laboratory organised in collaboration with the University of Padua—Department of Cultural Heritage. This laboratory was led by Michele Cupitò, Professor of European and Mediterranean Protohistory, with support from Giovanni Tasca and Vanessa Baratella (University of Padua) and Gaia Garosi (SE.Arch. Srl). The typo-chronological and functional study of this material is currently being completed by the same team.

Two significant fired material concentrations were documented, the function of which is still under investigation: the first (SU 5), characterised by decimetric and centimetric vessel fragments, had well-defined boundaries (Figure 8); the second (SU 22), more akin to a dump, was mostly characterised by significantly smaller fragments, with a chaotic arrangement and less-defined boundaries (Figure 9). An in-depth excavation below the aforementioned layer exposed a pyro-technological structure (Figure 10), with only the cut surviving. The collected data points to a small furnace, exhibiting a vague mushroom-like profile, characterised by a sub-circular combustion chamber (2.10 m diameter) and a long inlet channel/praefurnium (1.35 m in length).

**Figure 8.** Mira (Venice), Parco delle Giare. Concentration of fired clay fragments in Test Unit 2 (Photo: Se.Arch. Srl—Courtesy of SABAP VE-Lag).

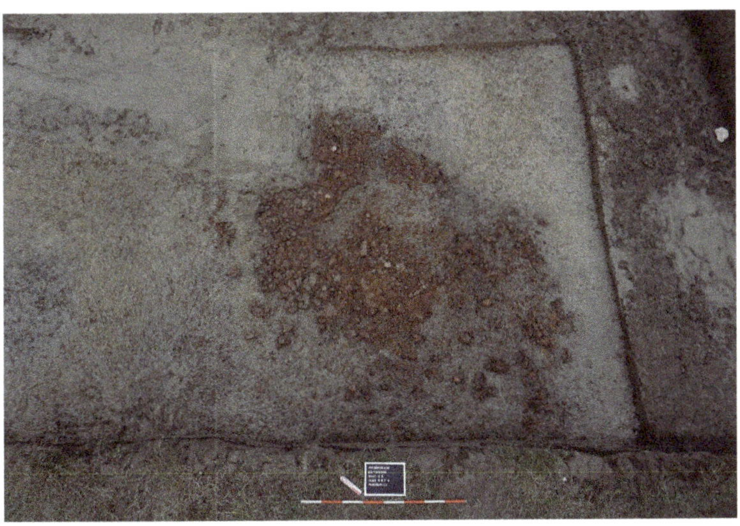

**Figure 9.** Mira (Venice), Parco delle Giare. Disordered deposit of fired clay fragments in Test Unit 2 (Photo: Se.Arch. Srl—Courtesy of SABAP VE-Lag).

**Figure 10.** Mira (Venice), Parco delle Giare. Structure cut identified in Test Unit 2 (Photo: Se.Arch. Srl—Courtesy of SABAP VE-Lag).

The structure's plan exhibits a characteristic configuration, comparable to a 'keyhole' or 'fungoid' shape, which finds a precise parallel in Type 3.1 as proposed by Agostino Sotgia [38] (pp. 308–309). The furnace classified as Type 3.1 corresponds to a two-chambered structure with a horizontal development, wherein the combustion and firing chambers are arranged according to a figure-eight planimetric configuration. This furnace typology emerged starting from the Middle Bronze Age and continued to be used through the early Iron Age. The combustion chamber housed the fuel, while the artefacts to be fired were placed in the firing chamber. This technical principle, based on the utilisation of heat rather than direct flame, facilitated a more homogeneous firing of the artefacts and greater potential for controlling the internal atmosphere of the firing chamber. Draft, therefore, occurred horizontally.

In the southern part of the archaeological test unit, a shallow channel/basin with stagnant water was identified. Due to its size, an in-depth transect was excavated (Figure 7). Below the alluvial deposit (SU 4), a primary fill (SU 25) was observed, consisting of a rather compact clay–silt deposit, of plastic consistency, 30 cm thick, rich in peat-like vegetal material but lacking inorganic inclusions. Beneath this level, a deposit (SU 36) with a weakly silty clay matrix was distinguished, poorly compacted and with varying thickness, reaching 5 cm in some areas and absent in others. Rare inclusions of wood, ceramics, bone and malacofauna were present. This level likely represents a wet ground surface within the channel/basin, possibly the visible bottom during the period of the site's activity. The layer's irregular morphology may have been caused by repeated animal trampling. Continuing downwards, a compact clay deposit (SU 37) is recognised, 22 cm thick, containing horizontally oriented brushwood. The disposition of this material suggests the presence of stagnant water, confirming a natural formation at the bottom of the channel/basin. The absence of water flow facilitated the accumulation and stagnation of vegetal debris.

Finally, the bottom of the channel/basin was composed of a light-grey, weakly sandy clay matrix deposit (SU 38) with diffuse calcium carbonate concretions. This layer, also found in core drillings and at the base of the three excavated pits, likely represents the original ground surface upon which the archaeological site was established.

In Test Unit 1 (Figure 4), the excavation uncovered a significant pyro-technological structure (2.5 m × 4.0 m), sub-rectangular in shape, with arched niches along its southern side. This structure cuts through an ancient ground surface (SU 505), consisting of a clay–silt matrix with fired clay fragments and sparse ceramics. Initially thought to be a furnace, the structure was found in good condition, with an intact perimeter curb and a side opening defined by curbs (Figure 11). A small in-depth excavation within the structure revealed a charcoal deposit of approximately 20 cm. Given the unique nature of the find, a survey technique was chosen to balance site excavation timelines with research needs. The documentation phase was therefore integrated with a high-definition 3D scan, necessary for future detailed studies and also usable for promotional purposes.

**Figure 11.** Mira (Venice), Parco delle Giare. Pyro-technological structure identified in Test Unit 1 (Photo: Se.Arch. Srl—Courtesy of SABAP VE-Lag).

Test Unit 3 (Figure 4), located near Test Unit 1, yielded no structural remains. However, fragments of ceramic and fired clay vessels, along with sparse millimetric charcoal inclusions, were found scattered throughout the layer (SU 605 = SU 505) at the same level as the structure in Test Unit 1.

## *4.2. Geophysical Results*

### 4.2.1. Electrical Resistivity Tomography

The results of the ERT measurements are shown in Figure 12. The three ERT sections were made to analyse the relationship of the archaeological evidence with the deeper-buried geomorphological system present in the area. The resistivity range of each section appeared very similar and consistent, with values ranging between a minimum of 0 and maximums lying between 25 and 40 Ohm*m. Starting from these ranges, a common resistivity range of 0–25 Ohm*m was adopted to evaluate similar behaviours in the three sections. The outcome of checking the direct and reciprocal measurements is that 95% of the data on all sections is saved, defining the absolute goodness of the dataset and the validity of the recorded signal. Line 1, according to the scheme of acquisition of geophysical measurements shown in Figure 5, appears outside the areas investigated with FDEM measurements. In contrast, Line 2 crosses NW area 2 investigated with FDEM measurements starting from metre 11 until the end of the line from south to north (yellow lines in Figure 5). Line 3 crossed area 1 investigated with FDEM measurements to the SE in the portion from metre 9 and for the entire line length (yellow lines in Figure 5). The analysis of the resistivity values obtained for the three sections, shown in Figure 12, depicts the presence of more-resistive sediments in the shallow part of all three sections, with values ranging from 25 Ohm*m to 11 Ohm*m, distributed differently both laterally and in depth for each section over a band affecting the first metre from ground level.

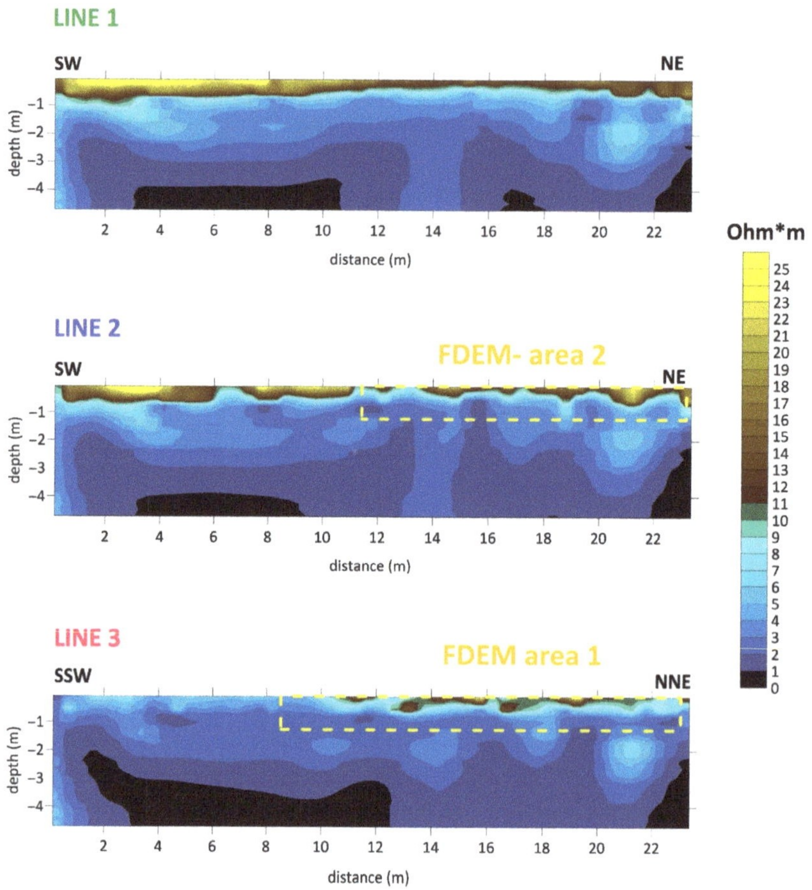


**Figure 12.** Results of the inversion processing of the three ERT lines (green, blue and pink lines in Figure 5).

In particular, where this more-surface-resistive level appears continuous, horizontal and homogeneous in Line 1, in Line 2, this layer appears irregular throughout the whole section. In contrast, in Line 3, this resistive layer seems to affect only the second part of the section, showing a more conductive zone even at the surface in the first portion of the line. The three deep sections show an expected increase in electrical conductivity (inverse of resistivity), albeit uneven, in the deepest part of all sections, with varying distributions for each. The most interesting area, however, seems to be the one with a thickness between 1 m and 3 m in depth above ground level (Figure 12). In fact, in this area, we note the presence of a series of isolated bodies varying in colour between cyan and light blue, apparently related to similar structures (e.g., deposits of materials with larger grain sizes than the host system), considering the resistivity values and electrically conductive behaviour compared to the background. Lateral and vertical resistivity trends and the total number of these structures for each ERT section appear variable. Note that the apparent vertical effect of some anomalies, such as those visible between 10 m and 16 m in Line 1, around 14 m in Line 2 and less markedly between 12 m and 22 m in Line 3, is provided by the impossibility, within the maximum depth of investigation obtainable here, of fully resolving the geometry of deeper, not necessarily vertical, structures such as those described here.

#### 4.2.2. Frequency-Domain Electromagnetic Measurements

The FDEM measurements produced three different electrical conductivity maps and three maps of the magnetic behaviour of the system, corresponding to three different depth ranges, as shown in the images in Figures 13 and 14.

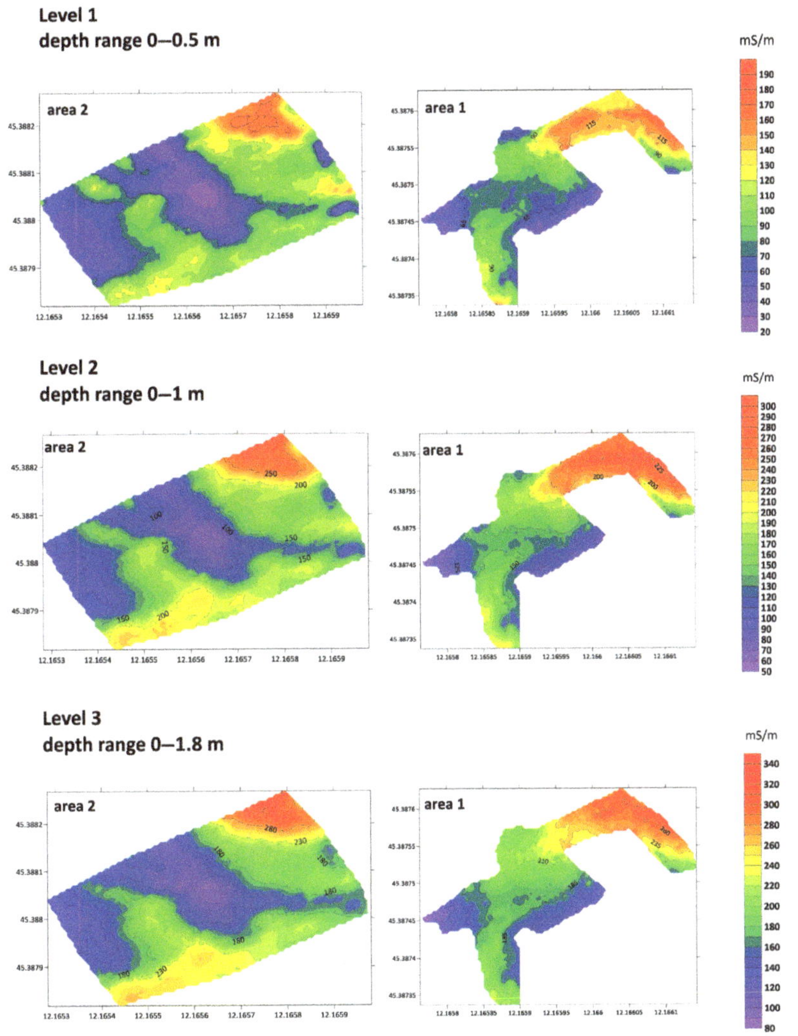


**Figure 13.** Results of FDEM measurements corresponding to the electrical conductivity at the three different investigated depth range levels.

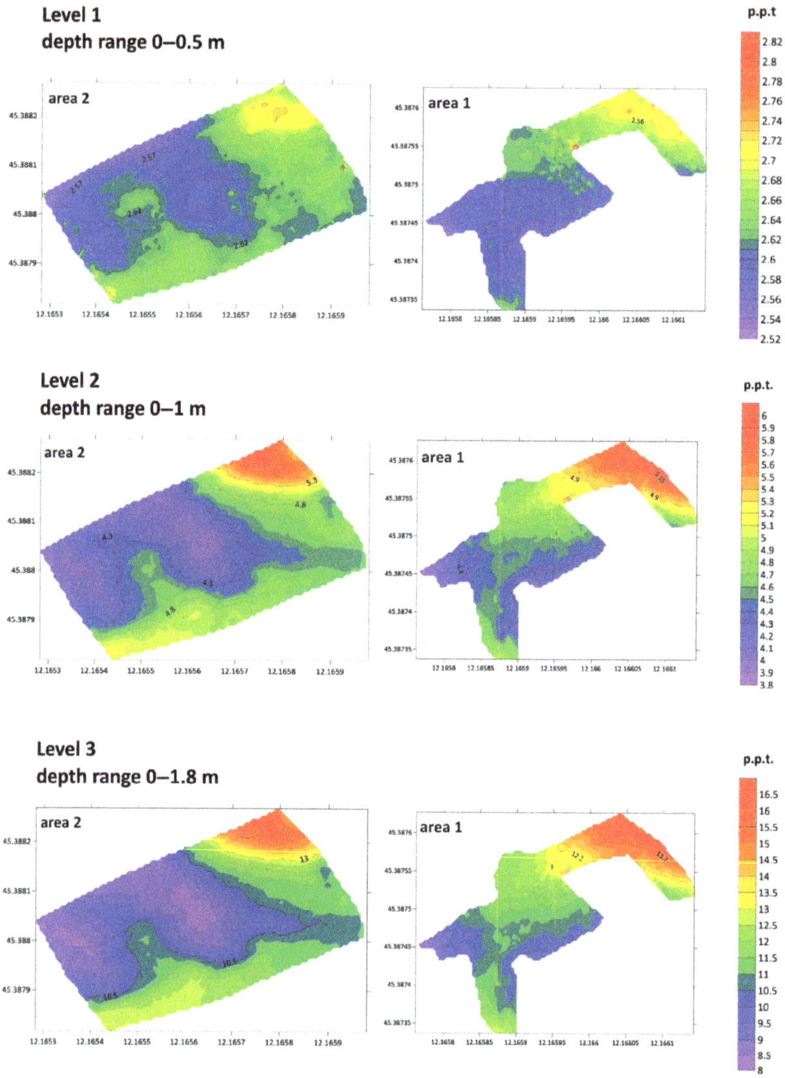


**Figure 14.** Results of FDEM measurements corresponding to the magnetic susceptibility at the three different investigated depth range levels.

For each depth range analysed, we decided to adopt a common range of values of the EM field component related to the electrical and magnetic behaviour of the system in the specific layer. The pattern of electrical conductivity in the three depth ranges investigated is consistent, with a significant increase in electrical conductivity values in the NE zone at all levels for both areas (red zones in the images in Figure 13).

In the images related to the variation in the magnetic behaviour detectable through the in-phase component of the EM field in the same depth ranges considered for the electrical response, it is interesting to note that the pattern is consistent across levels. Also, for the in-phase maps, we adopted a common scale for both areas investigated for each level considered. In an analysis of the resulting maps, shown in Figure 14, it is of extreme interest to note that the values of the parameter under investigation increased in the same areas (red) where we registered an increase in the electrical conductivity of the system (Figure 13).

*4.3. Core-Sample Analysis*

Hard and compacted buried soil with a silty clay, grey matrix and many centimetre-size carbonatic nodules were found in the lower part of most cores. The cross-section of Figure 15 shows the lateral extension of this soil, on which the archaeological deposits lie. Above the archaeological deposits and the buried soil, there are clay sediments that are topped by a peat layer. The sedimentary succession is closed at the top by an extensive unit of silty sand. These latter sediments are connected with a sandy palaeochannel located in the central portion of the section, whose erosive base cuts older sedimentary units.

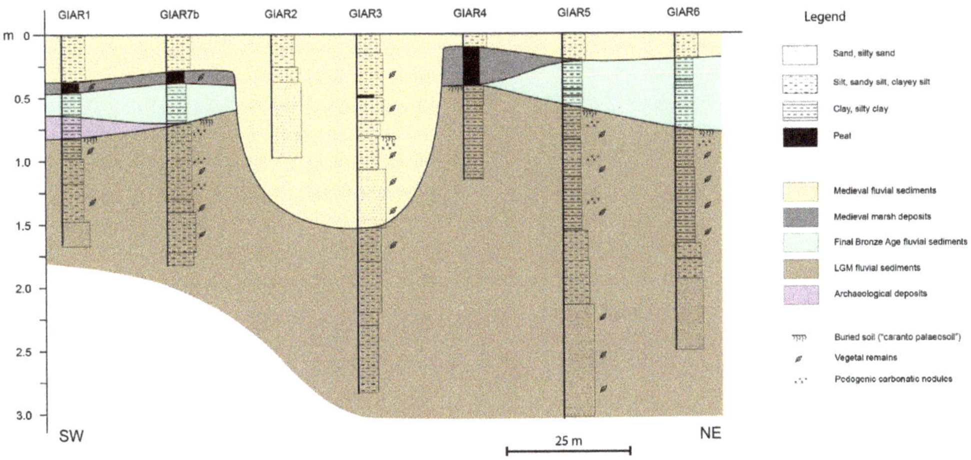


**Figure 15.** Geological section (for location of the cores, see Figure 6). The archaeological deposits at the SE end of the section correspond to the excavated archaeological site.

## 5. Discussion

Geographically, the site is situated within a liminal zone between the mainland and the lagoon basin, exhibiting conditions of hydro-geological instability and marginality vis-à-vis the principal settlement agglomerations thus far identified predominantly within the northern sector of the Venice lagoon. This location, coupled with the good state of preservation of the archaeological levels and the absence of significant post-depositional disturbances, has enabled the recognition of a complex stratigraphic palimpsest, yielding material evidence that offers significant new insights on human presence and productive activities in the central sector of the Venice lagoon during the protohistoric period.

The totality of the material and stratigraphic data allows for the assignment of the site to a phase dating between the late Recent Bronze Age 2 (first half of the 12th century BCE) and the early Final Bronze Age 1–2 (1175/50-1000 BCE). The recovered evidence—including two pyro-technological structures, extensive concentrations of fired clay materials and three potential basins containing refined clays—clearly indicates the artisanal and productive nature of the site, likely specialised in ceramic manufacture or the transformation of raw materials (possibly also lagoon resources such as fish or salt, a hypothesis currently under analytical verification). The typo-chronological and functional study of artefacts and structures is currently ongoing, and thus, both hypotheses remain open at this time.

The eminently productive connotation of the structural evidence uncovered during excavation, the scarcity of domestic pottery and the marked disproportion of the latter in relation to the large quantity of collected fired clay materials underscore the non-residential nature of the site. The settlement associated with the unearthed artisanal site was likely located in the vicinity, at a distance easily traversable on foot, allowing for the convenient monitoring of ongoing production processes and an easy return to dwellings at the conclusion of work activities, with minimal (or no) on-site consumption of meals.

ERT measurements conducted through three different 2D vertical sections at the Giare site around the two excavated areas provide interesting data for evaluating the ancient lagoon hydrologic system on which the archaeological site is set. In fact, in addition to a more shallow resistive system that seems to affect the first metre of depth below the present ground level, the ERT sections show the presence of an evident lateral and vertical variability between 1 m and 3 m depth, identifying selected areas with different resistivity values as zones of the accumulation of materials with different grain sizes (e.g., palaeochannels).

FDEM measurements combined with ERT measurements provided a clear and expected correlation with the more shallow palaeochannel system, also visible in traces from satellite imagery and likely corresponding to recent historical ages, given the maximum depth of investigation obtained here of about 1.5 m from ground level. The EM field's quadrature and in-phase component maps agree, respectively, in detecting anomalous increases in the electrical conductivity and magnetic behaviour of the system over two specific areas. The high conductivity supports the possibility of a buried layer with an anomalous salt concentration in a specific zone in these areas. On the other hand, fire clay materials and specific remains associated with pyro-technological activities represent a potential perfect magnetic target, producing anomalies in the in-phase component of the EM field, thus revealing their possible concentration in the actual subsoil.

The buried soil identified in the cores (Figure 15) is interpreted as being the caranto palaeosoil, developed on top of the LGM succession of the Brenta megafan. The topographic surface on which the productive site was settled corresponds to the top of the soil. From a palaeoenvironmental point of view, the pronounced soil evolution shows that the site was a relict, stable, relatively well-drained alluvial plain. The site was lying at a distance of about 3.7 km from the Bronze Age lagoon margin and 3.5 km from the then-active Brenta River channel (Figure 2).

The archaeological site was buried by fine-grained floodplain deposits. These deposits are related to distal overbank flooding by a major river, presumably the nearby Bronze Age Brenta River. The good preservation of the archaeological structure buried by fluvial sediments suggests that the site was abandoned not long before sedimentation started.

The peat layer can tentatively be correlated with the peat that was found in cores carried few hundred metres NW of the study area, and that was radiocarbon-dated to the Middle Ages [39]. This implies that peat deposition corresponds to a rather long period comprising the Iron Age, Roman times and early part of the Middle Ages. The final burial of the peat by the sandy silty deposits is attributed to proximal overbank deposition by the Brenta palaeochannel that was crossing the area in medaieval times.

## 6. Conclusions and Future Perspectives

The intervention of rescue archaeology, planned during water restoration works in the Giare–Mira area between 2023 and 2024, opened an important window into human activities in the Bronze Age at the edge of the central-southern Venice lagoon. The approach to studying this context was undoubtedly conditioned by the logistical limitations and time constraints imposed by the water restoration works. However, the fortunate possibility of collecting geophysical and geomorphological information, in addition to the important and unexpected archaeological structures found in the only three excavation assays carried out in the area, makes it possible to formulate new preliminary hypotheses on the occupation of this site since the Bronze Age.

To further investigate the site, various samples were collected for planned laboratory analyses during the excavation process. Specifically, to reconstruct the ancient environment, palynological analysis of soil samples and archaeobotanical analysis of plant macro-remains (including combusted remnants from the suspected furnaces) are scheduled. To refine the chronological data derived from vessel typologies, radiocarbon dating of organic materials is planned. To evaluate the site's potential primary function in processing lagoon products (fish/salt), biochemical analyses will be conducted on the internal surfaces of fired clay artefacts to identify markers of relevant biological products. Lastly, archaeometric analyses of both fired clay and ceramic vessels are slated to determine the artefacts' firing and usage temperatures.

To delineate the spatial extent of the habitation likely affiliated with the exposed protohistoric production site, a comprehensive, multi-scalar investigation of the encompassing area (c. 860 ha) is programmed, with initial emphasis on its detection. This analysis will proceed through sequential tiers of inquiry. The primary stage will entail the examination of historical cartographic sources and the application of aerial photographic interpretation to identify anthropogenic anomalies, leading to the creation of a predictive archaeological sensitivity map. Informed by the resultant dataset, a subsequent phase of systematic surface reconnaissance will be implemented, concentrating on the on-the-ground verification of the most archaeologically significant anomalies. The outcomes of the surface surveys will inform the subsequent in-depth phase, comprising a programme of geophysical prospection and mechanical coring aimed at the precise definition of the identified archaeological sites, determining their horizontal extent and the depth of archaeological horizons relative to the present ground level.

**Author Contributions:** Conceptualisation, C.R., G.A.G., P.M. (Paolo Mozzi) and S.P.; methodology, C.R. and G.A.G.; software, L.P., I.B., P.M. (Pietro Maniero) and S.R.; validation, C.R., R.D., P.M. (Paolo Mozzi) and S.P.; investigation, R.D., I.B., L.P., G.A.G., S.R., P.M. (Pietro Maniero) and A.d.L.; resources, C.R.; data curation, I.B. and L.P.; writing—original draft preparation, C.R., G.A.G., P.M. (Paolo Mozzi) and S.P.; writing—review and editing, C.R., G.A.G., P.M. (Paolo Mozzi) and S.P.; supervision, C.R., G.A.G., P.M. (Paolo Mozzi) and S.P.; project administration, C.R. and S.D.S.; funding acquisition, C.R. All authors have read and agreed to the published version of the manuscript.

**Funding:** This research was funded by Italian Ministry of Culture (MiC)—Soprintendenza Archeologia, Belle Arti e Paesaggio per il Comune di Venezia e Laguna (archeologia preventiva nella Laguna Sud—Programmazione triennale dei lavori pubblici 2024–2026. D.M. n. 327 del 20 ottobre 2023).

**Data Availability Statement:** The data will be made available on request to authors.

**Acknowledgments:** This research was supported by the World Class Research Infrastructure (WCRI) **SYCURI—SY**nergic strategies for **CU**ltural heritage at **RI**sk—and fully funded by the University of Padova.

**Conflicts of Interest:** Gaia Alessandra Garosi, Alessandro de Leo and Stefano Di Stefano are employed by the Se.Arch. Srl. The authors declare no conflicts of interest. The funders had no role in the design of the study; in the collection, analyses, or interpretation of data; in the writing of the manuscript; or in the decision to publish the results.

## References

1. Cupitò, M.; Leonardi, G. Il Veneto tra Bronzo Antico e Bronzo Recente. In *Preistoria e Protostoria del Veneto*; Leonardi, G., Tiné, V., Eds.; Istituto Italiano di Preistoria e Protostoria: Florence, Italy, 2015; pp. 201–239, ISBN 978-88-6045-056-2.
2. Cupitò, M.; Lotto, D.; Facchin, A. Dinamiche di popolamento e modelli di organizzazione del territorio nella bassa pianura veneta compresa tra Adige e Tagliamento durante l'Età del bronzo. In *Preistoria e Protostoria del Veneto, Studi di Preistoria e Protostoria, 2*; Leonardi, G., Tiné, V., Eds.; Istituto Italiano di Preistoria e Protostoria: Rome, Italy, 2015; pp. 295–306.

3. Bagolan, M.; Leonardi, G. Il bronzo finale nel Veneto. In *Il Protovillanoviano al di qua e al di là dell'Appennino", Biblioteca di Athenaeum 38*; Harari, M., Pearce, M., Eds.; Athenaeum Editrice: Pavia, Italy, 2000; pp. 13–44.
4. Bianchin Citton, E. Il sito umido di S. Gaetano—Casa Zucca. In *La Protostoria tra Sile e Tagliamento. Antiche Genti tra Veneto e Friuli, Catalogo della Mostra*; Concordia Sagittaria, 14 settembre–10 novembre 1996; Pordenone, 23 novembre 1996–8 gennaio 1997; Esedra: Padua, Italy, 1996; pp. 175–185.
5. Fontana, A.; Vinci, G.; Tasca, G.; Mozzi, P.; Vacchi, M.; Bivi, G.; Salvador, S.; Rossato, S.; Antonioli, F.; Asioli, A.; et al. Lagoonal settlements and relative sea level during Bronze Age in Northern Adriatic: Geoarchaeological evidence and paleogeographic constraints. *Quat. Int.* **2017**, *439*, 17–36. [CrossRef]
6. Malizia, A. La raccolta paletnologica del Museo Archeologico di Altino (Venezia). *Archeol. Veneta* **1985**, *8*, 125–148.
7. Citton, E. L'abitato del Bronzo finale e degli inizi dell'età del ferro di Portegrandi nella tenuta I Marzi. In *Altino Antica. Dai Veneti a Venezia*; Tirelli, M., Ed.; Marsilio: Venice, Italy, 2011; pp. 50–51.
8. Barbero, S.R.; Albani, A.; Zecchetto, S. Palaeoenvirnmental significance of a benthis foraminifera fauna from and archaeological excavation in the Lagoon of Venice. *Palaeogeogr. Palaeoclimatol. Palaeoecol.* **1997**, *136*, 27–34. [CrossRef]
9. Ammerman, A.J.; McClennen, C.E.; De Min, M.; Housley, R. Sea level change and the archaeology of early Venice. *Antiquity* **1999**, *73*, 303–312. [CrossRef]
10. Bondesan, A.; Meneghel, M. *Geomorfologia della Provincia di Venezia: Note Illustrative della Carta Geomorfologica della Provincia di Venezia*; Esedra: Padua, Italy, 2004.
11. Moine, C.; Corrò, E.; Primon, S. *Paesaggi Artificiali a Venezia. Archeologia e Geologia nelle terre del Monastero di Sant'Ilario tra alto Medioevo ed Età Moderna*; All'Insegna del Giglio: Florence, Italy, 2017; ISBN 9788878147768.
12. Mozzi, P.; Bini, C.; Zilocchi, L.; Becattini, R.; Mariotti Lippi, M. Stratigraphy, palaeopedology and palinology of late Pleistocene and Holocene deposits in the landward sector of the lagoon of Venice (Italy), in relation to caranto level. *Alp. Mediterr. Quat.* **2003**, *16*, 193–210.
13. Donnici, S.; Serandrei Barbero, R.; Bini, C.; Bonardi, M.; Lezziero, A. The Caranto Paleosol and Its Role in the Early Urbanisation of Venice. *Geoarchaeology* **2011**, *26*, 514–543. [CrossRef]
14. Tosi, L.; Rizzetto, F.; Bonardi, M.; Donnici, S.; Serandrei Barbero, R.; Toffoletto, F. *Note Illustrative della Carta Geologica d'Italia alla Scala 1:50.000*; 148–149—Chioggia-Malamocco; APAT, Dipartimento Difesa del Suolo, Servizio Geologico d'Italia: Rome, Italy, 2007.
15. Pacciarelli, M. Dal villaggio alla città. La svolta protourbana del 1000 a.C. nell'Italia tirrenica. In *Grandi Contesti e Problemi della Protostoria Italiana*; All'Insegna del Giglio: Florence, Italy, 2001; Volume 4.
16. Cardarelli, A. The collapse of the Terramare Culture and growth of new economic and social systems during the Late Bronze Age in Italy. *Sci. Antich.* **2009**, *15*, 449–520.
17. Leonardi, G. Il Bronzo finale nell'Italia nord-orientale. Proposte per una suddivisione in fasi. In *Il Bronzo Finale in Italia: Atti della XXI Riunione Scientifica*; Istituto Italiano di Preistoria e Protostoria: Florence, Italy, 1979; pp. 155–188.
18. Narkhov, E.; Muravyev, L.; Sergeev, V.; Fedorov, A.; Denisov, A.Y.; Shirokov, A.; Sapunov, V. Applications of modern high-precision Overhauser magnetometers. In *AIP Conference Proceedings, Proceedings of the International Conference on Astronomy & Geophysics in Mongolia, Ulaanbaator, Mongolia, 20–22 July 2017*; AIP Publishing LLC: Melville, NY, USA, 2017; Volume 1886, p. 20075. [CrossRef]
19. Shaar, R.; Tauxe, L.; Ron, H.; Ebert, Y.; Zuckerman, S.; Finkelstein, I.; Agnon, A. Large geomagnetic field anomalies revealed in Bronze to Iron Age archeomagnetic data from Tel Megiddo and Tel Hazor, Israel. *Earth Planet. Sci. Lett.* **2016**, *442*, 173–185. [CrossRef]
20. von der Osten-Woldenburg, H.; Chaume, B.; Reinhard, W. Magnetic imaging of a late Bronze Age tumulus in France before and during excavation. *Lead. Edge* **2002**, *21*, 465–466. [CrossRef]
21. Kuna, M.; Křivánek, R.; Chvojka, O.; Šálková, T. A quantitative approach to magnetometer survey data: The case of the Late Bronze Age site of Březnice. *J. Archaeol. Sci.* **2021**, *126*, 105298. [CrossRef]
22. Benech, C. Contribution of geophysics to the research on city planning in ancient and classical Near East. In Proceedings of the 11th International Conference for Archaeological Prospection, Warsaw, Poland, 15–19 September 2015; pp. 161–165.
23. Fedorova, N.; Noskevich, V. Geophysical Investigations of the Bronze Age Archaeological Site in the Trans-Urals, Russia. In *Recent Research on Geomorphology, Sedimentology, Marine Geosciences and Geochemistry*; Springer: Cham, Switzerland, 2022. [CrossRef]
24. Lowe, K.; Fogel, A.; Sneddon, A. Archaeological geophysical survey of a Prehistoric Bronze Age site in Cyprus (Alambra Mouttes)—Applications and limitations. *Archaeol. Anthropol. Sci.* **2018**, *10*, 1971–1989. [CrossRef]
25. Persico, R.; Colica, E.; Zappatore, T.; Giardino, C.; D'Amico, S. Ground-Penetrating Radar and Photogrammetric Investigation on Prehistoric Tumuli at Parabita (Lecce, Italy) Performed with an Unconventional Use of the Position Markers. *Remote Sens.* **2022**, *14*, 1280. [CrossRef]

26. Ulrich, B.; Günther, T.; Rücker, C. Electrical Resistivity Tomography Methods for Archaeological Prospection. In *Layers of Perception, Proceedings of the 35th International Conference on Computer Applications and Quantitative Methods in Archaeology (CAA), Berlin, Germany, 2–6 April 2007*; Posluschny, A., Lambers, K., Herzog, I., Eds.; Dr. Rudolf Habelt GmbH: Bonn, Germany, 2007.
27. Parsi, M.; Linck, R.; Fassbinder, J.; Eitel, M. Highlighting the Potential of 3D ERT by Comparing its Results with GPR and the Excavation Map of a Roman Building. *ArchéoSciences* **2021**, *45-1*, 183–186. [CrossRef]
28. Russolillo, A.; Martin, F.F.; Merico, A.; Sapia, V.; Talamo, P.; Materni, V.; Pischiutta, M.; De Vita, S.; Furlani, S.; Targia, D.; et al. Unveiling a hidden fortification system at "Faraglioni" Middle Bronze Age Village of Ustica Island (Palermo, Italy) through ERT and GPR prospections. *J. Appl. Geophys.* **2023**, *220*, 105272. [CrossRef]
29. De Smedt, P.; Van Meirvenne, M.; Saey, T.; Baldwin, E.; Gaffney, C.; Gaffney, V. Unveiling the prehistoric landscape at Stonehenge through multi-receiver EMI. *J. Archaeol. Sci.* **2014**, *50*, 16–23. [CrossRef]
30. Mele, M.; Cremaschi, M.; Giudici, M.; Lozej, A.; Pizzi, C.; Bassi, A. The Terramare and the surrounding hydraulic structures: A geophysical survey of the Santa Rosa site at Poviglio (Bronze Age, northern Italy). *J. Archaeol. Sci.* **2013**, *40*, 4648–4662. [CrossRef]
31. Deiana, R.; Vicenzutto, D.; Deidda, G.P.; Boaga, J.; Cupitò, M. Remote Sensing, Archaeological, and Geophysical Data to Study the Terramare Settlements: The Case Study of Fondo Paviani (Northern Italy). *Remote Sens.* **2020**, *12*, 2617. [CrossRef]
32. Cassiani, G.; Bellizia, E.; Fontana, A.; Boaga, J.; D'Alpaos, A.; Ghinassi, M. Geophysical and Sedimentological Investigations Integrate Remote-Sensing Data to Depict Geometry of Fluvial Sedimentary Bodies: An Example from Holocene Point-Bar Deposits of the Venetian Plain (Italy). *Remote Sens.* **2020**, *12*, 2568. [CrossRef]
33. Benech, C.; Lombard, P.; Rejiba, F.; Tabbagh, A. Demonstrating the contribution of dielectric permittivity to the in-phase EMI response of soils: Example from an archaeological site in Bahrain. *Near Surf. Geophys.* **2016**, *14*, 337–344. [CrossRef]
34. Simpson, D.; Meirvenne, M.; Lück, E.; Bourgeois, J.; Joerg, R. Prospection of two circular Bronze Age ditches with multi-receiver electrical conductivity sensors (North Belgium). *J. Archaeol. Sci.* **2010**, *37*, 2198–2206. [CrossRef]
35. De Smedt, P.; Meirvenne, M.; Meerschman, E.; Saey, T.; Bats, M.; Court-Picon, M.; De Reu, J.; Zwertvaegher, A.; Antrop, M.; Bourgeois, J.; et al. Reconstructing Palaeochannel Morphology with a Mobile Multicoil Electromagnetic Induction Sensor. *Geomorphology* **2011**, *130*, 136–141. [CrossRef]
36. De Smedt, P.; Saey, T.; Lehouck, A.; Stichelbaut, B.; Meerschman, E.; Islam, M.M.; Van De Vijver, E. Exploring the potential of multi-receiver EMI survey for geoarchaeological prospection: A 90ha dataset. *Geoderma* **2013**, *199*, 30. [CrossRef]
37. LaBrecque, D.J.; Morelli, G.; Daily, W.; Ramirez, A.; Lundegard, P. Occam's inversion of 3D ERT data. In *Three-Dimensional Electromagnetics*; Spies, B., Ed.; SEG: Tulsa, OK, USA, 1999; pp. 575–590.
38. Sotgia, A. Fornaci per ceramica ed aree produttive in Italia tra Età del Bronzo ed Età del Ferro. *IpoTESI Preistoria* **2019**, *12*, 301–318. [CrossRef]
39. Bondesan, A.; Primon, S.; Bassan, V.; Vitturi, A. *Le unità Geologiche della Provincia di Venezia*; Cierre Grafica, Report; Cierre Grafica: Sommacampagna, Italy, 2008.

land

MDPI

Article

# Conservation for Whom? Archaeology, Heritage Policy, and Livelihoods in the Ifugao Rice Terraces

**Stephen Acabado [1,2,*], Adrian Albano [3] and Marlon Martin [4]**

[1] Department of Anthropology, University of California (UCLA), Los Angeles, CA 90095, USA
[2] Partido State University, Goa 4422, Camarines Sur, Philippines
[3] College of Forestry and Natural Resources, University of the Philippines-Los Baños, Los Baños 4031, Laguna, Philippines; alalbano@up.edu.ph
[4] Save the Ifugao Terraces Movement, Inc., Kiangan 3604, Ifugao, Philippines; marlon.martin12@yahoo.com
* Correspondence: acabado@anthro.ucla.edu

**Abstract**

Heritage landscapes endure not through the preservation of fixed forms but through the capacity to adapt to changing social, political, economic, and environmental conditions. Conservation policies that privilege static ideals of authenticity risk undermining the very systems they aim to protect. This paper advances a model of shared stewardship that links conservation of heritage to support for livelihoods, functional flexibility, and community authority in decision-making. Using the Ifugao Rice Terraces of the Philippine Cordillera as a case study, we integrate archaeological, ethnographic, spatial, and agricultural economic evidence to examine the terraces as a dynamic socio-ecological system. Archaeological findings and oral histories show that wet-rice agriculture expanded in the 17th century, replacing earlier taro-based systems and incorporating swidden fields, managed forests, and ritual obligations. Contemporary changes such as the shift from heirloom *tinawon* rice to commercial crops, the impacts of labor migration, and climate variability reflect long-standing adaptive strategies rather than cultural decline. Comparative cases from other UNESCO and heritage sites demonstrate that economic viability, adaptability, and local governance are essential to sustaining long-inhabited agricultural landscapes. We thus argue that the Ifugao terraces, like their global counterparts, should be conserved as living systems whose cultural continuity depends on their ability to respond to present and future challenges.

**Keywords:** heritage conservation; rice terraces; cultural values; UNESCO; Ifugao; Philippines

## 1. Introduction

Heritage landscapes should be understood as dynamic systems whose continuity depends on their ability to adapt to changing social, environmental, political, and economic conditions. Conservation frameworks that privilege fixed ideals of authenticity and integrity, such as those emphasized by UNESCO [1–3], risk overlooking the historical fluidity and adaptive strategies embedded in these landscapes [4–7]. The Nara Document on Authenticity [8] reinforces this view, arguing that authenticity should be defined within the specific cultural and historical contexts of the communities concerned, and that its expression may include change, adaptation, and the continuation of living traditions. To remain meaningful, heritage conservation must move beyond static representations and instead recognize the processes of change that sustain both material structures and the communities that manage them. In this sense, the approach outlined here follows the

Nara Document on Authenticity's determination that heritage is most meaningful when its continuity reflects the evolving values, practices, and priorities and needs of the community from which it originates.

In the Philippine context, heritage conservation is guided by the National Cultural Heritage Act of 2009 [9], which mandates the protection, preservation, and conservation of cultural properties, including historic sites and traditional cultural landscapes. The law recognizes the importance of safeguarding both tangible and intangible heritage and directs local government units to maintain cultural heritage inventories and formulate conservation plans. These provisions are particularly relevant to the Ifugao Rice Terraces (Figure 1), which are designated as National Cultural Treasures. The Act also emphasizes community involvement, requiring consultation with stakeholders in the management of heritage sites. While the legislation provides a framework for protection, its implementation often confronts the same challenges noted in global heritage discourse, including tensions between preservation of form and the adaptive practices necessary for sustaining living landscapes. This tension is precisely what the Nara Document warns against, urging heritage managers to balance material conservation with the evolving practices that keep heritage alive [8].

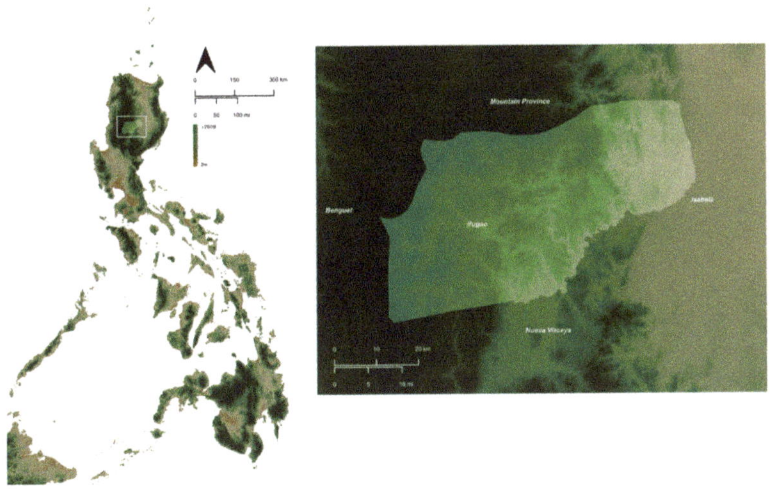


**Figure 1.** Ifugao Province, Cordillera Administrative Region, Philippines.

The Ifugao Rice Terraces of Northern Luzon, Philippines (Figure 2) illustrate this principle. Inscribed as a World Heritage Site in 1995 and celebrated as a "living cultural landscape" that exemplifies the synergy between human ingenuity and mountain ecology [10], the terraces are often framed through their aesthetic and symbolic value. Yet their significance lies equally in their role as a demonstration of Indigenous engineering and socio-ecological adaptation. Oral histories and ethnographic studies show the complex interplay of ritual practices, agricultural cycles, and environmental stewardship that underpins the terraces' management [11–14].

**Figure 2.** The Batad Rice Terraces, one of the five UNESCO clusters.

Archaeological and ethnographic evidence shows that land use in Ifugao has always been flexible, shifting between swidden fields, pond-field rice terraces, house platforms, and reforested or fallow zones [15–17]. Changes in terraces' function have historically responded to variations in labor availability, environmental pressures, and socio-political organization [18]. Contemporarily, terraces are sometimes repurposed for the production of vegetables, converted into residential plots, or left to regenerate as forest [19]. Such transformations reflect the productivity-focused decision-making that has long characterized Ifugao agricultural strategies.

Agricultural economic analyses confirm that decisions to maintain, abandon, or convert terraces are influenced by yield, household income, and labor requirements. In many cases, vegetable farming offers higher returns than wet-rice cultivation, making it a rational choice under current market conditions [20]. Archaeological surveys and excavations parallel these findings, documenting a history of terraces' reuse that includes conversion to sweet potato or taro fields, transformation of house platforms into gardens, and reoccupation of previously fallow plots. These data challenge the perception of the terraces as an unchanging tradition and instead depict them as part of a long record of continuous adaptation.

This perspective is consistent with settlement archaeology, which interprets landscapes as the product of evolving subsistence strategies, social relationships, environmental interactions, and ritual traditions. Recent research underscores this view [21,22]. For instance, in studies of Mesolithic landscapes, scholars argue that detailed reconstructions of settlement and mobility arise most fruitfully when grounded in the specific subsistence practices that formed social–environmental relationships, rather than generalized economic models [23]. Similarly, ecological niche modeling applied to long-term archaeological data has demonstrated how changes in environmental conditions actively shape settlement behaviors and landscape trajectories [24].

Thus, insights from landscape archaeology, historical ecology, and critical heritage studies further emphasize that landscapes are historically contingent and socially negotiated [25,26]. For Ifugao, this means that conservation strategies must begin with an understanding of dynamic patterns of land use and social organization.

Integrating archaeological evidence with agricultural economic analysis underscores the central question for heritage policy: what is to be conserved, for whom, and under what conditions? Effective conservation must address the economic realities of contemporary farming, the labor demands of terraces' maintenance, and the social values that local communities attach to land. By adopting this approach, heritage in Ifugao can be reframed as a living, adaptive system rather than a static representation of the past.

*Dynamic Landscapes and Ethics of Stewardship*

Settlement and landscape archaeology show the Ifugao Rice Terraces as dynamic systems, not static monuments. Terraces, irrigation, forests, and ritual spaces form an integrated landscape shaped by shifting social, economic, and ecological conditions [27,28]. Archaeological evidence shows the land use developed, from swidden to pond-field systems, residential zones, or fallow, demonstrating ongoing adaptation [29,30].

Historical ecology emphasizes that landscapes record negotiated human–environment interactions over time [25,26]. In Ifugao, centuries of Indigenous engineering and ritual practice produced resilient, flexible systems [31,32]. Yet heritage management often privileges static notions of authenticity, risking erasure of these adaptive histories [10,33].

In addition, intangible heritage such as songs, rituals, and ecological knowledge is embedded in cultivation. Hudhud chants and the punnuk ritual link farming to memory and identity [34,35]. As farming shifts, these traditions erode.

Moreover, ethical conservation requires acknowledging the political dimensions of heritage and asking who decides what is worth preserving and why [36,37]. Shared stewardship offers an alternative by supporting local agency, redistributing benefits, and sustaining both tangible and intangible heritage.

This approach resonates with emerging models of living heritage that center adaptability and community participation. Rather than preserving idealized forms, conservation should invest in practices that allow heritage to persist through change by supporting livelihoods, transmission of knowledge, and autonomy in heritage-making [38].

## 2. Methods

This study integrates archaeological, agricultural economics, and ethnographic approaches within a socio-ecological systems and political ecology framework to examine the Ifugao Rice Terraces as a dynamic landscape and to evaluate implications for heritage conservation [39,40]. The interdisciplinary design enables a diachronic perspective that connects historical land-use patterns to contemporary livelihood decision-making, linking empirical evidence with policy-relevant insights. Archaeological fieldwork was conducted at multiple sites representing diverse land-use contexts, including swidden fields, pond-field terraces, residential platforms, and managed forests. Sites were selected to capture spatial variability across the province, incorporating both UNESCO-inscribed clusters and non-listed areas (Figure 3). Selection criteria were informed by preliminary surveys, oral histories, prior research records, and spatial analysis of terraces' distribution, following recommendations to define the research universe explicitly through culturally grounded boundaries and mapped extents [41].

Archaeological excavation strategies, such as stratigraphic profiling to identify occupation layers, radiocarbon dating to establish chronological sequences, and analysis of artifacts to infer subsistence practices and social organization [42], were planned with an understanding that material signatures could later be interpreted alongside agricultural profitability trends. For example, archaeobotanical and zooarchaeological analyses—using soil flotation to recover charred plant remains and faunal assemblages [42,43]—provided evidence of past cultivation and domestication strategies, which were then compared to present-day crop profitability patterns to detect long-term continuities or disruptions in economic logic. This integration of datasets draws from Holmelin's [39] emphasis on tracing adaptive strategies through historical and contemporary material records to show how local actors balance environmental constraints with market incentives.

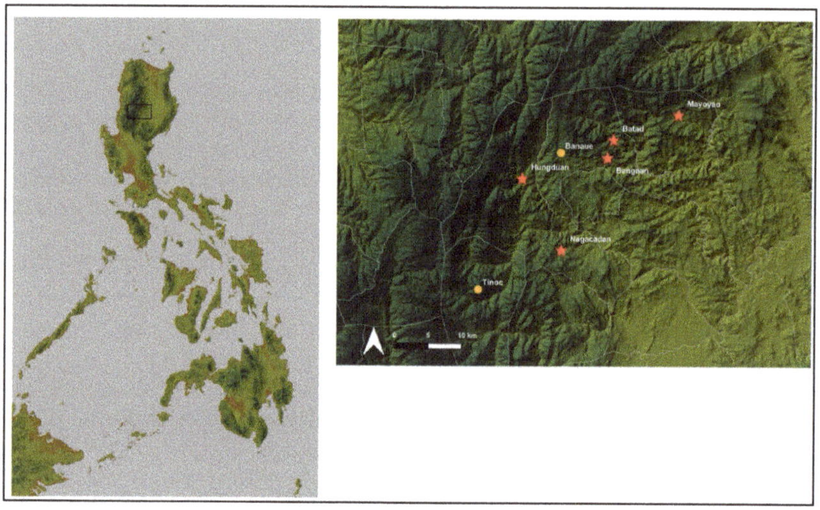


**Figure 3.** The five UNESCO heritage clusters (denoted by star points) and sources of agricultural economics data (denoted by circle points).

Spatial recording relied on precision GPS units and total stations, complemented by remote sensing data. Historical aerial imagery offered an early-20th century baseline, while LiDAR datasets exposed microtopographical features obscured by vegetation, enabling detection of abandoned terraces, modified irrigation routes, and shifts in cultivated area [15,44,45]. These spatial reconstructions, in addition to being archaeological records, informed calculations of potential agricultural yields under different land-use regimes. Qingwen [40] underscores that such geographic and infrastructural mapping can be used to model changes in productivity, labor distribution, and vulnerability to climate hazards—making it possible to assess the economic consequences of landscape change over time.

Participatory mapping workshops, conducted with farmers and elders, identified irrigation lines, muyong boundaries, and culturally significant landscape features, applying principles of community-based archaeology [46] and ensuring that local environmental knowledge was embedded in both heritage and economic analyses [39]. Ethnographic data from semi-structured interviews and focus group discussions with farmers, elders, and municipal officials in multiple municipalities provided qualitative detail on crop selection criteria, perceptions of climate variability, and the role of rituals in agricultural decision-making. These accounts informed the interpretation of both archaeological subsistence data and contemporary cost–benefit calculations, following Qingwen's [40] observation that governance and cultural practice are deeply intertwined in land-use choices.

To analyze the economic logic of ongoing crop transitions, secondary data were gathered on tinawon rice (local varieties), inbred rice (government-introduced), and commercially grown vegetables such as tomato and cabbage. These were drawn from municipal agriculture offices for municipalities where each crop type is primarily cultivated—Banaue for tinawon and inbred rice, Tinoc for tomato and cabbage—and included average yields per area, selling prices, and production costs (seeds, fertilizers, pesticides, labor, transportation, marketing) [47,48]. Net income per half-hectare crop land area was calculated, allowing for comparison of profitability and vulnerability to shocks such as sudden price drops or climate-induced crop loss. By embedding these economic measures within the archaeological–historical frame, the analysis could identify whether current shifts towards high-input, high-risk crops represent historical continuities in adaptive intensification or novel departures from traditional resilience strategies [41]. In line with Nan et al.'s [41] methodological emphasis, resilience indicators such as diversification potential and dependence on external inputs were incorporated to capture both material and socio-cultural dimensions of sustainability.

This integrated methodology ensures that archaeological, spatial, ethnographic, and economic datasets are analyzed not as parallel lines of evidence but as interdependent perspectives on the same socio-ecological system. Techniques and instruments in each domain were selected and applied with cross-domain utility in mind: archaeological mapping informs yield modeling; ethnographic narratives contextualize economic data; and economic calculations provide a contemporary counterpoint to historical subsistence reconstructions. Such cross-fertilization allows for an evaluation of Ifugao land-use strategies over the long term, grounded in both empirical measurement and culturally embedded interpretation.

## 3. The Ifugao Rice Terraces as an Archaeological Landscape

The UNESCO World Heritage inscription identifies five clusters: Nagacadan in Kiangan, the entire municipality of Hungduan, the municipality of Mayoyao, and the Batad and Bangaan terraces in Banaue (see Figure 3). However, many terraces are not included in the UNESCO list. For example, the Banaue Viewpoint terraces, though excluded from the UNESCO inscription, are designated as National Cultural Treasures. Across these sites, whether internationally listed or nationally recognized, the terraces transform steep mountain slopes into productive agricultural fields through stone and mud walls, gravity-fed irrigation systems, and continual upkeep. For generations, Ifugao communities have sustained this landscape through deep knowledge of ecology, hydrology, and land management, transmitted through oral traditions and customary practice.

Archaeological investigations challenge long-standing assumptions that the terraces are over 2000 years old. Excavations and chronometric dating at multiple Ifugao sites suggest a later and more dynamic formation, with significant expansion beginning in the 17th century [42,49]. This re-dating places the terraces' intensification in the context of Spanish colonial incursions into Northern Luzon. The terraces are better understood not as remnants of deep historic agriculture but as strategic responses to shifting political, economic, and demographic pressures. They served as both refuge and resistance by enabling an intensified food production that facilitated the consolidation of economic and political resources and helped reorganize social and ritual life.

Archaeological evidence indicates that prior to the adoption of wet-rice agriculture, Ifugao subsistence systems centered on irrigated taro pond fields, swidden plots, and managed forest zones. Excavated materials from the Old Kiyyangan Village (OKV) site, including carbonized plant remains and faunal assemblages, show that taro (Colocasia esculenta) was the staple crop until around 1650 CE, when wet-rice (*Oryza sativa*) cultivation emerged [42,43]. This agricultural shift coincided with increasing population and colonial pressures, prompting a reconfiguration of the landscape that allowed greater food production and territorial control. Rather than a complete replacement, this transition added to existing land-use systems, producing a mosaic of terraces (payo), swidden fields (uma), house platforms (boble), and private woodlots (muyong or pinugo) [17,44] (Figure 4).

**Figure 4.** An artist's rendition of Ifugao terrace ecology and landscape pattern (by N.J. Roxas).

Spatial modeling and GIS analyses from the OKV site also show that between 1570 and 1800 CE, the local population doubled, terrace expansion accelerated, and forested areas were preserved despite intensification [44]. Archaeological and ethnographic data show that land-use flexibility played a central role in Ifugao's adaptation. Communities shifted between swidden and terrace cultivation, allowed fallowing and replanting, and alternated forest regeneration with reclamation. Some abandoned terraces were later reclaimed for subsistence or commercial crops. This system illustrates that the Ifugao landscape is not a static heritage form but a living and adaptive socio-ecological complex.

A critical aspect of this complex is the muyong or pinugo system. These managed forest zones are maintained by extended families and provide wood, fruits, medicinal plants, and water for irrigation. The forests also stabilize slopes, reduce erosion, and support biodiversity [50–52]. The muyong exemplifies traditional ecological knowledge and upland resource management. It connects forest conservation with agricultural productivity and has proven resilient to demographic and environmental shifts [53]. Despite its importance, contemporary conservation strategies often overlook this component and focus narrowly on the rice terraces. Figures 5 and 6 show terrace abandonment and subsequent forest regrowth occur more frequently, as exposed by LiDAR scans of the Nagacadan Terraces in Kiangan, Ifugao. A historical photograph taken by Roy F. Barton in 1912—its exact location still being identified—is also included to illustrate the expansive distribution of terraces more than a century ago, when entire mountain slopes were carved into rice fields, a landscape no longer widely visible today. Some of these abandoned terraces are now being reclaimed for cash crop production. Contemporary conservation strategies often neglect this component, focusing narrowly on the visual icon of rice terraces while overlooking the broader socio-ecological system that sustains them.

**Figure 5.** The historical photo on the left, taken by R.F. Barton in 1912, shows the expansive nature of terracing in Ifugao. The image on the right, from the 2025 LiDAR coverage with bare-earth DEM, shows eroded terraces hidden beneath the forest canopy.

**Figure 6.** Bare-earth DEM and two-dimensional images of the terrace system shown above (map generated by the IAP and EJ Hernandez).

Labor organization and tenure systems are equally central to the maintenance of the terraces. Historically, large-scale tasks such as repairing terrace walls or clearing irrigation channels were completed through kin-based reciprocal labor exchanges (ubbu). Access to land was governed by customary systems of inheritance, alliance, and ritual participation. These practices supported both economic production and social cohesion. Rituals, including the *Hudhud* chants performed during harvest, reinforced these networks and transmitted genealogies, moral teachings, and ancestral memory [34]. From an archaeological perspective, the terraces reflect both material infrastructure and social relations.

Processes of abandonment and reuse are also part of the archaeological record. Without regular upkeep, terrace walls collapse, irrigation lines silt up, and grasses take over. Abandonment is increasing today due to labor migration, market pressures, and climate variability. However, archaeological and historical data indicate that cycles of expansion, maintenance, and disuse have long characterized the Ifugao landscape. Recognizing these patterns challenges idealized views of the terraces as timeless and affirms their dynamic history.

Archaeology brings a long-term perspective to heritage conservation [54,55]. The terraces have survived not by remaining unchanged but by being continually adapted. Conservation should not aim to restore a fixed past but should support communities in managing change while maintaining their heritage's value. Understanding the terraces as an archaeological landscape means recognizing them as living systems that require ongoing, locally rooted stewardship. This perspective informs how we address contemporary challenges such as economic transformation, environmental stress, and the erosion of intangible heritage.

### 3.1. Intangible Heritage and Knowledge Systems

The Ifugao Rice Terraces are also important ritual landscapes embedded in Ifugao's cosmology, social life, and identity. Every phase of the rice cycle—field preparation, planting, harvest—is marked by ritual offerings to ancestral spirits, affirming the interconnection of humans, land, and the spirit world. These practices turn the landscape into a material archive of memory and meaning, where technical maintenance is inseparable from social and spiritual obligations.

A key dimension of this heritage is gendered knowledge. Women, particularly grandmothers, have long served as custodians of seed selection, drawing on detailed ecological observations to determine planting strategies. Ethnographic interviews demonstrate that over 20 tinawon rice varieties were used as recently as the 1970s, each selected for drought tolerance, pest resistance, or ceremonial purposes (Table 1). This system offered ecological resilience through diversity. Today, however, commercial seeds' use, often promoted by external actors, has disrupted this intergenerational transmission. Elder women warn that without support, ancestral seed knowledge may disappear.

**Table 1.** Known and named commercial and local rice varieties cultivated in Kiangan and Hungduan Municipalities, gathered by Jacy Moore-Miller in 2014 [56].

| Commercial Varieties | *Tinawon* Varieties |
|---|---|
| 52 | Binogon |
| 82 | Botnol |
| 222 | Iggamay |
| C-12 | Imbannig |
| C-2 | Imbuukan |
| C-4 | Madduli |
| C-4 red | Mayawyaw |
| Diamond | |
| Halaylay | |
| Ingaspar | |
| Ingaspi | |
| Korean | |
| Migapas | |
| Minmis | |

**Table 1.** *Cont.*

| Commercial Varieties | *Tinawon* Varieties |
|---|---|
| Mukoz | |
| Mulmug | |
| Munoz | |
| NSCI-208 | |
| Pakulsa | |
| Oakland | |
| Oklan | |
| Oklan Minaangan | |
| Pangasinan Variety | |
| PJ-27 | |
| PJ-7 | |
| RC-218 | |
| RI-152 | |
| RI-238 | |
| Romelia | |
| RP 224 | |
| Super 60 | |
| Taiwan | |
| Thunder | |

Terrace farming also served as a context for transmitting cultural values. Through work in the fields, children learned how to maintain canals, repair terraces, and participate in communal labor. With the decline in tinawon cultivation and youth migration, these teaching contexts are vanishing. The result is a loss of collective memory and erosion of the values that once structured Ifugao social life.

UNESCO's recognition of elements like the *Hudhud* epic underscores the global importance of Ifugao's intangible heritage. Yet such listings can create tension. While raising awareness, they risk turning living practices into performances. Some rituals, once integral to farming and feasting, are now staged for tourists, severed from their agricultural roots. Conservation cannot rely solely on documentation or display. It must support living practices, especially those tied to terrace cultivation and community-based labor.

By examining these dimensions, we see that intangible heritage is not peripheral but foundational to the Ifugao landscape. Effective conservation requires engaging with the knowledge, values, and social practices that animate the terraces, ensuring that preservation efforts remain grounded in community agency and lived experience.

### *3.2. Economic Change and Livelihood Strategies*

The Ifugao Rice Terraces were historically constructed not solely for subsistence but for prestige-oriented production. Heirloom *tinawon* rice, cultivated once a year, was central to Ifugao ritual and social life. Its value lay less in caloric yield and more in its ceremonial role in feasting, kinship building, and status negotiation [16,57]. These practices were embedded in customary tenure systems, reciprocal labor exchanges, and ritual cycles [17].

Contemporary economic conditions, however, have altered livelihood strategies. Expanded access to education, healthcare, and markets has increased cash needs. As Ifugao households shift toward a mixed economy—including off-farm work, remittances, tourism, and commercial agriculture—land use has adapted accordingly. Tinawon's low yield and labor intensity make it less viable for income generation. In contrast, crops like inbred rice, cabbage, and tomato offer faster returns, multiple harvests, and easier marketability (Figure 7; Table 2).

**Figure 7.** Converted agricultural terraces in Banaue, Ifugao (photo by the IAP and EJ Hernandez).

**Table 2.** Comparison between income from cash crop vis-à-vis tinawon cultivation (in Philippine Pesos). (A figure of 0.5 ha is used here for comparison purposes. In practice, the average farm hectarage for rice is 0.25 ha, while 0.5 ha is for commercial vegetables such as tomato.)

| | **Rice** | | **Vegetable** | |
|---|---|---|---|---|
| | **Tinawon (Milled)** | **Inbred (Palay)** | **Cabbage** | **Tomato** |
| Ave. Yield (kg. per cropping) (Cropping period: Tinawon 7 months, inbred (irik) 5 months, cabbage 3 mos, tomato 4–6 mos, with weekly harvesting over a span of 1–2 months) | 1200 | 3000 | 10,000 | 30,000 |
| Average price (PHP per kg)<br>Price Range (min–max) (Unlike rice, price of vegetables is highly volatile. Data sources: Ifugao PAENRO, Tinoc OMAg (Office of the Municipal Agriculturist), key informant interviews) | 100.00<br>90–110 | 17<br>14–20 | 25<br>5–70 | 20<br>4–90 |
| Gross Income | 120,000 | 51,000 | 250,000 | 600,000 |
| *Less Cost* | 59,500 | 34,000 | 214,000 | 435,500 |
| Farm Input | | | | |
| Seed<br>Fertilizer<br>Pesticide<br>Others | 4500 | 1500<br>6000<br>1500 | 16,000<br>45,000<br>17,000<br>6000 | 5500<br>50,000<br>60,000<br>60,000 |

**Table 2.** *Cont.*

| | Rice | | Vegetable | |
|---|---|---|---|---|
| | **Tinawon (Milled)** | **Inbred (Palay)** | **Cabbage** | **Tomato** |
| Farm Input | | | | |
| Labor<br>Transportation<br>Marketing | 55,000 | 25,000 | 100,000<br>20,000<br>10,000 | 170,000<br>60,000<br>30,000 |
| Net Income | | | | |
| Normal Income/0.5 ha | PHP 60,500.00 | PHP 17,000.00 | PHP 36,000.00 | PHP 164,000.00 |
| Maximum Income | PHP 72,500.00 | PHP 26,000.00 | PHP 486,000.00 | PHP 2,264,500.00 |
| Minimum Income (Maximum Loss) | PHP 48,500.00 | PHP 8000.00 | (PHP 164,000.00) | (PHP 315,500.00) |

These shifts reflect adaptation, not cultural erosion. Diversification is an economic necessity for many families, one that responds to changing household needs and labor availability. The terraces remain culturally significant, but their use now encompasses both ceremonial and commercial production.

Labor, Migration, Tourism, and Climate Stress

The movement away from tinawon is also shaped by changing labor dynamics. Terrace farming demands intensive maintenance of walls, canals, and soil fertility. Yet rural labor pools are shrinking as youth migrate for education and jobs. Remittances support household expenses but reduce the labor available for farming, particularly for older community members.

Commercial cropping offers some labor efficiencies. Though requiring external inputs, these crops demand shorter growing seasons or less intensive upkeep. For families with reduced human labor, this is a practical tradeoff.

Environmental stress compounds these choices. Farmers report abandoning rice terraces damaged by typhoons or landslides, citing the difficulty of repair [58]. Climate change increases risk and diminishes the appeal of labor-intensive rice farming. Local narratives describe tinawon as "for consumption only, not for selling" and explain that vegetables are planted "because they have expenses" [58]. These decisions reflect pragmatism, not cultural disinterest.

The terraces are designed to manage water, soil, and slope stability in a mountainous environment. Gravity-fed irrigation channels bring water from springs and streams to the fields, while the muyong forest system above the terraces regulates hydrology and slope stability [53,59,60]. This system reflects centuries of adaptive management but depends on delicate balances now threatened by climate change [61].

Climate projections for the Cordillera region indicate rising temperatures and more intense rainfall. Local farmers already observe drying springs and more frequent landslides. Torrential rains damage terrace walls, while erratic dry spells leave paddies parched. A 2024 risk assessment [58] highlights how the terraces, once adapted to stable rainfall, are increasingly vulnerable to climate extremes.

Landslides and erosion present immediate threats. Saturated soils weaken structural integrity, collapsing walls and burying fields. Soil loss depletes fertility, and damaged irrigation channels disrupt entire networks. Repairing terraces requires coordinated labor, but when abandonment follows damage, the reciprocal labor systems that sustained terraces' maintenance can unravel.

Drought compounds these risks. Tinawon rice requires steady water throughout its long growing cycle. Water scarcity leads many farmers to leave terraces fallow or shift to vegetable crops, as in Poblacion, Tinoc, where water shortages have driven the full replacement of rice with vegetables [19]. Others adopt lowland rice varieties that demand external inputs, further disrupting local seed systems and traditional practices.

These ecological changes ripple through cultural life. Rituals like the *Hudhud* chants and the *punnuk* tug of war are tied to the rice cycle. When terraces are not planted, ceremonies do not take place, weakening the transmission of intergenerational knowledge. As one farmer put it, "How can we teach the children if we have no rice to plant?" Such losses erode collective identity and the cultural biography of the landscape.

Despite these pressures, Ifugao communities continue to adapt. The muyong system exemplifies long-standing climate resilience through forest management, erosion control, and water regulation. Current initiatives by groups like the Save the Ifugao Terraces Movement (SITMo) integrate traditional practices with climate science, including forest restoration, irrigation repair, and blending customary water sharing with modern monitoring. These efforts show that adaptation is ongoing and locally driven.

Still, successful adaptation requires external support. Resources, infrastructure, and policies must align with Indigenous knowledge systems to sustain these landscapes under changing environmental conditions.

Settlement archaeology provides valuable parallels. Across the Andes and the Mediterranean, terraces have expanded, been reorganized, or been abandoned in response to economic and demographic shifts. The Ifugao case mirrors this pattern. Previous transformations occurred during colonial incursions and demographic growth; today's changes reflect similar adaptive cycles.

Conservation narratives often call for a return to traditional tinawon cultivation, defining authenticity in static terms. Yet Ifugao farmers have always made strategic land-use decisions. Viewing terraces' changes through an archaeological perspective highlights continuity in adaptation rather than rupture.

Ethnographic research demonstrates that economic, demographic, and climate pressures drive decision-making. Supporting farmers as heritage stewards means addressing the material conditions of their lives. This includes creating viable incentives for traditional rice cultivation, supporting premium markets for heirloom rice, and improving infrastructure to mitigate climate vulnerability.

Without such support, conservation risks echoing colonial dynamics, prescribing land use without addressing community needs. An ethical heritage policy must support locally defined strategies that balance cultural practices with survival and dignity.

Tourism has also grown into a significant industry in Ifugao Province. Before the COVID-19 pandemic, the province averaged around 71,000 tourists annually (2017–2019), with nearly half being foreign visitors [62]. Banaue, Batad, and surrounding areas developed as prominent tourism hubs, offering homestays, guided hikes, cultural performances, and viewpoints showcasing the terraces.

Tourism revenues are substantial. In 2019 alone, visitor spending in the province was estimated at almost USD 18 million [62]. This income supports local businesses such as hotels, restaurants, souvenir shops, and transportation services. For the local government, tourism taxes and fees represent important budget contributions.

From a heritage management perspective, tourism provides both opportunities and challenges. On the one hand, tourism revenue can theoretically fund conservation, generate employment, and promote cultural awareness. On the other hand, without equitable distribution, tourism can reproduce social inequalities and extract value from local heritage while returning minimal benefits to those who maintain it.

## 4. Discussion and Comparison

The Ifugao Rice Terraces share important characteristics with other long-inhabited cultural landscapes across the globe where heritage recognition frequently intersects with the realities of evolving land use. Examining comparative cases from Bali, the Peruvian Andes, Engaruka in Tanzania, and the Hani Rice Terraces in China underscores the universality of adaptive landscape management and provides specific analytical insights that can inform heritage policy in Ifugao. These cases indicate that the dynamics of agricultural change, socio-political transformation, and environmental variability are not unique to the Cordillera. They are recurring features of human–environmental interaction in highland and terraced landscapes globally.

In Bali, "the Cultural Landscape of Bali Province: the Subak System as a Manifestation of the Tri Hita Karana Philosophy", inscribed on the UNESCO World Heritage List in 2012, is often cited as an exemplary integration of spiritual, social, and agricultural management. The subak system is a complex, community-based irrigation network linked to water temples that mediate both resource allocation and ritual obligations [63]. Its heritage framing emphasizes ritual continuity and the philosophical ideal of harmony between people, nature, and the divine. However, rapid urban expansion, tourism's development, and generational disengagement from rice farming have reduced active participation in the subak. Hauser-Schäublin [64] characterizes this process as the "ritualization of agriculture," where symbolic preservation of ceremonies can mask the decline in agricultural productivity. The Balinese example parallels the Ifugao experience in that heritage recognition risks emphasizing performance and aesthetics over function. Without explicitly linking heritage designation to tangible livelihood benefits, both cases demonstrate how conservation can inadvertently transform a working agricultural system into a staged representation for external audiences.

The Peruvian Andes offer another instructive example. Agricultural terraces in the Sacred Valley and around Cusco are celebrated as emblems of national identity and as tangible expressions of Inca and pre-Inca ingenuity. They are central to Peru's heritage tourism sector and are maintained in part for their visual and symbolic value [18]. Yet in many areas, these terraces no longer function as productive agricultural systems. Heritage conservation in this context often prioritizes the physical preservation of terrace walls and surrounding landscapes while overlooking their agronomic rehabilitation. This creates a form–function disconnect that is also evident in Ifugao when terraces are restored primarily for their appearance, without addressing the economic viability of traditional cultivation. The Andean case reinforces the point that physical restoration must be coupled with support for agricultural productivity if heritage landscapes are to remain both culturally and materially relevant.

Engaruka, in Northern Tanzania, while not inscribed as a UNESCO World Heritage site, is frequently cited in heritage and landscape studies as an important example of historical land-use complexity in Africa. Built and intensively used between the 15th and 18th centuries, Engaruka's terrace and irrigation systems were later abandoned. Yet the site has seen episodic reuse by Maasai and other groups responding to environmental pressures and shifting social needs [65]. This case highlights a crucial insight: abandonment is not necessarily indicative of cultural decline. It can represent a strategic adaptation to new ecological, demographic, or political realities. This cyclical pattern of use, abandonment, and reclamation is directly comparable to Ifugao, where terraces have historically shifted between rice cultivation, vegetable production, and forest regeneration depending on labor availability, market access, and environmental stability.

The Hani Rice Terraces of Yunnan, China, provide an additional valuable comparison. Inscribed on the UNESCO World Heritage List in 2013, the terraces extend over more than

16,000 hectares and have been cultivated for over 1300 years by the Hani people. The Hani system integrates forests, villages, terraces, and water systems into an interdependent ecological and social framework, similar to the Ifugao muyong–terrace complex [66–68]. As in Ifugao, the terraces rely on intricate water management derived from forested catchments, with agricultural cycles deeply embedded in ritual and cultural practice. However, heritage recognition in the Hani case has coincided with growing pressures from tourism's development, market integration, and youth outmigration.

Studies indicate that while UNESCO inscription has increased international visibility, it has also contributed to the commodification of cultural practices and a reorientation of agricultural priorities toward visitor expectations rather than local needs [67,69]. These pressures mirror those faced in Ifugao, where conservation narratives can valorize traditional forms while neglecting the socio-economic mechanisms that sustain them. The Hani example further illustrates that long-term viability depends on maintaining the integrity of the whole agro-ecological system, ensuring that tourism revenues and conservation investments support the communities who actively manage the terraces.

These cases highlight several important themes for the Ifugao context. First, economic viability is a critical determinant of conservation's success. Agricultural heritage systems endure when they generate income or food security for local communities. Without this material basis, even well-intentioned conservation efforts may falter. Second, heritage's sustainability depends on flexibility rather than fixity. In all four comparative examples, communities adapted land use in response to changing conditions, yet heritage frameworks sometimes penalize such adaptations as "inauthentic" [4,7]. In Ifugao, recognizing adaptive change as a legitimate expression of heritage could help reconcile conservation policy with local livelihood strategies. Third, the integration of local governance and decision-making into heritage policy strengthens outcomes by ensuring they align with community priorities. The subak councils in Bali, Andean village cooperatives, Maasai communal management systems in Engaruka, and local Hani administrative committees all demonstrate that participatory governance is key to balancing heritage goals with evolving socio-economic needs.

The comparative perspective suggests that Ifugao's conservation framework should explicitly link heritage recognition to the enhancement of livelihoods, agricultural flexibility, environmental sustainability, and community authority in decision-making. In practice, this means designing policies and programs that not only maintain the visual integrity of the terraces but also support diverse forms of production, reward environmental stewardship, provide economic incentives, and sustain highland farming. By embracing change as an intrinsic element of cultural continuity, heritage policy can move beyond preservation of form toward stewardship through function, a principle that is as relevant in Ifugao as it is in Bali, the Andes, Engaruka, and the Hani Rice Terraces.

In the Philippine context, these goals are reinforced by the National Cultural Heritage Act of 2009 [9], which provides the legal basis for protecting traditional cultural landscapes and requires community involvement in their management. However, as in the comparative cases, achieving these objectives in practice will require aligning the law's provisions with adaptive, livelihood-centered strategies that recognize the dynamic nature of living landscapes like the Ifugao Rice Terraces.

## 5. Conclusions: Conservation as Shared Stewardship

This paper set out to examine the Ifugao Rice Terraces as a living archaeological landscape, to interrogate how heritage policy addresses (or fails to address) their dynamic nature, and to propose a framework for conservation that foregrounds local livelihoods, intangible heritage, and adaptive strategies. Through the integration of archaeological, spatial, ethnographic, and economic evidence, we have shown that the terraces are not static relics but evolving socio-ecological systems. Their continuity lies in their capacity for change, a pattern visible in both the long-term archaeological record and the rapid transformations documented through ethnography and remote sensing. This reflects the position of the Nara Document on Authenticity, which holds that authenticity must be understood within cultural contexts and that adaptation can be a legitimate and necessary expression of heritage [8].

Insights from other long-inhabited agricultural landscapes reinforce this perspective. Across global contexts, economic viability, adaptive land-use strategies, and community governance emerge as the conditions that sustain heritage landscapes over the long term. For Ifugao, this means that conservation must be grounded in livelihoods' security, flexibility in agricultural practices, and local authority in decision-making.

Although landscapes might appear stable to the casual observer, in fact they are in constant motion. Archaeological and historical records, combined with modern remote sensing data, show that land-use systems can shift dramatically within human lifetimes. If LiDAR datasets indicate that terraced landscapes can be abandoned and reforested within 50 to 100 years, then many of the patterns and transitions visible in archaeological contexts may also be unfolding in contemporary times. What we see as "traditional" landscapes today may be snapshots in a long continuum of change rather than fixed remnants of the past.

These patterns challenge the way heritage policy and popular narratives often frame agricultural systems. Conservation discourse often privileges the image of stability, fields frozen in an idealized past, over the lived, adaptive practices that sustain them. The persistence of a landscape in cultural memory, photography, or tourism marketing can mask the reality that the forms and functions of that landscape are continually renegotiated. Recognizing that change is the norm, not the exception, creates space for more nuanced, community-driven approaches to heritage management.

The terraces illustrate this tension between perceived permanence and actual dynamism. They are often viewed as timeless symbols of cultural resilience, yet archaeological and ethnographic research documents a long trajectory of transformation. Before wet-rice agriculture dominated the slopes, Ifugao farmers cultivated taro in pond fields and practiced swiddens' cultivation in surrounding uplands. Over centuries, these systems gave way to wet-rice terraces, later complemented or replaced by vegetable farming, and, in some areas, eventual forest regeneration. The historical record shows repeated adjustments to meet changing socio-political, economic, and ecological conditions. These shifts are part of a continuing pattern in which socio-political upheavals, the introduction of new crops, altered trade networks, pest outbreaks, changing rainfall patterns, or soil degradation required technical and organizational responses. Livelihood strategies, whether driven by subsistence needs or opportunities in wage labor and tourism, also shaped land-use choices [70].

Heritage frameworks, especially those influenced by UNESCO, have often privileged visual permanence over lived practice [4,7]. In Ifugao, the terraces were inscribed on the World Heritage List in 1995 as a "living cultural landscape", because they represent continuity. But this continuity is maintained through flexibility, not fixity. Historical aerial imagery, complemented by LiDAR and other remote sensing datasets, shows that thousands of hectares of terraces have already been reclassified as forestlands. This process can take place within 50 to 100 years, well within a single human lifespan. Thus, abandoned terraces can become indistinguishable from natural forest cover in a short period of time [14,45].

The speed of these transitions underscores the vulnerability of agricultural landscapes when labor, water, and market incentives shift. It also shows that terraces' abandonment is not necessarily a sign of cultural decline. In some cases, it reflects strategic decision-making to reduce labor demands, avoid land-rights conflicts, or take advantage of new economic opportunities. For others, migration or engagement in off-farm employment makes the maintenance of labor-intensive terraces less viable. These realities complicate preservationist visions that emphasize form over function. Conservation policies that focus solely on retaining the physical appearance of terraces risk ignoring the drivers of abandonment, such as unpaid or undercompensated labor, inequitable tourism revenue, climate variability, and shifting demographic patterns. Without addressing these underlying causes, preservation becomes an exercise in maintaining symbols while undermining the social and economic foundations that sustain them.

As Cameron [71] notes, living heritage must be "responsive to contemporary realities," not immobilized by aesthetic ideals. Khalaf [7] similarly warns that "unexamined assumptions about authenticity can render adaptation illegitimate," marginalizing the very communities whose practices gave rise to the heritage in question. In Ifugao, this risk is real: terraces actively managed and repurposed may be viewed as less "authentic" than those maintained for their traditional appearance, even if the latter are no longer fully integrated into local livelihood systems. The Nara Document cautions against such reductive interpretations, emphasizing that authenticity resides in the cultural values and practices of the community, even when these result in material change [8].

The comparative cases examined in this paper reinforce these conclusions. Bali's subak system, the terraces of the Peruvian Andes, and the Engaruka fields in Tanzania each reveal that long-inhabited agricultural landscapes survive not through rigid adherence to past forms but through the ability to adapt to changing ecological, social, and economic circumstances [64,65]. Across these examples, economic viability, functional flexibility, and local governance emerge as the central determinants of conservation's success. Where heritage frameworks acknowledge and support these factors, landscapes remain both culturally meaningful and materially productive. Where they do not, conservation risks producing hollow monuments to an imagined past.

For Ifugao, the key takeaway is that the future of the terraces depends on shifting from preservation of form to stewardship through function. The National Cultural Heritage Act of 2009 [9] offers a framework that can support this approach by mandating the conservation of traditional cultural landscapes and requiring stakeholder consultation. Leveraging this law to operationalize shared stewardship would align national policy with on-the-ground realities, ensuring that conservation strategies protect both the terraces' cultural significance and the livelihoods that sustain them.

A conservation model grounded in shared stewardship places local communities at the center of decision-making, directs conservation funding toward sustaining livelihoods, and strengthens the ability of landscapes to adapt to change. This can include initiatives such as supporting agricultural diversification, improving market access for highland products, and compensating communities for the ecosystem services provided by reforested terraces. What requires protection is not only the physical form of the terraces but also the social, ecological, and historical processes that give them meaning. Because these processes are dynamic, their vitality depends on the capacity to adapt. For heritage to remain meaningful in the 21st century, it must recognize change as an essential component of cultural continuity.

**Author Contributions:** Conceptualization, S.A. and M.M.; methodology, S.A.; validation, S.A., M.M. and A.A.; formal analysis, S.A., M.M. and A.A.; investigation, S.A., M.M. and A.A.; resources, S.A. and M.M.; data curation, S.A.; writing—original draft preparation, S.A.; writing—review and editing, S.A., M.M. and A.A.; visualization, S.A.; supervision, S.A.; project administration, S.A.; funding acquisition, S.A. Archaeological and ethnographic information were collected by S.A. and M.M., while A.A. conducted the agricultural economics research. All authors have read and agreed to the published version of the manuscript.

**Funding:** This research was supported by funding from the US National Science Foundation (Award #1460665), the Henry Luce Foundation, the National Geographic Society, and the Cotsen Institute of Archaeology at UCLA.

**Data Availability Statement:** The original contributions presented in this study are included in the article. Further inquiries can be directed to the corresponding author.

**Acknowledgments:** We are grateful to the three anonymous reviewers for their thoughtful comments, which greatly improved this work. Our deepest thanks go to the people of Ifugao for their generosity in welcoming and allowing the Ifugao Archaeological Project to learn and work with them over the past two decades. We are also thankful to NJ Roxas and EJ Hernandez for their creativity in preparing the maps and images.

**Conflicts of Interest:** The authors declare no conflicts of interest.

## References

1. UNESCO. *Operational Guidelines for the Implementation of the World Heritage Convention*; CC-77/CONF.001/8Rev; UNESCO: Paris, France, 1977; Available online: http://whc.unesco.org/archive/opguide77b.pdf (accessed on 7 August 2025).
2. UNESCO. *Final Report: Informal Consultation of Intergovernmental and Non-Governmental Organizations in the Implementation of the Convention Concerning the Protection of the World Cultural and Natural Heritage, Morges, Switzerland, 19–20 May 1976*; CC-76/WS/25; UNESCO: Paris, France, 1976; Available online: http://unesdoc.unesco.org/images/0002/000213/021374eb.pdf (accessed on 7 August 2025).
3. UNESCO. *Operational Guidelines for the Implementation of the World Heritage Convention*; WHC.05/2; UNESCO: Paris, France, 2005; Available online: https://whc.unesco.org/archive/opguide05-en.pdf (accessed on 7 August 2025).
4. Meskell, L. *A Future in Ruins: UNESCO, World Heritage, and the Dream of Peace*; Oxford University Press: Oxford, UK, 2018.
5. Su, M.M.; Wang, M.; Yu, J.; Wall, G.; Jin, M. Measuring tourism impacts on community well-being at the Hani Rice Terraces GIAHS site, Yunnan Province of China. *Soc. Nat. Resour.* **2023**, *36*, 796–820. [CrossRef]
6. Winter, T. Clarifying the critical in critical heritage studies. *Int. J. Herit. Stud.* **2012**, *19*, 532–545. [CrossRef]
7. Khalaf, M.C. Women and heritage: The politics of gender, authenticity and preservation. *Int. J. Herit. Stud.* **2020**, *26*, 489–504.
8. ICOMOS. *The Nara Document on Authenticity. Adopted at the Nara Conference on Authenticity in Relation to the World Heritage Convention, Nara, Japan, 1–6 November 1994*; International Council on Monuments and Sites: Paris, France, 1994; Available online: https://landscapes.icomos.org/wp-content/uploads/1994-NARA-document-on-authenticity-International-Council-on-Monuments-and-Sites-1.pdf (accessed on 7 August 2025).
9. An Act Providing for the Protection and Conservation of the National Cultural Heritage, Strengthening the National Commission for Culture and the Arts (NCCA) and Its Affiliated Cultural Agencies, and for Other Purposes. Republic Act No. 10066; Official Gazette of the Republic of the Philippines: Manila, Philippines, 2009. Available online: https://ncca.gov.ph/republic-act-no-10066/ (accessed on 7 August 2025).
10. UNESCO World Heritage Centre. *Rice Terraces of the Philippine Cordilleras [World Heritage List]*; UNESCO: Paris, France; Available online: https://whc.unesco.org/en/list/722/ (accessed on 7 August 2025).
11. Brosius, J.P. Significance and social being in Ifugao agricultural production. *Ethnology* **1988**, *27*, 97–110. [CrossRef]
12. Codamon-Dugyon, E. Indigenous knowledge in traditional production of rice: Impact on food security in the upland households in Ifugao, Philippines. *Plant Sci. Today* **2023**, *10*, 27–33. [CrossRef]
13. Aclibon Puguon, K.; Arriola, J.L. Beyond sustenance: Exploring the cultural and narrative significance of rice in five Ifugao narratives. *Divers. J.* **2025**, *10*, 0884–0905. [CrossRef]
14. Martin, M.; Acabado, S.B.; Macapagal, R. Hongan di Pa'ge: The sacredness and realism of terraced landscape in Ifugao culture, Philippines. In *Indigenous Perspectives on Living with Cultural Landscapes*; Liljeblad, J., Verschuuren, B., Eds.; Routledge: Abingdon, UK, 2019; pp. 167–179.
15. Avtar, R.; Tsusaka, K.; Herath, S. Assessment of forest carbon stocks for REDD+ implementation in the muyong forest system of Ifugao, Philippines. *Environ. Monit. Assess.* **2020**, *192*, 571. [CrossRef]
16. Acabado, S.B.; Martin, M. *Indigenous Archaeology in the Philippines: Decolonizing Ifugao History*; University of Arizona Press: Tucson, AZ, USA, 2022.
17. Conklin, H.C. *Ethnographic Atlas of Ifugao: A Study of Environment, Culture, and Society in Northern Luzon*; Yale University Press: New Haven, CT, USA, 1980.

18. Murphy, B. Terracing, Land Management and Agricultural Soils in the Andagua Valley of the Southern Peruvian Andes. Master's Thesis, University of Denver, Denver, CO, USA, 2017.
19. Albano, A. From rice to vegetable terraces: Agricultural transition and sustainability in western Ifugao, Philippines. *Southeast Asian Stud.* **2025**, *14*, 37–65.
20. Estacio, I.; Basu, M.; Sianipar, C.P.; Onitsuka, K.; Hoshino, S. Dynamics of land cover transitions and agricultural abandonment in a mountainous agricultural landscape: Case of Ifugao rice terraces, Philippines. *Landsc. Urban Plan.* **2022**, *222*, 104394. [CrossRef]
21. Flannery, K.V. *The Early Mesoamerican Village, Updated ed.*; Academic Press: Cambridge, MA, USA, 2019.
22. Smith, M.E. *The Comparative Archaeology of Complex Societies*; Cambridge University Press: Cambridge, UK, 2011.
23. Taylor, B. Subsistence, environment and Mesolithic landscape archaeology. *Camb. Archaeol. J.* **2018**, *28*, 493–510. [CrossRef]
24. Wilson, K.M.; McCool, W.C.; Brewer, S.C.; Zamora-Wilson, N.; Schryver, P.J.; Lamson, R.L.F.; Huggard, A.M.; Coltrain, J.B.; Contreras, D.A.; Codding, B.F. Climate and demography drive 7000 years of dietary change in the Central Andes. *Sci. Rep.* **2022**, *12*, 2026. [CrossRef] [PubMed]
25. Crumley, C.L. (Ed.) *Historical Ecology: Cultural Knowledge and Changing Landscapes*; School of American Research Press: Santa Fe, NM, USA, 1994.
26. Balée, W.; Erickson, C.L. (Eds.) *Time and Complexity in Historical Ecology: Studies in the Neotropical Lowlands*; Columbia University Press: New York, NY, USA, 2006.
27. Fletcher, R. Settlement archaeology: World-wide comparisons. *World Archaeol.* **1986**, *18*, 59–83. [CrossRef]
28. Hayashida, F.M. Archaeology, ecological history, and conservation. *Annu. Rev. Anthropol.* **2005**, *34*, 43–65. [CrossRef]
29. Guimbatan, R.; Baguilat, T., Jr. Misunderstanding the notion of conservation in the Philippine rice terraces–cultural landcapes. *Int. Soc. Sci. J.* **2006**, *58*, 59–67. [CrossRef]
30. Dizon, J.T.; Calderon, M.M.; Sajise, A.J.U.; Andrada, R.T., II; Salvador, M.G. Youths' perceptions of and attitudes towards the Ifugao Rice Terraces. *J. Environ. Sci. Manag.* **2012**, *15*, 52–58.
31. Acabado, S.B. The Ifugao agricultural landscapes: Complementary systems and the intensification debate. *J. Southeast Asian Stud.* **2012**, *43*, 500–522. [CrossRef]
32. Acabado, S.B.; Martin, M.M. Older is not necessarily better: Decolonizing Ifugao history through the archaeology of the rice terraces. *Land* **2024**, *13*, 237. [CrossRef]
33. Martin, M.M. The rice terraces of Ifugao province, Philippines. *J. World Herit. Stud.* **2017**, 1–5.
34. Dulay, M.J. The values of the Ifugao Hudhud. *Int. J. Sci. Res.* **2015**, *4*, 2776–2779.
35. Stanyukovich, M.V. A living shamanistic oral tradition: Ifugao Hudhud, the Philippines. *Oral Tradit.* **2003**, *18*, 249–251. [CrossRef]
36. Smith, L. *Archaeological Theory and the Politics of Cultural Heritage*; Routledge: Abingdon, UK, 2004.
37. Meskell, L. UNESCO's World Heritage Convention at 40: Challenging the economic and political order of international heritage conservation. *Curr. Anthropol.* **2013**, *54*, 483–494. [CrossRef]
38. Smith, L. *Uses of Heritage*; Routledge: Abingdon, UK, 2006.
39. Holmelin, N.B. National specialization policy versus farmers' priorities: Balancing subsistence farming and cash cropping in Nepal. *J. Rural Stud.* **2021**, *83*, 71–80. [CrossRef]
40. Qingwen, M. Agri-cultural heritage: An interdisciplinary field with development prospects. *J. Resour. Ecol.* **2021**, *12*, 437–443. [CrossRef]
41. Nan, M.A.; Lun, Y.; Qingwen, M.; Keyu, B.; Wenhua, L. The significance of traditional culture for agricultural biodiversity—Experiences from GIAHS. *J. Resour. Ecol.* **2021**, *12*, 453–461. [CrossRef]
42. Acabado, S.B.; Koller, J.M.; Liu, C.-H.; Lauer, A.; Farahani, A.; Barretto-Tesoro, G.; Martin, J.A.; Peterson, J.A. The short history of the Ifugao Rice Terraces: A local response to the Spanish conquest. *J. Field Archaeol.* **2019**, *44*, 195–214. [CrossRef]
43. Eusebio, M.S.; Ceron, J.R.; Acabado, S.B.; Krigbaum, J. Rice pots or not? Exploring ancient Ifugao foodways through organic residue analysis and paleoethnobotany. *Nat. Mus. J. Cult. Herit.* **2015**, *1*, 11–20.
44. Findley, D.M.; Bankoff, G.; Barretto-Tesoro, G.; Hamilton, R.; Kay, A.U.; Acabado, S.B.; Amano, N.; Kaplan, J.O.; Roberts, P. Land use change in a pericolonial society: Intensification and diversification in Ifugao, Philippines between 1570 and 1800 CE. *Front. Earth Sci.* **2022**, *10*, 680926. [CrossRef]
45. Avtar, R.; Sawada, H.; Watanabe, T.; Ahamad, F. The impact of REDD+ implementation on Indigenous communities in Southeast Asia. *Land* **2021**, *10*, 283. [CrossRef]
46. Atalay, S. *Community-Based Archaeology: Research with, by, and for Indigenous and Local Communities*; University of California Press: Berkeley, CA, USA, 2012.
47. PAENRO (Provincial Agriculture, Environment and Natural Resources Office), Ifugao. *Cost of Production of Tinawon and Inbred Rice, Banaue*; PAENRO: Lagawe, Philippines, 2025.
48. OMAg (Office of the Municipal Agriculturist), Municipality of Tinoc, Ifugao. *Cost of Production of Tomato and Cabbage*; OMAg: Tinoc, Philippines, 2025.
49. Maher, R.F. Archaeological investigations in central Ifugao. *Asian Perspect.* **1973**, *16*, 39–70.

50. Camacho, L.D.; Gevaña, D.T.; Carandang, A.P.; Camacho, S.C. Indigenous knowledge and practices for the sustainable management of Ifugao forests in Cordillera, Philippines. *Int. J. Biodivers. Sci. Ecosyst. Serv. Manag.* **2016**, *12*, 5–13. [CrossRef]
51. Serrano, R.C.; Cadaweng, E.A. The Ifugao Muyong: Sustaining water, culture and life. In *Search of Excellence: Exemplary Forest Management in Asia and the Pacific*; Regional Community Forestry Training Center for Asia and the Pacific, Food and Agriculture Organization of the United Nations: Bangkok, Thailand, 2005; pp. 103–112.
52. Acabado, S.; Martin, M. The Ifugao agroecological system: Bridging culture and nature to enhance tropical biodiversity. In *Exploring Frameworks for Tropical Biodiversity: Integrating Natural and Cultural Diversity for Sustainability: A Global Perspective*; Sanz, N., Ed.; UNESCO Mexico Office: Mexico City, Mexico, 2018; pp. 228–253.
53. Eder, J.F. No water in the terraces: Agricultural stagnation and social change at Banaue, Ifugao. *Philipp. Q. Cult. Soc.* **1982**, *10*, 101–116.
54. Carman, J. *Archaeology and Heritage: An Introduction*; A&C Black: London, UK, 2003.
55. Comer, D.C. Conservation and preservation in archaeology in the twenty-first century. In *Encyclopedia of Global Archaeology*; Springer: Berlin/Heidelberg, Germany, 2014; pp. 1640–1646.
56. Acabado, S.B.; Martin, M. Between pragmatism and cultural context: Continuity and change in Ifugao wet-rice agriculture. In *Water and Heritage: Material, Conceptual and Spiritual Connections*; Willems, W., Van Schaik, H., Eds.; Sidestone Press Academic: Leiden, The Netherlands, 2015; pp. 275–297.
57. Acabado, S.B. Defining Ifugao social organization: "House," field, and self-organizing principles in the northern Philippines. *Asian Perspect.* **2013**, *52*, 161–189. [CrossRef]
58. Martin, M.; Laurice, J.; Paterno, M.C.; Megarry, W.; Hermann, V. *Climate Change Vulnerability Assessment of the Rice Terraces of the Philippine Cordilleras*; Project Report; ICOMOS Philippines: Manila, Philippines, 2023.
59. Ellison, D.; Futter, M.N.; Bishop, K. On the forest cover–water yield debate: From demand- to supply-side thinking. *Glob. Change Biol.* **2012**, *18*, 806–820. [CrossRef]
60. Dorren, L.; Schwarz, M. Quantifying the stabilizing effect of forests on shallow landslide-prone slopes. In *Ecosystem-Based Disaster Risk Reduction and Adaptation in Practice*; Advances in Natural and Technological Hazards Research; Renaud, F., Sudmeier-Rieux, K., Estrella, M., Nehren, U., Eds.; Springer: Berlin/Heidelberg, Germany, 2016; Volume 42, pp. 197–210. [CrossRef]
61. Soriano, M.A., Jr.; Diwa, J.; Herath, S. Local perceptions of climate change and adaptation needs in the Ifugao Rice Terraces (Northern Philippines). *J. Mt. Sci.* **2017**, *14*, 1455–1472. [CrossRef]
62. Lapniten, K. Sustainable Livelihood Offers a Lifeline to Philippines' Dying Rice Terraces. Mongabay. 2021. Available online: https://news.mongabay.com/2021/08/sustainable-livelihood-offers-a-lifeline-to-philippines-dying-rice-terraces/ (accessed on 21 June 2025).
63. Lansing, J.S. Balinese "water temples" and the management of irrigation. *Am. Anthropol.* **1987**, *89*, 326–341. [CrossRef]
64. Hauser-Schäublin, B. Land donations and the gift of water. On temple landlordism and irrigation agriculture in pre-colonial Bali. *Hum. Ecol.* **2011**, *39*, 43–53. [CrossRef]
65. Lang, C.; Stump, D. Geoarchaeological evidence for the construction, irrigation, cultivation, and resilience of 15th--18th century AD terraced landscape at Engaruka, Tanzania. *Quat. Res.* **2017**, *88*, 382–399. [CrossRef]
66. UNESCO World Heritage Centre. Cultural Landscape of Honghe Hani Rice Terraces. Available online: https://whc.unesco.org/en/list/1111/ (accessed on 8 August 2025).
67. Gu, H.; Jiao, Y.; Liang, L. Strengthening the socio-ecological resilience of forest-dependent communities: The case of the Hani Rice Terraces in Yunnan, China. *Forest Policy Econ.* **2012**, *22*, 53–59. [CrossRef]
68. Yuan, Z.; Lun, F.; He, L.; Cao, Z.; Min, Q.; Bai, Y.; Liu, M.; Cheng, S.; Li, W.; Fuller, A.M. Exploring the State of Retention of Traditional Ecological Knowledge (TEK) in a Hani Rice Terrace Village, Southwest China. *Sustainability* **2014**, *6*, 4497–4513. [CrossRef]
69. Su, X.; Song, C.; Sigley, G. The uses of reconstructing heritage in China: Tourism, heritage authorization, and spatial transformation of the Shaolin Temple. *Sustainability* **2019**, *11*, 411. [CrossRef]
70. Gonzalvo, C.M.; Maharjan, K.L.; Baggo, J.C.; Embate, J.M. Farmer perceptions of GIAHS: Analyzing farmer involvement and GIAHS benefits in the Ifugao Rice Terraces. *Agriculture* **2024**, *14*, 2305. [CrossRef]
71. Cameron, C. Heritage as process: Reframing conservation through engagement, not fixation. *Int. J. Herit. Stud.* **2020**, *26*, 210–223.

land

MDPI

*Article*

# Temporal and Spatial Distribution of Sanxingdui (3.7–3.1 ka BP) and Shi'erqiao (3.1–2.6 ka BP) Sites on the Chengdu Plain, Southwest China

**Hui Chao [1,2,*], Xiaolin Chang [2,3,4] and Changhao Xu [5]**

[1] School of Atmospheric Sciences, Chengdu University of Information Technology, Chengdu 610225, China
[2] Key Laboratory of Deep-Time Geography and Environment Reconstruction and Applications of Ministry of Natural Resources, Chengdu University of Technology, Chengdu 610059, China; changxiaolin10@cdut.edu.cn
[3] Institute of Sedimentary Geology, Chengdu University of Technology, Chengdu 610059, China
[4] Center for Scientific and Technological Archaeology, Chengdu University of Technology, Chengdu 610059, China
[5] Sichuan Metallurgical Geological Exploration Institute, Chengdu 610000, China; xuzhanghao@scnrig.com.cn
* Correspondence: chaohui@cuit.edu.cn

**Abstract**

Understanding the spatial distribution of ancient settlements is essential for reconstructing human–environment dynamics. The Chengdu Plain, a core region of the ancient Shu civilization, provides a key setting for examining early cultural development in Southwest China. This study applies GIS-based spatial analysis to 116 prehistoric sites, 27 Sanxingdui (3.7–3.1 ka BP) and 89 Shi'erqiao (3.1–2.6 ka BP), to evaluate their relationships with elevation, relief, slope, aspect, and distance from rivers. Sanxingdui settlements show a tightly constrained pattern, clustering at 450–500 m a.s.l. elevation, on gentle slopes (0–5°), within moderate relief (30–150 m), and almost entirely inside the 0–0.01 river buffer, reflecting strong dependence on stable alluvial landscapes. In contrast, Shi'erqiao sites occupy a broader environmental spectrum, spanning 400–650 m a.s.l. in elevation, concentrating in 30–60 m relief zones, and extending beyond the 0.02 hydrological buffer, indicating enhanced adaptability supported by improved water management and sociopolitical reorganization. The transition from Sanxingdui to Shi'erqiao thus represents a shift from selective environmental reliance to more flexible landscape engagement under mid–late Holocene climatic variability. These results underscore the dynamic nature of prehistoric human–landscape interaction on the Chengdu Plain and provide new insights into cultural resilience in the upper Yangtze River basin.

**Keywords:** Sichuan Province; geographic information systems; cultural development; environmental adaption

## 1. Introduction

During the Holocene, the intensification of human activities produced increasingly complex interactions between societies and their surrounding environments [1]. Understanding the relationship between environmental change, regional cultural evolution, and human–land dynamics has therefore become a central focus of environmental archaeology, attracting sustained scholarly attention both within China and internationally [2,3]. The Yangtze River Basin, one of the major population centers and cultural heartlands, offers an important context for examining these long-term interactions [4]. Within this region, the Chengdu Plain, located in the upper reaches of the basin, stands out as a principal core area of early Chinese civilization. Archaeological investigations have documented a long and continuous cultural sequence extending from early agrarian settlements to increasingly complex Bronze Age societies [5]. High-quality archaeological datasets now enable more refined reconstructions. These include models of human habitation patterns, cultural transmission pathways, migration trajectories, and broader processes of social development [6]. The Shu cultural sequence developed from the Baodun culture, flourished at Sanxingdui, and transitioned into Shi'erqiao, forming a continuous trajectory of cultural evolution in the upper Yangtze River [7,8]. Among these cultural stages, the Sanxingdui and Shi'erqiao periods are particularly significant for understanding the interplay between cultural transformation and environmental setting [9–12]. A growing body of research demonstrates that the evolution of ancient settlements and cultural systems was closely tied to both natural and social environmental conditions [1–3]. Moreover, key environmental variables, such as topography, hydrology, and geomorphology, have been recognized as decisive factors shaping cultural development, land-use strategies, and spatial organization over millennia [1,13]. Consequently, analyzing the spatial logic of prehistoric settlement systems provides essential insights into how ancient communities adapted to and modified their local environments [14].

The Sanxingdui culture represents a highly centralized and materially distinctive Bronze Age society on the Chengdu Plain, characterized by monumental ritual architecture, urban-scale fortified settlements, and an advanced bronze industry that relied on the region's rich agricultural and metallurgical resource base [5,7,9]. Archaeological studies suggest that the remarkable spatial concentration of Sanxingdui sites around the Guanghan area may reflect both the presence of a strong political ritual center and potential research biases arising from the preferential discovery and excavation of large, architecturally prominent sites [11,12]. Following the decline of this urban-focused system, the succeeding Shierqiao culture exhibits a markedly different spatial pattern, typified by smaller, more dispersed settlements distributed across the Chengdu Plain, as well as diversified subsistence and land-use strategies [1]. Although the two cultures differ significantly in settlement scale, spatial organization, and modes of environmental engagement, they form a continuous cultural sequence from Baodun through Sanxingdui to Shierqiao, reflecting a long-term sociopolitical transformation within the ancient Shu region [7,15]. The contrast between their centralized and dispersed spatial patterns provides a crucial framework for examining how environmental factors, resource availability, and cultural dynamics jointly shaped the evolution of prehistoric settlement systems on the Chengdu Plain.

Over the past decades, archaeological discoveries related to the Sanxingdui–Shi'erqiao cultural sequence have attracted wide scholarly attention, providing new insights into early state formation, sociopolitical organization, and human–environment interactions in prehistoric China [1,5,8]. Nevertheless, several key challenges remain. First, most previous research has concentrated on dwelling features and artifact typology, whereas systematic, quantitative assessments of settlement distribution and environmental preferences remain limited [16,17]. Second, the nature of the cultural transition from Sanxingdui to Shi'erqiao, whether abrupt collapse, gradual transformation, or adaptive relocation, continues to be debated [5,7]. Third, the specific role of climatic fluctuations, particularly extreme climate events, in shaping settlement strategies and sociopolitical reorganization has yet to be fully clarified [18–20]. Recent advances in Geographic Information Systems (GIS) offer powerful tools for modeling settlement spatial logic and quantifying environmental determinants of site selection [21,22]. GIS-based analyses of elevation, slope, relief, aspect, and hydrological proximity have proved effective for identifying the geo-environmental drivers behind prehistoric land-use choices [23–26]. By integrating these analytic approaches with archaeological datasets, researchers can reconstruct diachronic trajectories of environmental adaptation and cultural evolution [27].

This study applies GIS spatial analysis to 116 prehistoric sites on the Chengdu Plain, including 27 Sanxingdui and 89 Shi'erqiao sites, to examine their distribution in relation to five key environmental variables. Although a modern Digital Elevation Model (DEM) was used, numerous geomorphological studies and borehole records indicate that the Chengdu Plain has experienced minimal vertical displacement (<10 m) since the mid-Holocene [2,5], allowing modern elevation and slope data to serve as reasonable proxies for paleo-topographic conditions. The research aims to characterize the environmental preferences of Sanxingdui and Shi'erqiao settlements, identify diachronic changes in settlement–environment relationships, and interpret how these spatial patterns reflect broader processes of cultural transformation and environmental adaptation. The findings contribute to ongoing discussions of prehistoric human–landscape dynamics in southwestern China and offer a replicable methodological framework for spatial analysis in heritage science.

## 2. Study Area

The Chengdu Plain, situated in the western Sichuan Basin of Southwest China, provides an exemplary setting for examining long-term human–environment interactions [28,29]. Flanked by the Longmen Mountains to the northwest and the Longquan Mountains to the east, the plain extends approximately 166 km from north to south and 192 km from east to west, covering ~12,400 km$^2$, about 2.6% of Sichuan Province [5]. Formed largely by overlapping alluvial fans of the Min and Tuo rivers, which descend from the mountainous sectors of the northwestern Sichuan Plateau [13], the region comprises one of the largest and most continuous alluvial lowlands in Southwest China. Its flat terrain, well-developed dendritic river network, and humid subtropical monsoon climate (mean annual precipitation 1000–1200 mm, concentrated in summer) have supported continuous human occupation since the Neolithic [20,28]. These environmental conditions facilitated early agricultural intensification, dense settlement development, and sustained cultural evolution. Historically, the Chengdu Plain served as the political core of the ancient Shu state and hosted several major prehistoric urban centers, including the Sanxingdui complex in the Guanghan area (Deyang City) and the Shi'erqiao cultural center in modern Chengdu [30] (Figure 1).

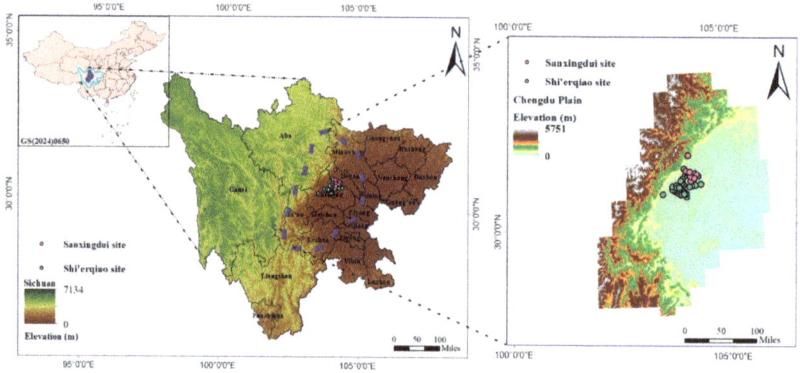


**Figure 1.** Location of Sanxingdui and Shi'erqiao site in Chengdu Plain, SW China.

Sanxingdui and Shi'erqiao cultures represent two sequential yet distinct modes of settlement organization on the Chengdu Plain [31]. The Sanxingdui culture is characterized by its large urban center, advanced bronze technology, and monumental ritual architecture, and broadly contemporaneous with Erlitou Phase IV and Yinxu Phase II in North

China [32] (Figure 2A,B). Current archaeological research widely agrees that the Shi'erqiao culture constitutes a direct cultural continuation of the Sanxingdui. After the decline of the Sanxingdui urban core, the political and ceremonial center of the ancient Shu state shifted from the Guanghan area (Deyang City) to the Chengdu area, where the Shi'erqiao culture developed as its immediate successor [9,10]. Shi'erqiao culture is a continuation of the Sanxingdui culture. It is represented by the Shi'erqiao site in Jinniu District and the Jinsha site in Qingyang District (Figure 2C,D). It was a period of regional transition, contemporaneous with the early Western Zhou to the early Spring and Autumn Period in the Central Plains, characterized by a shift toward more dispersed and diverse settlement forms [33].

**Figure 2.** (**A**) "Behind the Masks: Discovering Sanxingdui and Jinsha", University of Oxford; (**B**) Plan of the Qinguanshan No. 1 Large Building Foundation at the Sanxingdui Site in Guanghan, Sichuan (Sichuan Provincial Institute of Archaeological Research); (**C**,**D**) The Excavation Site at the Shi'erqiao Site (Sichuan Provincial Institute of Archaeological Research), Chengdu.

The chronological framework adopted in this study is grounded in published radiocarbon evidence (Table 1) and follows the established cultural sequence of the ancient Shu region, encompassing the Sanxingdui (3.7 ka BP) and Shi'erqiao (3.1–2.6 ka BP) periods [29]. This framework provides the temporal foundation for examining long-term changes in settlement patterns and human–environment interactions across the two cultural phases.

**Table 1.** $^{14}C$ radiocarbon dating data of the Sanxingdui and Shi'erqiao period.

| Period | Age (Minimum) | Lab | Age (Maximum) | Lab |
| --- | --- | --- | --- | --- |
| Sanxingdui | 3.15–2.97 ka BP [34] | BA210001, Peking University | $3.615 \pm 0.080$ ka BP [35] | Laboratory of the Institute of Archaeology, Chinese Academy of Sciences |
| Shi'erqiao | $2.567 \pm 0.078$ ka BP [13] | Peking University | 3.15–2.97 ka BP [34] | BA210001, Peking University |

## 3. Materials and Methods

The archaeological site data used in this study are derived from published sources, including the journals *Archaeology*, *Sichuan Cultural Relics*, and *the report Archaeological Discoveries in Chengdu*. A total of 116 prehistoric sites were compiled from published survey reports, excavation bulletins, and official heritage databases. These records were generated by government-authorized archaeological institutions through systematic surveys, scientific and rescue excavations, and standardized registration, representing the most authoritative dataset for the Chengdu Plain. Site detection follows a dual-track system: rescue excavations preferentially target large or highly visible sites, while survey techniques have evolved from early pedestrian methods to recent remote sensing and geophysical prospection. Despite these biases, cross-checking among multiple sources ensures that the dataset provides the most reliable baseline for regional prehistoric settlement analysis. All archaeological points used in this study represent confirmed habitation or multi-functional settlement sites, each characterized by well-defined cultural layers documented in published excavation or survey reports. These cultural deposits provide clear evidence of sustained or recurrent human occupation, allowing their unequivocal identification as settlements.

This study compiled data from 116 archaeological sites on the Chengdu Plain, including 27 from the Sanxingdui period and 89 from the Shierqiao period. For each site, recorded attributes include its name, cultural type, precise location, and elevation. These data were organized in a spreadsheet for GIS analysis. All sites were georeferenced using WGS84 UTM Zone 48N coordinates and cross-validated against topographic base maps. Data normalization and error correction were applied using ArcGIS 10.2 to ensure spatial accuracy. Five environmental variables were selected to evaluate the geomorphological and hydrological constraints influencing the spatial distribution of the Sanxingdui and Shi'erqiao settlements: elevation, relief degree, slope, aspect, and Euclidean distance to rivers. These indicators are widely recognized in archaeological landscape studies as key determinants of settlement suitability because they jointly reflect terrain stability, flood risk, resource accessibility, and environmental habitability [36]. Elevation provides the general altitudinal setting and helps assess flood avoidance or strategic visibility. Relief degree, calculated as the elevation range within a 3 × 3 pixel neighborhood (approximately 90 m × 90 m for a 30 m DEM), represents local terrain variability, which affects construction feasibility and agricultural potential [36]. Because even 30–50 m of relief in this low-gradient alluvial setting delineates discrete Holocene geomorphic surfaces with different hydrological behaviours, our elevation, local-relief, and slope metrics are designed to capture microtopographic suitability rather than absolute terrain variation, ensuring robust differentiation of settlement distributions across cultural phases.

Slope was derived from the DEM using the Spatial Analyst "Slope" tool and reflects terrain steepness, a critical factor in evaluating accessibility, defense, and suitability for habitation [37]. In the low-gradient alluvial context of the Chengdu Plain, even slight slope differences (<1–2°) measurably influence surface drainage behaviour and the propensity for seasonal waterlogging. Consequently, the slope metric here is intended to characterise microtopographic hydrological performance rather than broader regional contrasts in terrain. This approach enables a more reliable assessment of fine-scale settlement suitability across the Sanxingdui and Shi'erqiao cultural phases. Aspect, generated using the "Aspect" tool and categorized into eight directional classes, captures insolation and microclimate differences that influence agricultural productivity [38]. Using a 30 m resolution river network dataset, we computed the Euclidean distance to rivers to evaluate water accessibility, essential for daily consumption, agriculture, and transportation. Although river size varies, previous studies confirm that this proximity measure is a robust proxy for hydrological suitability in prehistoric contexts [36].

All variables were derived from a 30 m resolution DEM provided by the Geospatial Data Cloud Platform, reprojected to a unified coordinate system. The 3 × 3 pixel moving window is sufficiently large to capture meaningful microtopographic variation around each site while remaining small enough to avoid overgeneralizing the terrain; although it does not cover the entire extent of large settlements, it reliably represents the immediate environmental conditions relevant to prehistoric habitation. Terrain anomalies such as cliffs or incised valleys are inherently captured through elevation gradients, relief degree, and slope values, while areas of nearly imperceptible slope variation are appropriately reflected in low-gradient outputs. Together, these variables provide an integrated representation of the topographic and hydrological factors that most likely shaped prehistoric settlement choices on the Chengdu Plain.

## 4. Results

### 4.1. Elevation

In the ArcGIS software, the DEM of Sichuan province was projected and symbolized. Subsequently, the spatial data of ancient sites spanning from the Sanxingdui period to the Shi'erqiao period were overlaid onto the DEM. Ultimately, the distribution patterns of these ancient sites across varying elevations were visualized through mapping. Furthermore, we determined the elevation of each ancient site and calculated the frequency of sites at different elevation intervals (Figure 3).

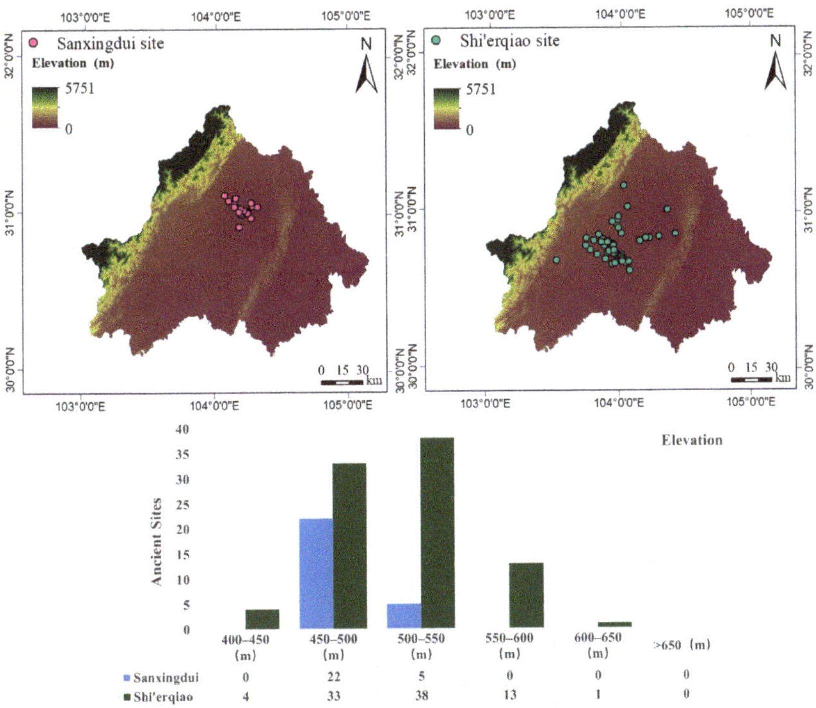


**Figure 3.** Distribution of Sanxingdui and Shi'erqiao sites at different elevations in the Chengdu Plain.

The altitudinal distribution of all sites from both phases in the study area falls within the range of 400–650 m a.s.l. Sanxingdui-period sites are predominantly clustered within a narrow band of 450–500 m a.s.l.; specifically, 22 of the 27 total sites fall within the 400–500 m a.s.l. range, with the remaining 5 located between 500 and 550 m a.s.l. These sites are situated near the junction of the piedmont and mid-fan zones, a region characterized by gently undulating relief and abundant water resources.

In comparison, Shi'erqiao period sites exhibit a significantly broader spatial distribution, spanning the full spectrum from 400 to 650 m a.s.l., with a notable concentration between 450 and 550 m a.s.l. The 500–550 m a.s.l. interval hosts the highest density of remains (38 sites), followed by 37 sites at 400–500 m a.s.l., 13 sites at 550–600 m a.s.l., and 1 site extending to 600–650 m a.s.l. This pattern indicates a more advanced capacity for exploiting diverse micro-topographic settings in site selection and land use.

Although the overall elevational range is relatively constrained, it corresponds to distinct geomorphic zonation on the Chengdu Plain. Even modest altitudinal differences here reflect transitions from floodplains to low terraces and gentle foothills. Consequently, the tight clustering of Sanxingdui sites at 450–500 m a.s.l. aligns with stable floodplain-terrace surfaces, while the wider elevational span of Shi'erqiao sites demonstrates an adaptation to more varied terrain.

### 4.2. *Relief Degree*

Relief degree is a fundamental geomorphological metric that captures elevation variability within a defined area and serves as a key proxy for surface heterogeneity, exerting strong control over prehistoric settlement placement [36]. In this study, we derived relief values from a 30 m-resolution DEM using a standard classification scheme and mapped their spatial distribution across the Chengdu Plain. The locations of Sanxingdui and Shi'erqiao sites were then superimposed onto the relief map to assess topographic preferences (Figure 4). Elevation in the study area increases from the low-lying alluvial plain toward the piedmont zone and decreases again toward the Min River terraces; accordingly, relief was categorized into six classes: 0–30 m, 30–60 m, 60–90 m, 90–150 m, 150–180 m, and >180 m. The number of sites falling within each relief class was quantified to evaluate differences between the two cultural phases, enabling a comparative analysis of how variability in microtopography influenced settlement distribution.

The results show that overall relief variability across the study area is relatively low, yet clear differences emerge between the two cultural phases. Sanxingdui sites are distributed across a broader relief spectrum, with no sites found in the lowest (0–30 m) category. Their distribution is as follows: 18 sites within the 30–150 m classes, 2 sites in the 150–180 m range, and 7 sites in areas exceeding 180 m.

In contrast, Shi'erqiao sites display a pronounced concentration in low-relief settings. More than 60% (64 sites) occur within the 30–60 m class. The remaining locations are distributed as follows: 15 sites in the 0–30 m class, 9 sites within 60–150 m, and only one site in the 150–180 m category. No Shi'erqiao sites are located in high-relief areas (>180 m). As illustrated in Figure 3, Sanxingdui sites reflect a broader tolerance for topographic variability and a more dispersed spatial configuration, suggesting flexible settlement strategies. Conversely, the Shi'erqiao sites are tightly clustered within low-relief alluvial plains characterized by minimal surface undulation. This shift indicates a morphological evolution in settlement organization, from the more outward-expanding and topographically diverse pattern of the Sanxingdui phase to the more inward-focused and spatially intensive configuration characteristic of the Shi'erqiao period.

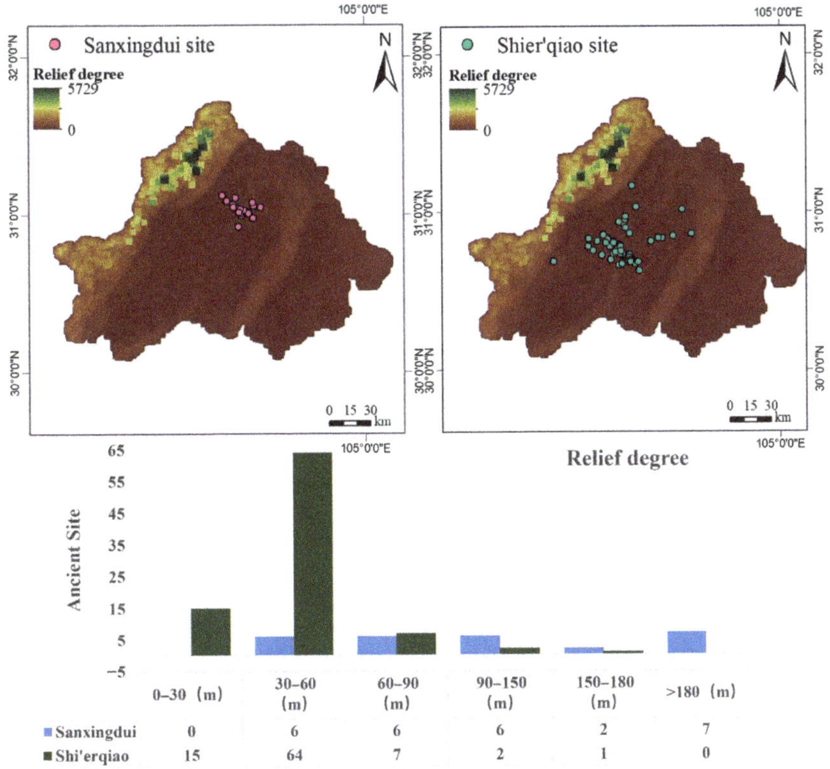


**Figure 4.** Distribution of Sanxingdui and Shi'erqiao sites at different relief degrees in the Chengdu Plain.

### 4.3. Slope

Slope is a key geomorphological parameter that strongly shapes settlement suitability, influencing construction feasibility, land-use intensity, water retention, erosion risk, and transportation efficiency [37]. Although elevation data portray the Chengdu Plain as broadly level, fine-scale slope variations reveal subtle terrain differences that may have conditioned early habitation choices. Slope analysis, integrated with the spatial distribution of archaeological sites, demonstrates that both Sanxingdui and Shi'erqiao settlements are overwhelmingly concentrated within gentle-slope zones (0–5°), reflecting the inherently low-gradient landscape of the region (Figure 5).

Slope extraction analysis shows that both the Sanxingdui and Shi'erqiao settlements overwhelmingly occupy gentle-slope zones (0–5°), though with distinct patterns of variability. Sanxingdui sites are primarily concentrated within the 0–5° range (19 sites), but a smaller number extend into higher slope categories, including 5 sites in the 6–10° range, 2 sites in the 11–15° range and 1 site in the 16–25° range. No Sanxingdui sites occur on slopes exceeding 15°. In contrast, Shi'erqiao settlements display an even stronger preference for flat terrain: 79 sites fall within the 0–5° zone, accounting for more than 80% of all occurrences, and only 10 sites lie within the 6–10° category. No Shi'erqiao sites are identified on slopes above 10°. Overall, Sanxingdui exhibits a slightly broader slope tolerance, whereas the Shi'erqiao settlements are tightly clustered within the flattest parts of the plain, reflecting an increasing reliance on low-relief environments during the later cultural phase.

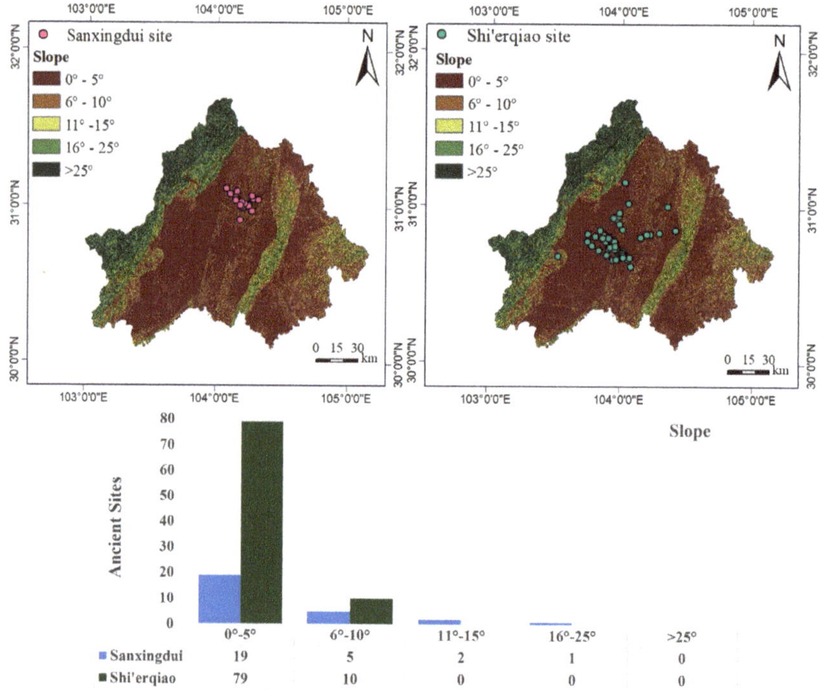


**Figure 5.** Distribution of Sanxingdui and Shi'erqiao sites at different slopes in the Chengdu Plain.

### 4.4. Aspect

Aspect, defined as the compass direction of slope orientation, is a key geomorphic parameter influencing micro-environmental conditions such as solar radiation, evapotranspiration, wind exposure, and vegetation growth [38]. These factors can shape local habitability and thus affect ancient settlement choices. In this study, the aspect of the Chengdu Plain was categorized into nine classes (Figure 5): Flat terrain (−1), North (0–22.5°), Northeast (22.5–67.5°), East (67.5–112.5°), Southeast (112.5–157.5°), South (157.5–202.5°), Southwest (202.5–247.5°), West (247.5–292.5°), and Northwest (292.5–337.5°), with the final interval returning to North (337.5–360°). Sanxingdui sites are primarily concentrated within the central portion of the study area, whereas the Shi'erqiao sites exhibit a broader spatial distribution across multiple aspect categories, reflecting more extensive use of diverse micro-environments.

The aspect analysis reveals a relatively even distribution of Sanxingdui sites across multiple orientation categories. Specifically, 2 sites occur on flat terrain, while north-, east-, southeast-, south-, and west-facing slopes each host between 4 and 6 sites, with smaller occurrences in the northeast, southwest, and northwest categories. In contrast, the Shi'erqiao sites exhibit a more concentrated pattern. A total of 11 sites are situated on flat terrain, and the majority are distributed across east-, southeast-, south-, northeast-, and west-facing slopes (9–13 sites each), with no sites recorded on northwest-facing slopes. This indicates a marked preference for flat surfaces and east-facing orientations during the Shi'erqiao period. The histogram in Figure 6 further highlights these contrasts, showing the broad, nearly uniform aspect distribution of Sanxingdui sites versus the strong clustering of Shi'erqiao settlements in low-relief, east-facing environments.

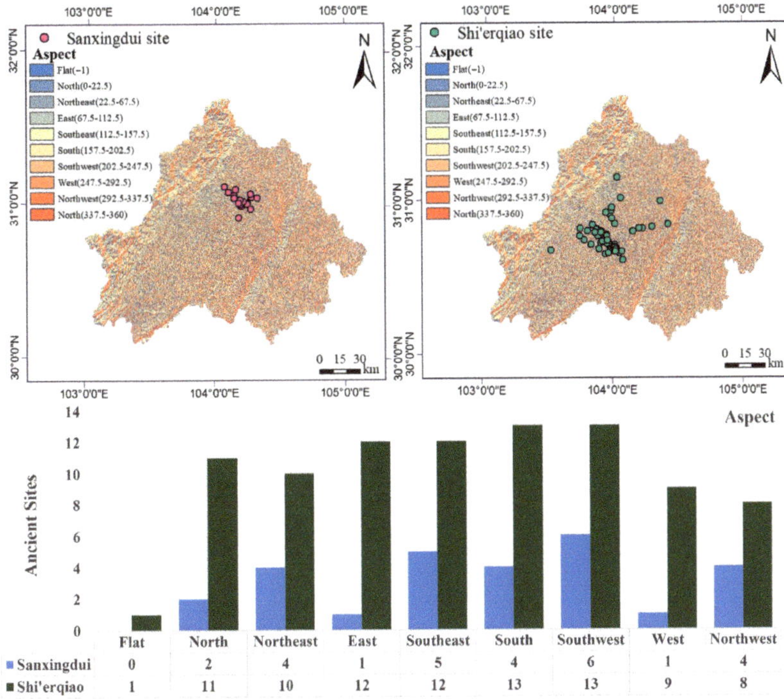


**Figure 6.** Distribution of Sanxingdui and Shi'erqiao sites at different aspects.

### 4.5. Euclidean Distance to Rivers

Hydrological systems are a fundamental component shaping the formation and sustainability of human settlements, influencing access to drinking water, irrigation, transportation, daily activities, and defensive planning [21]. To assess the spatial relationship between settlements and water availability, this study generated proximity buffers around the regional river network, and overlaid the Sanxingdui and Shi'erqiao sites' distributions onto a Euclidean distance-to-river surface (Figure 6). This approach allows for a comparative assessment of hydrological dependency and settlement location preferences between the two cultural phases [39].

Figure 7 shows that Sanxingdui sites are predominantly concentrated within the 0–0.01 distance interval, indicating strong spatial clustering in areas immediately adjacent to river channels. Most sites occur along higher-order rivers (Classes 3–4), which represent larger, more stable waterways. This pattern underscores the marked hydrological dependence of Sanxingdui settlements and highlights the central role of major river systems in sustaining their habitation and socio-economic development.

In contrast, the Shi'erqiao sites exhibit a more dispersed pattern of hydrological proximity. Although many sites remain within 0.05 units of the nearest river, a notable proportion extends to greater distances (up to 0.2578 units), indicating reduced spatial dependence on major waterways. As shown in Figure 8, Sanxingdui settlements are strongly clustered in the nearest interval (0–0.01) with 11 sites, followed by progressively fewer sites in the subsequent distance classes (5 in 0.01–0.02, 4 in 0.02–0.03, 5 in 0.03–0.04, 1 in 0.04–0.05, and 1 in 0.05–0.06), with none beyond 0.06.

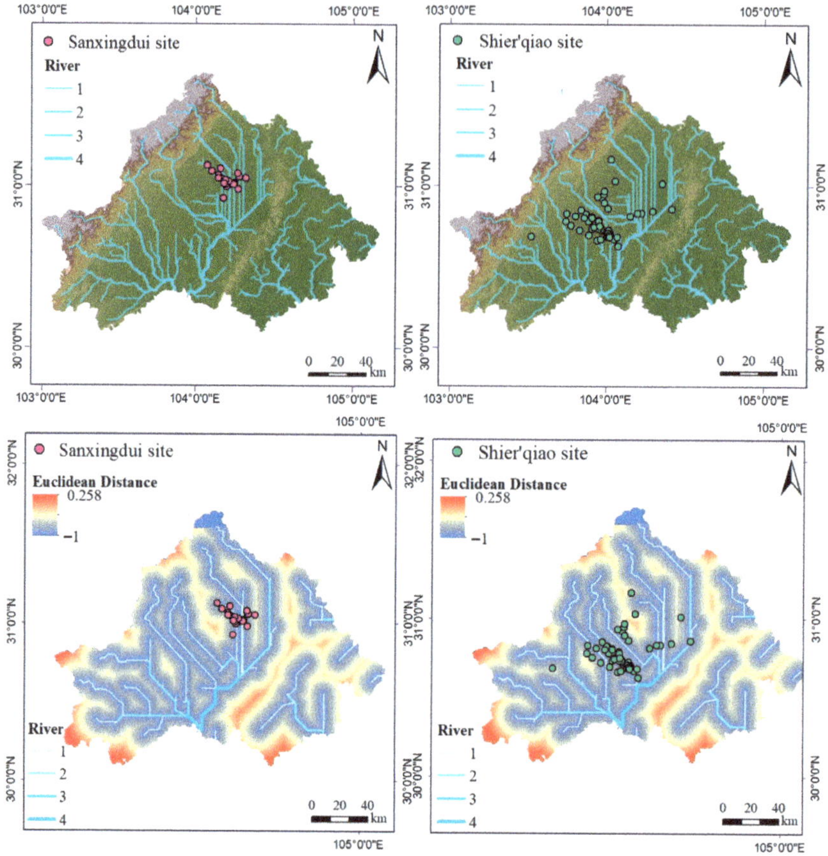


**Figure 7.** The Euclidean distance from the rivers of the Sanxingdui and Shi'erqiao sites in the Chengdu Plain.

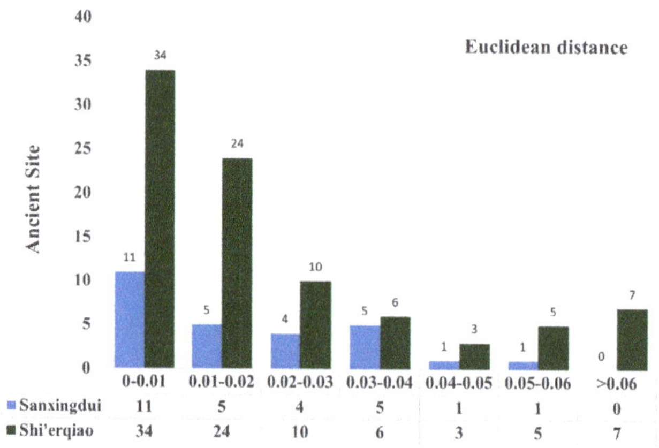


**Figure 8.** Distribution of Sanxingdui and Shi'erqiao sites at different Euclidean distances from rivers.

Shi'erqiao sites display a markedly different pattern: the highest frequency occurs in the 0.01–0.02 interval (34 sites), followed by 24 sites in the 0–0.01 interval. Site numbers decline gradually with increasing distance (10 in 0.02–0.03, 6 in 0.03–0.04, 3 in 0.04–0.05, and 5 in 0.05–0.06), yet 7 sites occur beyond 0.06. This distribution suggests a broader range of hydrological adaptability during the Shi'erqiao period. These contrasting patterns indicate a clear diachronic shift from the Sanxingdui culture's strong dependence on major river channels to the more diversified hydrological strategies of the Shi'erqiao period.

## 5. Discussion

### *5.1. Geomorphic Evolution and Settlement Adaptation*

Rivers, particularly unstable channels in alluvial proluvial plains, are highly sensitive to environmental change, and the development of a river system records variations linked to tectonic activity, paleoclimatic conditions, and other factors [40]. The Chengdu Plain is a tectonically conditioned depression undergoing sustained uplift along the eastern flank of the Longmen Mountains, where active orogenesis and consequent fluvial incision have long shaped regional landscape evolution [2,5,41]. Increased relief contrasts enhanced sediment flux from the mountain front, particularly along the Minjiang and its distributary channels emerging from the Longmen Shan foothills [42]. This promoted frequent avulsion of proximal channels such as the paleo-Minjiang and early Jinsha–Qingshui tributaries, facilitating the progradation of fantoe lobes along the northwestern Chengdu Plain [42,43]. Sustained river incision driven by tectonic uplift caused the abandonment of older alluvial-fan surfaces adjacent to the Minjiang and Qingshui River corridors, leading to the dissection of mid- to late Holocene fan deposits and the formation of inset fluvial terraces [41–43]. Differential uplift along the mountain front induced subtle tilting of the foreland surface toward the southeast. This tectonically forced reorientation produced an asymmetric pattern of fan development, with more deeply entrenched channels in the northwest and broader, low-gradient fan surfaces to the southeast [41]. Collectively, the Minjiang River and its tributary systems have progressively lowered regional base levels, expanded drained areas, and promoted the formation of stepped river terraces since the mid-Holocene [42,43]. As a result, the plain developed a dense dendritic drainage system incorporating the Minjiang, Jinsha, Qingshui, and the Xi River, creating a heterogeneous mosaic of floodplains, paleo-channels, and low-relief foothills [13] (Figure 9).

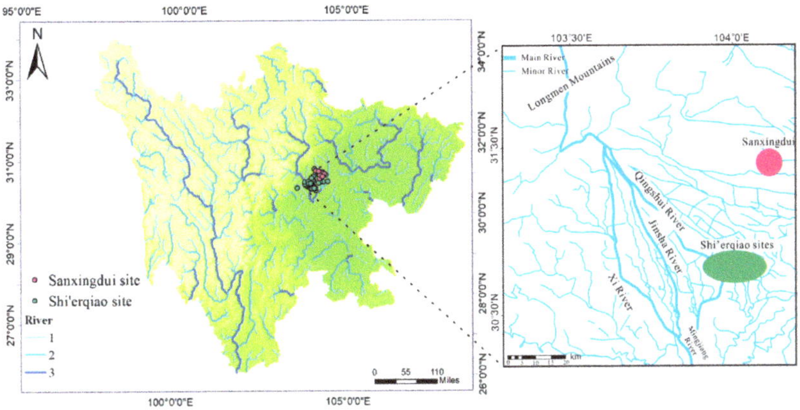


**Figure 9.** Location of the Sanxingdui and Shi'erqiao sites on river system map: (**left**) River system background of the study area in Sichuan Province; (**right**) Cultural sites on a small-scale modern river system map.

During the mid–late Holocene, the Chengdu Plain was shaped by an actively adjusting fluvial regime dominated by lateral accretion, channel avulsion, and episodic overbank deposition [42]. These processes produced a tiered geomorphic surface composed of low-relief floodplains (30–90 m), aggradational terrace margins (60–150 m), and abandoned channel depressions. Consistent with this framework, Sanxingdui settlements are concentrated in the northeastern plain, where moderate relief, natural levees, and high hydraulic connectivity offered favorable conditions for irrigation, drainage, and soil productivity [44]. Portions of the Minjiang floodplain have since undergone extensive alluvial burial and modern land modification, suggesting that some prehistoric sites may remain undiscovered; the present dataset therefore reflects only the identifiable archaeological record [43].

Sanxingdui sites show strong clustering within the 0–0.01 river-distance buffer and a marked preference for relief values > 90 m, indicating avoidance of flood-prone basin floors while maintaining direct access to perennial water. To reduce the impact of later hydraulic engineering, the hydrological analysis focuses on geomorphically persistent river corridors and tributary sectors supported by mid-Holocene sedimentary evidence. Their tight proximity to rivers underscores a high dependence on stable surface water typical of early agrarian systems [44], while the tendency toward elevated microtopography suggests intentional selection for improved drainage and reduced flood risk [45]. By contrast, Shi'erqiao settlements occupy a broader geomorphic spectrum, extending southward toward the Jinsha drainage and into higher microbasins (up to 550–600 m a.s.l.) (Figures 3 and 4). Their greater representation in the 0.01–0.03 buffer zones (Figure 7) reflect decreased reliance on major channels and expanded use of secondary water sources. This diversification is consistent with archaeological evidence for small-scale hydraulic installations, canals, ponds, and weirs, that enabled irrigation and facilitated settlement expansion into more dispersed terrain [46,47]. These features contrast sharply with the Sanxingdui period, where only simple earthen ditches have been documented, and demonstrate a clear technological transition toward engineered water redistribution systems [48]. Such installations would have reduced dependence on proximity to major channels and enabled reliable irrigation in more distant or geomorphically varied settings, consistent with the observed increase in site frequency within the 0.01–0.03 hydrological buffer zones [49]. Both cultural phases exhibit a preference for low slopes (0–5°) and low-to-moderate relief (<90 m), but the broader topographic tolerance of Shi'erqiao communities indicates gradually diminished geomorphic constraints and improved adaptability to varied fluvial landscapes [50]. Although the absolute differences in elevation and slope across the Chengdu Plain are small, these parameters retain functional significance in a low-gradient alluvial environment. Even modest elevation steps (30–50 m) and microslope variations (<1–2°) can substantially influence surface drainage efficiency, seasonal waterlogging, and the stability of cultivated landforms. These microtopographic controls operate at scales far below regional geomorphic contrasts; thus, the observed differences in settlement preferences cannot be attributed simply to the Sanxingdui and Shi'erqiao sites occupying different subregions of the plain. After controlling for regional distribution patterns, both cultural phases display internally consistent elevation and slope tendencies, Sanxingdui sites clustering on slightly elevated, better-drained levee–ridge margins, and Shi'erqiao sites extending into lower, gentler, more hydrologically varied surfaces. This persistence of microtopographic preference across spatially distinct areas indicates that elevation and slope reflect genuine environmental selection behaviours rather than artefacts of uneven site locations. Accordingly, the shift from Sanxingdui concentration on modestly higher, well-drained micro-relief to the broader topographic tolerance of the Shi'erqiao period is best understood as a cultural adaptation facilitated by advances in water-management capacity rather than a by-product of regional terrain differences.

Agricultural development played a significant role in this transition. During the Sanxingdui–Shi'erqiao sequence, settlements increasingly occupied flatter terrain (0–60 m), particularly within the 30–60 m band, aligning with areas offering stable irrigation potential and arable land [29,50]. Sanxingdui communities favored mid-elevation (400–550 m a.s.l.) zones with a stable water supply, well-suited for diversified early agriculture [51]. By the Shi'erqiao period, agricultural practices expanded further, with rice cultivation supplemented by foxtail millet, broomcorn millet, barley, and wheat [29,52]. These developments strengthened water-management capacity and increased tolerance for more distant hydrological settings. Spatially, 41% of Sanxingdui settlements fall within the 0–0.01 hydrological buffer, whereas this proportion decreases to 38% for the Shi'erqiao period (Figure 8). In contrast, the proportion in the 0.01–0.03 interval increases to 38%, reflecting reduced dependence on primary river channels and increasingly flexible land-use strategies enabled by hydraulic innovation [53,54]. Overall, the agricultural intensification and technological advances of the Shi'erqiao period supported settlement dispersal into a wider range of geomorphic contexts, marking a shift from river-dominated settlement logic to a more diversified and resilient land-use system.

Shi'erqiao culture, therefore, displays a broader environmental tolerance, encompassing higher elevations, lower-relief plains, and more distant positions from major rivers. This dispersed pattern suggests increased landscape flexibility likely driven by technological innovation, demographic expansion, or emerging sociopolitical organization [55]. The transition from tightly clustered riverine settlements in the Sanxingdui period to more topographically dispersed sites in the Shi'erqiao period reflects an evolving strategy of environmental adaptation supported by improved water management and expanding agricultural capacity [56,57]. Furthermore, the emergence of formalized burial zones, standardized ceramic forms, and organized production areas at Late Shi'erqiao sites supports interpretations of increased political coordination and community-level planning [11,58]. Together, these lines of evidence substantiate the argument that both hydraulic innovation and rising sociopolitical complexity contributed to broader settlement tolerance and more flexible land-use strategies in the Shi'erqiao period.

### 5.2. Impacts of Climatic Conditions on the Distribution of Archaeological Sites

The formation, development, and exchange of human cultures are often influenced by geographic environments, as evidenced by the effects of climate change and extreme weather events on human migration and cultural turnover [59,60]. The settlement behaviors of the Sanxingdui and Shi'erqiao cultures occurred during a transitional period of the Holocene climate. Globally, the Middle to Late Holocene as characterized by a gradually weakening Asian summer monsoon (ASM), declining temperatures, and regional aridification in parts of inland Asia and northern China [13,61].

Throughout the Holocene, the Sichuan Basin experienced a long-term shift from warm–humid conditions to a hotter phase and then to cooler regimes, with climate variability intensifying after the mid-Holocene [62,63]. During the Sanxingdui period, this transition manifested as a decline in mean annual temperature from ~+1 °C above modern values to ~0.5–0.8 °C below present, accompanied by a ~15% decrease in precipitation and a weakening ASM around 3.2 ka BP. In contrast, the subsequent Shi'erqiao occupation unfolded under persistently cooler and drier conditions, with climatic archives revealing centennial-scale oscillations of ±0.7 °C, >100 mm $yr^{-1}$ precipitation variability, and two notable ASM "weak–strong–weak" cycles at ~2.7 and ~2.4 ka BP [62] (Figure 10A,B). Based on extensive research on the ASM, stalagmite $\delta^{18}O$ is widely used as a proxy for the influence of solar radiation fluctuations on the ASM over orbital timescales. Lighter (more negative) $\delta^{18}O$ values indicate a stronger summer monsoon, while heavier (less negative) values indicate a weaker one (Figure 10C) [18,63]. As shown in Figure 10C, from 4.2 to 2.2 ka BP, $\delta^{18}O$ values exhibited a long-term trend toward lighter values (decreasing from −6.13‰ to −7.57‰), suggesting a strengthening ASM forcing at the orbital scale [64].

However, this trend was superimposed on substantial variability, and regional precipitation proxies for the Chengdu Plain tell a more complex story. Some records indicate an overall trend towards drier conditions [64]. While others show periods of increased rainfall, potentially linked to factors such as the transition from El Niño to La Niña conditions during the 4.2 ka BP event, which can lead to anomalous precipitation increases in the Yangtze River Basin [62,63,65] (Figure 10D). These hydroclimatic stresses coincided with the terminal stage of the Sanxingdui culture, marked by settlement contraction and large-scale ritual bronze deposition [63,65,66]. While the Sanxingdui period (3.7–3.1 ka BP) initially developed under a relatively warm and humid regime [67,68]. The ensuing amplified late-Holocene instability meant that Shi'erqiao communities faced more frequent and severe environmental fluctuations, likely prom [69].

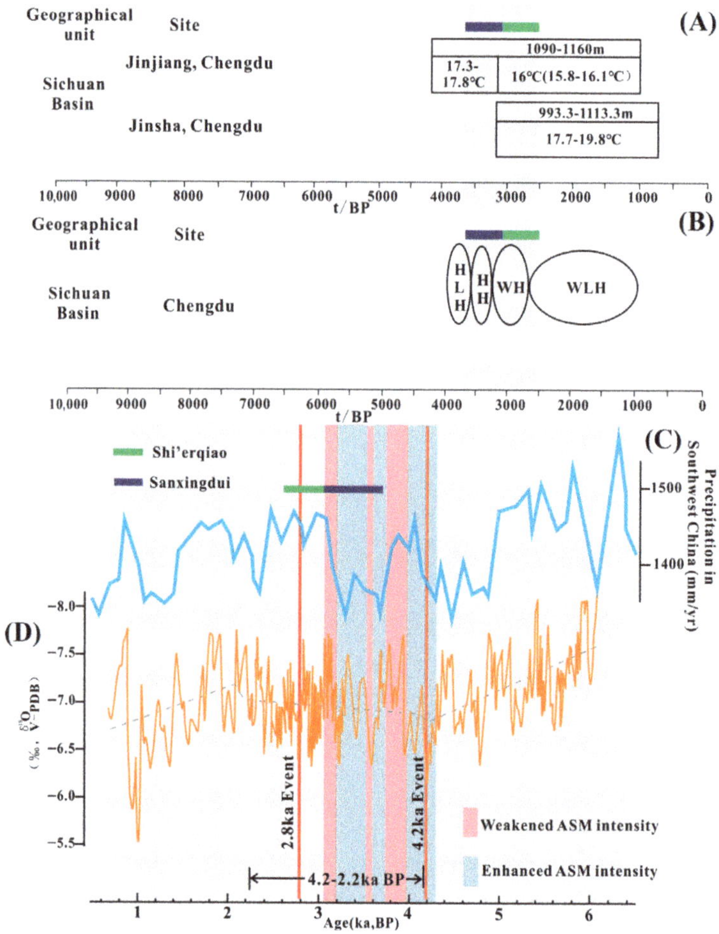


**Figure 10.** (**A**) Quantitative and (**B**) Qualitative reconstruction of climate change since Holocene in the Sichuan Basin, SW China [64]. HLH: Hot and little humid; HH: Hot and humid; WH: Warm and humid; WLH: Warm and little humid; (**C**) Reconstructed past precipitation in Southwest China [62]; (**D**) The Stalagmite oxygen isotope ($\delta^{18}O$) record from 6.2 to 0.69 ka BP in Southwest China. The gray dashed lines indicate the variation trends of the $\delta^{18}O$. The red lines indicate the extreme climatic events, including the 2.8 and 4.2 ka event [65].

Against this backdrop of progressively intensified late-Holocene climate variability, the settlement preferences of the Sanxingdui and Shi'erqiao cultures exhibit clear geomorphic signatures that closely track these hydroclimatic transitions. During the warm–humid and hydrologically stable Sanxingdui phase (3.7–3.1 ka BP), communities preferentially occupied extremely east-southeast-facing aspects (Figure 11A), low slopes (0–5°) (Figure 11B), and floodplain-proximal locations. These settings maximized biomass productivity, reliable hydrology, and radiative advantages favorable for intensive rice agriculture [70–72]. This pattern aligns with relatively high temperatures and abundant moisture before the weakening of the ASM at ~3.2 ka BP, as reflected in declining precipitation, weakening monsoon intensity, and changes in $\delta^{18}O$ values (Figure 10) [65]. Hydroclimatic instability intensified around 3.3–3.0 ka BP, marked by cold–wet anomalies, floods, and tectonically induced disturbances [13,65]. A 20–50 cm layer of coarse sand and fine gravel (0.5–2 cm, well-rounded) in the upper cultural deposits at Sanxingdui provides clear evidence of a high-energy flood event. This interpretation is reinforced by a 1.2 m grayish-white sandy gravel lens at the base of the cultural layer at the Shi'erqiao site (Lanyuan locality), dated to 3080 ± 30 BP, which aligns closely with the same flood phase [73–75]. The terminal Sanxingdui stage thus experienced environmental stress, settlement contraction, and large-scale ritual bronze deposition [65,69].

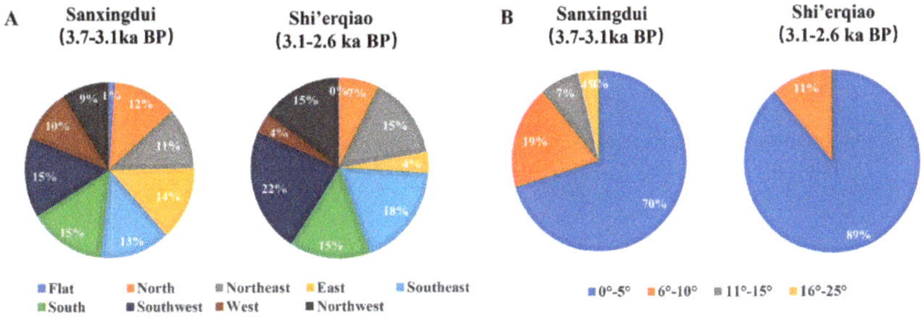


**Figure 11.** The statistics of Sanxingdui and Shi'erqiao sites at (**A**) aspect and (**B**) slope.

In contrast, the Shi'erqiao period (2.8–2.3 ka BP), coinciding with cooler, drier, and more variable climatic conditions linked to the 2.8 ka event and a weakened ASM, shows increasingly concentrated settlement distributions: 89% of sites lie on 0–5° slopes, yet slope-aspect preferences broaden to include north- and northwest-facing orientations (Figure 11) [65,76]. This shift indicates enhanced adaptive capacity and refined land–energy strategies under heightened climate variability [77]. Significantly, hydraulic installations, including wood-pile wharves, reservoirs, covered culverts, composite stone–timber floodwalls, and paddy-ditch systems, first appear or reach their greatest density during this period, whereas Sanxingdui features only simple irrigation ditches [78,79]. These facilities cluster on north-to northwest-facing slopes and gently rolling terrain of 5–10°, matching the expanded aspect range characteristic of Shi'erqiao settlements. This spatial correspondence provides material confirmation of the causal chain in which improved water-management technology enabled greater locational flexibility and resilience under increasingly unstable late-Holocene climatic conditions [80,81]. Together with the broader influence of the 4.2 ka and 2.8 ka events on regional hydroclimatic stability, these patterns underscore a marked contrast: warm–humid stability during the Sanxingdui period fostered centralized settlements and intensive agriculture, whereas the more variable climate of the Shi'erqiao period promoted settlement decentralization, diversified land use, and heightened topographic adaptability [68,82].

### 5.3. Responses of Environment Evolution to Culture

The close relationship between environmental change and cultural transformation in the Chengdu Plain provides an essential context for interpreting the evolution from the Sanxingdui to the Shi'erqiao stage of ancient Shu civilization [83]. Archaeological and chronological evidence demonstrates that these two cultural phases form a continuous developmental sequence rather than separate cultural entities [82,84]. Stratified excavations at Sanxingdui and Shi'erqiao show uninterrupted ceramic traditions, shared jade-working methods, and stable bronze typologies, particularly in mask-making, anthropomorphic imagery, and high-tin ritual vessels, indicating a deep lineage continuity across the transition [84,85]. During the Sanxingdui phase, settlement patterns and material assemblages reveal a highly centralized polity whose power was expressed through monumental ritual complexes, large-scale sacrificial pits, and specialized bronze workshops concentrated near the core site [4,5]. Their spatial clustering on low-relief, hydrologically accessible terrain reflects a landscape strategy dependent on predictable water supply and state-directed labor mobilization [86].

Around 3.1 ka BP, weakening of the Asian summer monsoon produced drier climatic conditions, while tectonically driven reorganization of the Minjiang and its distributaries reshaped the hydrological framework of the Chengdu Plain [34]. This environmental inflection point coincides with marked cultural adjustments recorded at Sanxingdui: abrupt burning remains, large-scale backfilling, and the termination of centralized bronzeworking within the ritual core [16]. Meanwhile, new occupational surfaces and cultural layers at Jinsha and Shierqiao reveal the emergence of continuously inherited ceramic forms, duplicated ritual motifs, and relocated craft-production loci, showing that cultural authority did not disappear but shifted spatially within the same cultural tradition [5,85,87]. The succeeding Shi'erqiao period, corresponding broadly to the early Western Zhou, exhibits greater settlement dispersal, diversified craft zones, and increased functional mixing of domestic, ritual, and production activities [58]. The rise of secondary centers and small-scale villages, together with the expansion of bead-making, gold-crafting, and sandstone-implement workshops, reflects a maturing sociopolitical and economic system that had moved beyond the earlier highly centralized model [85]. These changes indicate not cultural rupture but adaptive decentralization, a process that allowed communities to buffer hydrological uncertainty and broaden their ecological footprint [88].

In this broader context, the Sanxingdui–Shi'erqiao transformation mirrors synchronous patterns observed across Bronze Age Eurasia, where climatic instability and shifting river regimes frequently corresponded with political decentralization, innovation in land management, and restructured ritual practices [89–91]. Thus, the evolution from Sanxingdui to Shi'erqiao reflects both a deep cultural inheritance, evidenced by continuous ceramic lineages, bronze traditions, and ritual symbolism, and a strategic reorientation in human–landscape interaction driven by intertwined environmental pressures and sociopolitical recalibration.

## 6. Conclusions

This study demonstrates that the spatial and temporal patterning of Sanxingdui and Shi'erqiao cultural sites on the Chengdu Plain was jointly shaped by geomorphology, hydrology, and changing cultural–technological practices. Analysis of 116 confirmed settlement sites, 27 Sanxingdui and 89 Shi'erqiao reveals clear contrasts in land-use strategies between the two periods.

Sanxingdui settlements show strong environmental coherence, clustering at 450–500 m a.s.l. elevation, on gentle slopes, within moderate relief zones, and almost entirely inside the 0–0.01 river buffer. Their consistent orientation toward east- and southeast-facing slopes reflects dependence on fertile floodplains, stable river corridors, and microclimatic conditions that supported intensive rice agriculture and centralized sociopolitical organization. Shi'erqiao sites, in contrast, occupy a wider elevational and hydrological spectrum, with over one-third located beyond major river buffers and a marked preference for flat terrain and warm-aspect slopes. This broader distribution indicates reduced dependence on trunk rivers and greater reliance on small-scale irrigation, flexible water management, and diversified subsistence strategies.

The transition from Sanxingdui to Shi'erqiao therefore signals a shift from selective environmental preference to active environmental engagement, driven by both hydrogeomorphic adjustment and technological innovation. These findings highlight the dynamic interaction between human groups and the Chengdu Plain landscape and provide a framework for understanding cultural adaptation during the late Bronze Age. Future research integrating paleoenvironmental records and subsurface survey data will further clarify the mechanisms guiding settlement organization and cultural transformation in this region.

**Author Contributions:** Conceptualization, H.C. and X.C.; methodology, H.C., X.C. and C.X.; formal analysis, H.C. and C.X.; investigation, H.C. and C.X.; writing—original draft preparation, H.C., X.C. and C.X.; writing—review and editing, H.C. and X.C. All authors have read and agreed to the published version of the manuscript.

**Funding:** This research was funded by the Open Fund (DGERA20221101) of Key Laboratory of Deep-time Geography and Environment Reconstruction and Applications of Ministry of Natural Resources, Chengdu University of Technology, Talent Introduction Program of Chengdu University of In-formation Technology, (KYTZ202201).

**Data Availability Statement:** The original contributions presented in the study are included in the article, further inquiries can be directed to the corresponding author.

**Conflicts of Interest:** The authors declare no conflicts of interest.

## References

1. Zhao, C.; Mo, D. Holocene hydro-environmental evolution and its impacts on human occupation in Jianghan-Dongting Basin, middle reaches of the Yangtze River, China. *J. Geogr. Sci.* **2020**, *30*, 423–438. [CrossRef]
2. Zeng, M.X.; Zhu, C.; Song, Y.G.; Wu, Y.Q.; Wang, H.J.; Zhang, Y.W.; He, K.Y.; Chen, J.; Wu, Z.R.; Zeng, Q. Significance of fluvial landform evolution and river reorganization in settlement development over the past 50 ka: A case study of Baodun site on Chengdu Plain, SW China. *Geomorphology* **2022**, *409*, 108267. [CrossRef]
3. Wen, X.; Chen, M.; Feng, W.; Huang, C. Mid-late Holocene climatic changes recorded by loess deposits in the eastern margin of the Tibetan Plateau: Implication for human migrations. *Quat. Int.* **2017**, *441 Pt A*, 77–88. [CrossRef]
4. Chen, C.J.; Wu, Y.; Li, J.Y.; Zhang, H.W.; Cheng, H.; Shen, C.C.; Li, T.Y. Hydrological changes in the east asian monsoon region around 4.2 ka precisely reconstructed from multi-proxy stalagmites. *Quat. Sci. Rev.* **2025**, *357*, 109321. [CrossRef]
5. Huang, M. Investigation and research on environmental archaeology in Chengdu Plain. *China Cult. Herit.* **2017**, *6*, 87–93. (In Chinese)
6. Dong, G.; Li, T.; Zhang, S.; Ren, L.; Li, R.; Xiao, Y.; Wang, Z.; Chen, F. Precipitation in surrounding mountains instead of lowlands facilitated the prosperity of ancient civilizations in the eastern Qaidam Basin of the Tibetan Plateau. *Catena* **2021**, *203*, 105318. [CrossRef]
7. Yang, Q. A study on the correlation between the migration of major city sites of the ancient Shu civilization and climate change. *Sichuan Archit.* **2015**, *3*, 7–10.
8. Jiang, L.; Wu, Z.R. Human Activities Flexibility under Volatile Environment Conditions around 4000 a BP: Insights from the Jinsha Site in the Chengdu Plain, SW China. *Quat. Sci. Rev.* **2024**, *326*, 108612. [CrossRef]
9. Chen, D.A.; Chen, X.D. The Excavation of the No.1 Sacrificial Pit at Sanxingdui Site, Guanghan. *Cult. Relics* **1987**, *10*, 1–15+97–101. (In Chinese)
10. Chen, D.A.; Chen, X.D. The Excavation of the No.2 Sacrificial Pit at Sanxingdui Site, Guanghan. *Cult. Relics* **1989**, *5*, 1–20+97–103. (In Chinese)
11. Jiang, J.H.; Wang, Y.; Zhang, Q. A Preliminary Research on the Predynastic Qin Culture in the Chengdu Plain. *Acta Archaeol. Sin.* **2002**, *1*, 1–22. (In Chinese)

12. Fan, J. Research on Prehistoric Agriculture in the Chengdu Plain. *Forum Chin. Cult.* **2017**, *12*, 80–83. (In Chinese)
13. Wu, J.H.; Zeng, M.X.; Zhu, C.; Song, Y.G.; Wu, Y.Q.; Mao, X.Y.; Wang, N.J.; Wang, X.L. Drivers of prehistoric cultural evolution in the Chengdu Plain: Fire events and environmental changes during the middle and late Holocene. *Ecol. Indic.* **2024**, *169*, 112833. [CrossRef]
14. Zhu, L.; Li, Z.; Su, H.; Wang, X. Temporal and spatial distribution of ancient sites in Shaanxi Province using geographic information systems (GIS). *Herit. Sci.* **2021**, *9*, 1–10. [CrossRef]
15. Sun, H. Constitution and Origin of Wave Patterns on Bronze Ware: Issues of Sanxingdui/Shierqiao Culture and Wave Patterns of the Zhou Dynasty. *Sichuan Archit.* **2021**, *1*, 107–118. (In Chinese)
16. Zhao, D.Z. Basic Structure of the Sacrificial Activities at Sanxingdui Site: Altars, Temple and Sacrificial Pits. *Sichuan Cult. Relics* **2022**, *201*, 80–94. (In Chinese)
17. Wang, T.Y. The Identification of the Symbols Related to Bashu Sacred Tree—Taking the Yucong Sacred Tree Pattern Unearthed from Sanxingdui No.3 Pit as the Main Clue. *J. Chin. Cult.* **2024**, *4*, 113–123. (In Chinese)
18. Tan, M. Circulation Effect: Response of Precipitation $\delta^{18}O$ to the ENSO Cycle in Monsoon Regions of China. *Clim. Dyn.* **2014**, *42*, 1067–1077. [CrossRef]
19. Sun, Q.L.; Liu, Y.; Wünnemann, B.; Peng, Y.J.; Jiang, X.Z.; Deng, L.J.; Chen, J.; Li, M.T.; Chen, Z.Y. Climate as a factor for Neolithic cultural collapses approximately 4000 years BP in China. *Earth Sci. Rev.* **2019**, *197*, 102915. [CrossRef]
20. Huang, M.; Deng, Y.; Peng, H.; Wen, Z.; Shang, G.; Guan, H.; Ma, C. Hydroclimatic changes since the Last Glacial Maximum recorded in mountain peat deposit on the southwestern margin of the Sichuan Basin, China. *Front. Ecol. Evol.* **2022**, *10*, 1050429. [CrossRef]
21. Zhu, C.; Xu, J.J.; Huang, M.; Yang, Z.F.; Jiang, Z.H.; Bai, T.Y.; Lu, F.Z. Archaeological discoveries and research on the remains of an ancient flood event at the Majie Site in the Chengdu Plain. *Front. Earth Sci.* **2021**, *28*, 181–201. (In Chinese)
22. Hu, H.N. Study on the Distribution of Archaeological Sites from Yangshao to Shang periods in the Zhuozhang River Basin and Its Relationship with Physical Environment. Master's Thesis, Henan University, Henan, China, 2023. (In Chinese).
23. Feng, Y.; Wei, H.; Huang, Y.; Li, J.; Mu, Z.; Kong, D. Spatiotemporal evolution characteristics and influencing factors of traditional villages: The Yellow River Basin in Henan Province, China. *Herit. Sci.* **2023**, *11*, 97. [CrossRef]
24. Fang, Y.L.; Lu, H.Y.; Huang, Z.F.; Zhu, Z.G. Spatial and temporal distribution characteristics and influencing factors of traditional villages in China. *Econ. Geogr.* **2023**, *43*, 187–196.
25. Chen, W.; Yang, L.; Wu, J.; Wu, J.; Wang, G.; Bian, J.; Liu, Z. Spatio-temporal characteristics and influencing factors of traditional villages in the Yangtze River Basin: A Geodetector model. *Herit. Sci.* **2023**, *11*, 111. [CrossRef]
26. Zhang, H.P.; Liu, W.M.; Zhao, H.M.; Ye, Z.L.; Pan, M.H. The Paleolithic–Bronze age sites in Sichuan Province: Distribution and influencing factor. *Mar. Geol. Quat. Geol.* **2025**, *45*, 188–198. (In Chinese)
27. Cheng, J.N.; Zhao, G.X.; Zhang, Z.X.; Wang, S. Characterization of spatial pattern and geographic differentiation of traditional villages in Shandong Province. *J. Guilin Univ. Technol.* **2022**, *42*, 845–852. (In Chinese)
28. Zeng, M.X.; Ma, C.M.; Zhu, C.; Song, Y.G.; Zhu, T.X.; He, K.Y.; Chen, J.; Huang, M.; Jia, T.J.; Guo, T.H. Influence of climate change on the evolution of ancient culture from 4500 to 3700 cal. yr BP in the Chengdu Plain, upper reaches of the Yangtze River, China. *Catena* **2016**, *147*, 742–754. [CrossRef]
29. Yang, Y.; Li, H.; Lu, Y.; Xia, R.; James, N.; Chen, H.; Zhao, Y. The agricultural economy of the Sanxingdui culture (3700–3100 BP): Archaeological and historical evidence from the Chengdu Plain. *Land* **2024**, *13*, 787. [CrossRef]
30. Zhao, Z.Z.; Qiao, Y.S.; Wang, Y.; Fu, J.L.; Wang, S.B.; Li, C.Z.; Yao, H.T.; Jiang, F.C. Magnetostratigraphic and paleoclimatic studies on the red earth formation from the Chengdu Plain in Sichuan Province. *Sci. China Ser. Earth Sci.* **2007**, *50*, 927–935. [CrossRef]
31. Jiang, S.B. Preliminary study on pre-Qin agriculture in Chengdu Plain. *Agric. Archaeol.* **2002**, *3*, 26–34. (In Chinese)
32. Chen, X.D. The culture and chronology at Sanxingdui Site. *South. Ethnol. Archaeol.* **1990**, 213–231+312+318. (In Chinese)
33. Shi, J.S. Shi'erqiao sites and Shi'erqiao culture. *Archaeology* **2015**, *2*, 79–91+2. (In Chinese)
34. Xie, Z.B.; Xu, D.Y.; Han, Y.; Chen, M.; Wang, T.; Zhan, T.Y.; He, K.; Qiu, Y.W.; Li, G.H.; Wang, Q. Carbon-14 Dating Study of the No. 4 Sacrificial Pit at the Sanxingdui Site in Guanghan, Sichuan. *Sichuan Cult. Relics* **2021**, *2*, 117–120. (In Chinese)
35. Laboratory of the Institute of Archaeology, Chinese Academy of Social Sciences. Radiocarbon Dating Report (14). *Archaeology* **1987**, *7*, 658–662. (In Chinese)
36. Callow, J.N.; Van Niel, K.P.; Boggs, G.S. How does modifying a DEM to reflect known hydrology affect subsequent terrain analysis? *J. Hydrol.* **2007**, *332*, 30–39. [CrossRef]
37. Deng, C.; Liu, L.; Li, H.; Peng, D.; Wu, Y.; Xia, H.; Zhu, Q. A data-driven framework for spatiotemporal characteristics, complexity dynamics, and environmental risk evaluation of river water quality. *Sci. Total Environ.* **2021**, *785*, 147134. [CrossRef]
38. Liu, Y.; Cui, J.; Kong, X.; Zeng, C. Assessing suitability of rural settlements using an improved technique for order preference by similarity to ideal solution. *Chin. Geogr. Sci.* **2016**, *26*, 638–655. [CrossRef]
39. Askman, J.; Nilsson, O.; Becker, P. Why people live in flood-prone areas in Akuressa, Sri Lanka. *Int. J. Disaster Risk Sci.* **2018**, *9*, 143–156. [CrossRef]

40. Dewan, A.; Corner, R.; Saleem, A.; Rahman, M.; Haider, M.; Rahman, M.; Sarker, M. Assessing channel changes of the Ganges–Padma River system in Bangladesh using Landsat and hydrological data. *Geomorphology* **2017**, *276*, 257–279. [CrossRef]
41. He, Y. The preliminary discussion of the formation of Chengdu Basin. *Geol. Bull. China* **1987**, *2*, 75–82. (In Chinese)
42. Yan, L.; Li, Y.; Zhao, G.H.; Liang, M.J.; Yan, Z.K.; Shao, C.J.; Li, F.S.; Yun, K.; Ma, C.; Zhang, W. The uplift process and the geomorphological features of the rivers system in the Longmen Shan since Late Quaternary. *Quat. Sci.* **2018**, *38*, 232–246.
43. Lu, H.; Wang, C.; Cai, F.; Wu, C.; Bao, J.; Zheng, Y.; Cao, C. Morphological change and its response to human activities in the estuary of Minjiang River, China, since the last early century. *Earth Surf. Process. Landf.* **2023**, *48*, 1940–1955. [CrossRef]
44. Xu, J.; Jia, Y.; Ma, C.; Zhu, C.; Wu, L.; Li, Y.; Wang, X. Geographic distribution of archaeological sites and their response to climate and environmental change between 10.0–2.8 ka BP in the Poyang Lake Basin, China. *J. Geogr. Sci.* **2016**, *26*, 603–618. [CrossRef]
45. Gao, Y.; Li, J.; Di, B. Early human settlements and mobility patterns in the Sichuan–Chongqing region from the late Neolithic to the Bronze Age. *Quat. Int.* **2025**, *724*, 109726. [CrossRef]
46. Yu, M.Z.; Wu, C.M. Preliminary Discussion on the Early Remains of the Shierqiao Culture. *Acta Archaeol. Sin.* **2019**, *2*, 163–192. (In Chinese)
47. Wu, C.M. Preliminary Study of the Majiashan Type of the Shi'erqiao Culture. *Jianghan Archaeol.* **2023**, *5*, 82–92+121. (In Chinese)
48. Li, H.; Zuo, Z.; Cui, J.; Tian, J.; Yang, Y.; Yi, L.; Zhou, Z.; Fan, J. Bronze Production in the Ancient Chengdu Plains: A Diachronic Metallurgical Perspective on a Separate Cultural Region. *J. Cult. Herit.* **2020**, *43*, 26–36. [CrossRef]
49. Hosner, D.; Wagner, M.; Tarasov, P.E.; Chen, X.; Leipe, C. Spatiotemporal distribution patterns of archaeological sites in China during the Neolithic and Bronze Age: An overview. *Holocene* **2016**, *26*, 1576–1593. [CrossRef]
50. Li, Y.P.; Zhou, X.Y.; Li, J.Y.; Sun, N.; Zhao, K.L.; Chen, G.H.; Zhang, Y.P.; Li, X.Q. Evolution of agricultural structure and its adaptation to the environment in North China during the middle Holocene warm period. *Quat. Sci.* **2024**, *44*, 1433–1449. (In Chinese)
51. Ren, X.; Lemoine, X.; Mo, D.; Kidder, T.R.; Guo, Y.; Qin, Z.; Liu, X. Foothills and intermountain basins: Does China's fertile arc have "hilly flanks"? *Quat. Int.* **2016**, *426*, 86–96. [CrossRef]
52. Lü, Y.; Zhang, J.P.; Tang, M.; Yang, Z.F.; Huang, M.; Jiang, M.; Yan, X.; Shao, K.; Wen, C.H.; Ye, M.L.; et al. Phytolith Analysis Reveals the Agricultural Development and Environmental Background of the Pre-Qin Period in Chengdu Plain—A Case Study of Baodun and Sanxingcun Sites. *Quat. Sci.* **2021**, *41*, 1475–1488. (In Chinese)
53. Gregory, K.J. The Human Role in Changing River Channels. *Geomorphol.* **2006**, *79*, 172–191. [CrossRef]
54. Chen, T.; Ryves, D.; Wang, Z.; Lewis, J.; Yu, X. Mid- to Late Holocene Geomorphological and Hydrological Changes in the South Taihu Area of the Yangtze Delta Plain, China. *Palaeogeogr. Palaeoclimatol. Palaeoecol.* **2018**, *498*, 127–142. [CrossRef]
55. Zhang, S.; Zhang, D.D. Centralization or Decentralization? A Spatial Analysis of Archaeological Sites in Northern China during the 4.2 ka BP Event. *Front. Earth Sci.* **2023**, *11*, 1135395. [CrossRef]
56. Montgomery, D.R. *Dirt: The Erosion of Civilizations, with a New Preface*; Univsity of California Press: Oakland, CA, USA, 2012.
57. Zheng, Z.; Ma, T.; Tang, Y.J.; Rolett, B.V.; Zong, Y.Q.; Huang, K.Y.; Zhang, S.X. Holocene Environmental Changes and Prehistoric Agricultural Development along the Coastal Areas of South China to Southeast Asia. *Chin. Sci. Bull.* **2024**, *69*, 2553–2567. (In Chinese) [CrossRef]
58. Huo, W. Shi'erqiao Sites in Chengdu. *J. Tianfu* **2023**, *6*, 2+161. (In Chinese)
59. Staubwasser, M.; Weiss, H. Holocene Climate and Cultural Evolution in Late Prehistoric–Early Historic West Asia. *Quat. Res.* **2006**, *66*, 372–387. [CrossRef]
60. Shi, Z.; Zhou, P.; Li, X.; Cheng, H.; Sha, Y.; Xie, X.; Liu, X. Distinct Holocene Precipitation Trends over Arid Central Asia and Linkages to Westerlies and Asian Monsoon. *Quat. Sci. Rev.* **2021**, *266*, 107055. [CrossRef]
61. Zhao, K.; Wang, Y.; Edwards, R.L.; Cheng, H.; Kong, X.; Liu, D.; Shao, Q.; Cui, Y.; Huang, C.; Ning, Y.; et al. Late Holocene Monsoon Precipitation Changes in Southern China and Their Linkage to Northern Hemisphere Temperature. *Quat. Sci. Rev.* **2020**, *232*, 106191. [CrossRef]
62. Shen, J.; Xiao, X. Evolution of the South Asian Monsoon during the Last 20 ka Recorded in Lacustrine Sediments from Southwestern China. *Quat. Sci.* **2018**, *38*, 799–820.
63. Liu, B.; Sheng, E.; Yu, K.; Zhou, K.E.; Lan, J. Variations in Monsoon Precipitation over Southwest China during the Last 1500 Years and Possible Driving Forces. *Sci. China Earth Sci.* **2022**, *65*, 949–965. [CrossRef]
64. Wang, N.; Zhong, J.; Li, Y.; Bao, L.; Lu, T. Research Progress on Pollen-Inferred Holocene Climate Histories in Southwestern China. *Chin. J. Appl. Environ. Biol.* **2018**, *24*, 1. (In Chinese)
65. Chen, C.J. The Evolution of Asian Summer Monsoon and Its Impact on Human Activities During 2.9–8.2 ka BP. Ph.D. Thesis, Southwest University, El Paso, TX, USA, 2022. (In Chinese).
66. Zhang, X.; Zhang, H.; Wang, M.; Cheng, H. The 2.8 ka Abrupt Climate Event in East Asian Monsoon Region: High-Resolution Stalagmite Records from Southern China. In Proceedings of the EGU General Assembly 2025, Vienna, Austria, 27 April–2 May 2025; p. EGU25-8706.

67. Wu, L.; Lu, S.; Zhu, C.; Ma, C.; Sun, X.; Li, X.; Guo, Q. Holocene Environmental Archaeology of the Yangtze River Valley in China: A Review. *Land* **2021**, *10*, 302. [CrossRef]
68. Huang, M.; Ma, C.M.; He, K.Y.; Tang, M.; Xu, J.J.; Zhu, C. Environmental Changes and Human Activities Recorded by Pollen in the Middle–Late Holocene at Baodun Site, Chengdu Plain. *Quat. Sci.* **2022**, *4*, 1078–1093. (In Chinese)
69. Wang, F.; Fang, Q. The Locality and Adaptability of Early Settlement Construction in Shijiahe Settlement. *Jianghan Archaeol.* **2022**, *6*, 49–56+145. (In Chinese)
70. Zeng, M.X. The Variation of Vegetation and Climate and Its Impact on Human Activities from Late Deglacial Period in Western Sichuan. Ph.D. Thesis, Nanjing University, Nanjing, China, 2017. (In Chinese).
71. Barry, R.G.; Blanken, P.D. Top of Climatic Effects on Microclimate. In *Microclimate and Local Climate*; Cambridge University Press: Cambridge, UK, 2016; pp. 261–274.
72. Cheng, P.; Liu, H.; Zhang, L. Research on the Spatial–Temporal Distribution and Morphological Characteristics of Ancient Settlements in the Sichuan Basin. *Land* **2024**, *13*, 1622. [CrossRef]
73. Lin, A.; Wang, M. Great Earthquakes and the Fall of the Sanxingdui and Jinsha Civilizations in Central China. *Geoarchaeology* **2017**, *32*, 479–493. [CrossRef]
74. Zhu, Y.; Chen, G.; Wang, H.; Li, X. Holocene Fluvial Processes in the Chengdu Plain: Responses to Climate Change and Tectonic Activity. *Geomorphology* **2021**, *385*, 107742. (In Chinese)
75. Sichuan Provincial Institute of Cultural Relics and Archaeology. *Environmental Archaeology Research Report of the Sanxingdui Site*; Science Press: Beijing, China, 2019.
76. Luo, L.P.; Zhu, L.D.; Xiang, F.; Yang, W.G.; Han, A.K.; Luo, H.; Zhang, Q.; Zhu, Z.Y. Spore–Pollen Assemblage Age and Environmental Changes of the Chengdu Plain during the Late Holocene. *Acta Palaeontol. Sin.* **2008**, *47*, 195–202.
77. Jia, M.; Li, C.; Mao, X.; Zhang, D.; Wei, C.; Liu, H. Climate–Human–Environment Interactions in the Middle Yangtze Basin (Central China) during the Middle Holocene, Based on Pollen and Geochemical Records from the Sanfangwan Site. *Catena* **2021**, *204*, 105357. [CrossRef]
78. Huang, G.L.; Ma, C.M.; Zhu, C. Progress of the Mid–Late Holocene Environmental Archaeology in Chengdu Plain. *J. Paleogeography (Chin. Ed.)* **2017**, *19*, 1087–1098. (In Chinese)
79. Zhang, W.Y.; Wu, C.N.; Zhang, L.Y. A Review of Early Shi'erqiao Culture Remains. *Sichuan Cult. Relics* **2018**, *5*, 88–96. (In Chinese)
80. Huang, M. Prehistoric Floods and Water Control Documented at the Archaeological Sites in the Chengdu Plain. *J. Earth Environ.* **2024**, *15*, 193–206. (In Chinese)
81. Li, H.Y.; Wu, Y.; Li, T.Y.; Duan, R.; Wang, H.B.; Cui, J.H.; Li, J.Y. Variation in Climate and Hydrological Conditions in Southwest China during the Mid- to Late Holocene, Inferred from Stalagmite Multiple Proxies. *Palaeogeogr. Palaeoclimatol. Palaeoecol.* **2025**, *666*, 112831. [CrossRef]
82. Shi, J.S. Burial, Culture and Society in the Pre-Qin Period of Chengdu Plain. *Archaeol.* **2019**, *4*, 74–87. (In Chinese)
83. Wang, T. Research on Cultural Development and Man-Environment Interactions in Chengdu Plain during the Pre-Imperial Period. Ph.D. Thesis, Jilin University, Jilin, China, 2015. (In Chinese).
84. Huo, W. The Resource System and Exchange Network of the Sanxingdui Civilization. *J. Tsinghua Univ. (Philos. Soc. Sci.)* **2024**, *39*, 1–11+237. (In Chinese)
85. Lin, K.C. On Craft Production and the Settlement Pattern of the Jinsha Site Cluster on the Chengdu Plain. *Asian Perspect.* **2019**, *58*, 366–400. [CrossRef]
86. Xu, J.J. Mid- to Late Holocene Environmental Archaeology in Chengdu Plain. Ph.D. Thesis, Nanjing University, Nanjing, China, 2017. (In Chinese).
87. Wen, X.; Song, B.; Na, Z. Interruptions of the Ancient Shu Civilization: Triggered by Climate Change or Natural Disaster. *Int. J. Earth Sci.* **2013**, *102*, 933–947. [CrossRef]
88. Shi, J.S. On the Sanxingdui-Jinsha Culture. *Archaeol. Cult. Relics* **2020**, *5*, 65–72. (In Chinese)
89. Yan, Z.; Du, L.; Alipujiang, N.; Alimu, A.; Zhang, Z.; Qiu, M.; Dong, G. Spatio–Temporal Changes of Prehistoric Human Activities and Subsistence in Relation to Trans-Eurasian Exchange in the Inner Asian Mountain Corridor. *J. Geogr. Sci.* **2025**, *35*, 1667–1682. [CrossRef]
90. Giosan, L.; Clift, P.D.; Macklin, M.G.; Fuller, D.Q.; Constantinescu, S.; Durcan, J.A.; Tabrez, A.R. Fluvial Landscapes of the Harappan Civilization. *Proc. Natl. Acad. Sci. USA* **2012**, *109*, E1688–E1694. [CrossRef]
91. Molloy, B. Was There a 3.2 ka Crisis in Europe? A Critical Comparison of Climatic, Environmental, and Archaeological Evidence for Radical Change during the Bronze Age–Iron Age Transition. *J. Archaeol. Res.* **2023**, *31*, 331–394. [CrossRef]

MDPI AG
Grosspeteranlage 5
4052 Basel
Switzerland
Tel.: +41 61 683 77 34

*Land* Editorial Office
E-mail: land@mdpi.com
www.mdpi.com/journal/land

www.ingramcontent.com/pod-product-compliance
Lightning Source LLC
Chambersburg PA
CBHW061837210926
51850CB00006B/909
* 9 7 8 3 7 2 5 8 7 1 8 4 1 *